Domestic Dog Cognition and Behavior

Alexandra Horowitz
Editor

Domestic Dog Cognition and Behavior

The Scientific Study of *Canis familiaris*

 Springer

Editor
Alexandra Horowitz
Department of Psychology
Columbia University Barnard College
New York, NY
USA

ISBN 978-3-642-53993-0   ISBN 978-3-642-53994-7 (eBook)
DOI 10.1007/978-3-642-53994-7
Springer Heidelberg New York Dordrecht London

Library of Congress Control Number: 2014930349

© Springer-Verlag Berlin Heidelberg 2014
This work is subject to copyright. All rights are reserved by the Publisher, whether the whole or part of the material is concerned, specifically the rights of translation, reprinting, reuse of illustrations, recitation, broadcasting, reproduction on microfilms or in any other physical way, and transmission or information storage and retrieval, electronic adaptation, computer software, or by similar or dissimilar methodology now known or hereafter developed. Exempted from this legal reservation are brief excerpts in connection with reviews or scholarly analysis or material supplied specifically for the purpose of being entered and executed on a computer system, for exclusive use by the purchaser of the work. Duplication of this publication or parts thereof is permitted only under the provisions of the Copyright Law of the Publisher's location, in its current version, and permission for use must always be obtained from Springer. Permissions for use may be obtained through RightsLink at the Copyright Clearance Center. Violations are liable to prosecution under the respective Copyright Law.
The use of general descriptive names, registered names, trademarks, service marks, etc. in this publication does not imply, even in the absence of a specific statement, that such names are exempt from the relevant protective laws and regulations and therefore free for general use.
While the advice and information in this book are believed to be true and accurate at the date of publication, neither the authors nor the editors nor the publisher can accept any legal responsibility for any errors or omissions that may be made. The publisher makes no warranty, express or implied, with respect to the material contained herein.

Printed on acid-free paper

Springer is part of Springer Science+Business Media (www.springer.com)

Preface

"Dogs," the developmental psychologist Paul Bloom declared in 2004, "are the next chimpanzees." Bloom was right: while chimpanzees were the stars, and, with other primates, the primary subjects of comparative cognition research for decades, in the last 15 years their dominance has been challenged. In the late 1990s, research with dogs began to appear intermittently in major scientific journals of behavior and cognition; since that time, the number of scientific papers written about dog abilities has skyrocketed. "Dog labs" dedicated to studying the behavior of the species have been developed internationally, and dogs have been accepted as (and indeed are becoming) some of the most well-researched and interesting subjects of contemporary psychology and ethology.

The field of "animal cognition" is motivated by an interest in the description of animals' capacities and mental processes (including beliefs, desires, memories, and other cognitive content), especially insofar as it is explanatory of behavior. It began developing in the mid-twentieth century, arguably spurred by Donald Griffin's 1976 *The Question of Animal Awareness*, which legitimized questions of the mental experience of animals. By 1998, the field had grown sufficiently that the journal *Animal Cognition* (Springer) was launched. The current volume is concerned with a veritable subfield of animal cognition: "dog cognition." Here, too, investigation on dogs has quickly escalated: a 2002, 50-essay volume on animal cognition contained only essay using dogs as the main subject model (by a contributor in this volume, M. Bekoff).

While there are now many multi-author volumes dedicated to the study of nonhuman primate cognition and animal behavior and cognition in general, no edited academic volume on *Domestic Dog Cognition and Behavior* currently exists. Thus, the current volume represents the first compilation bringing together the writings and research of a number of the leading and most forward-thinking researchers in the field. The authors includes the head of the preeminent dog research program at Budapest's Eötvös University; the principal investigators at the University of Veterinary Medicine in Vienna; the heads of the University of Arizona and of Barnard College's dog cognition research groups; as well as some of the first biologists and researchers to study dogs and to write about their cognition; early, renowned investigators of the dog–human bond and effects of domestication and breed; and experts in olfaction, social cognition, and comparative canid work.

Though some research, and some of these researchers, emerge from a "comparative" perspective—which science is concerned with comparing the behavior and abilities of animals, especially human and nonhuman—others come from veterinary, cognitive science, and ethological backgrounds. Instead of being a corpus of comparative findings, therefore, this volume is designed to describe the results of the dog *qua* dog. The results of studies are used to draw a picture, inasmuch as it is possible, of the capabilities of dogs as subjects of interest themselves.

Dogs are not a new subject of study, but this interest in exploring the species' behavior for its own sake is new. Pavlov's dogs were used to demonstrate a form of learning which is seen in most animals (Pavlov 1927); Darwin, for his great personal interest in dogs, studied domesticated animals as a means to understand how artificial selection worked (Townshend 2009). Neither's research was intrinsically motivated by the dog: indeed, Pavlov's invasive work, including decerebration, could be seen as antithetical to developing an understanding of the behaving dog per se (Pavlov 1927).

My own arrival upon dogs as a research subject emerged from a comparative interest as well. When I began studying dog behavior, only 15 years ago, it was not dogs as a species which intrigued me. Instead, I was in search of any animal behavior which could give insights into the mental experience and cognitive understanding of the species, on average. In particular, I was interested in identifying behaviors from which one might be able to infer the presence or absence of a "theory of mind" in animals. Investigation of such "metacognitive" topics as theory of mind, or understanding of the intentionality of others' behavior, is notoriously intractable with nonverbal animals: most studies of human metacognition require verbal response or confirmation from subjects to confirm presence of an ability. Experimental metacognitive paradigms with nonhuman animals have usually led to ambiguous results, and even in cases where a species "passed" the test, alternative explanations for their behavior were easy to find (Shettleworth 1998). Given that any metacognitive ability in animals must have a function in ordinary intraspecific interactions, I went in search of a naturally occurring behavior tightly linked to the development of theory of mind. In human children, social and pretend play are implicated in development of metacognitive abilities (and an absence of play, in some children's difficulty with theory of mind) (see, e.g. Baron-Cohen et al. 1993). Thus I began to look at play in animals to see if it had any of the markers of consideration of others' states of mind. Domestic dogs regularly engage in intraspecific and interspecific play, up to one-third of their awake life as juveniles, and continuing into adulthood (Bekoff and Byers 1998; Fagen 1981; Horowitz 2002). Through detailed characterization of their behaviors in dyadic social rough-and-tumble play, I found that dogs used communicative play signals with sensitivity to the attentional state of a potential playmate, and used attention-getters suited to the level of inattention of the audience, often in order to gain attention before play-signaling (Horowitz 2002, 2009). This presaged the now myriad findings of the dogs' ability to identify and use human attentional states (e.g., Call et al. 2003; Schwab and Huber 2006).

In terms of metacognition, the main result of the play research was a suggestion that dogs may have a "rudimentary," inceptive theory of mind (Horowitz 2009); but the recognition of dogs as a viable and somewhat surprising research subject was the other practical result of the research.

The dog Domestic dogs are members of the Canidae family, along with social carnivores like the wolf, dingo, fox, coyote, and jackal (Serpell 1995). *Canis familiaris*[1] is the only canid species to be fully domesticated. Archeological evidence suggests that domestication of dogs, from wolves, *Canis lupus,* began at least 10–15,000 years ago, about the time that nomadic hunter-gathering humans settled into more agricultural societies (Clutton-Brock 1999), and perhaps at multiple locations (e.g. Boyko et al. 2009)—although mitochondrial DNA from wolves and dogs dates their divergence to 145,000 years ago (Vilà et al. 1997).

Dogs were also the first domesticated animal (Clutton-Brock 1999). This was surely due in part to their social nature, but also implicated must be their ancestors' willingness or ability to change behavior importantly in response to human behavior. Although humans have long explicitly bred animals for specific characteristics, domestication generally begins with a gradual association of a species with humans, whereby successive generations grower tamer and, finally, behaviorally and physiologically distinct from their wild ancestors. In particular, one speculated origin of domesticated dogs is that ancestral wolves began to exploit the new ecological niche that was trash-heaps, or "dumps," on the periphery of early human communities; these wolves may have been, tended to, occasionally eaten by, but generally tolerated by human (Coppinger and Coppinger 2001; Fuller and Fox 1969; Serpell 1995). Human selection began perhaps inadvertently by allowing those animals that were useful or pleasing to survive, while deterring or destroying those that were not (Hale 1969).

This earliest artificial selection would have favored an animal that was flexible in its behaviors, able to, to some extent, anticipate human (interspecific) behavior, and not strongly territorial. Belyaev's famous study creating a cadre of what he called "domesticated elite" foxes by selecting those who reacted to humans non-aggressively or fearfully (Belyaev 1979; Trut 1999) indicates that domestication may change behavioral thresholds (as of a fear response, predatory urge, or aggression).

Artificial selection for the characteristics that comprise today's current dog breeds began relatively recently, only 200 years ago, with the rise of dog "fancies." Breed "standards" appeared which listed the desired traits for members of a breed, which individuals were interbred to perfect the line (Garber 1996). The physical and behavioral diversity apparent in today's dog population arises from this intensive (and often destructive (Asher et al. 2009)) breeding practice.

---

[1] Throughout the volume, authors alternately use the Latin *Canis familiaris* and *Canis lupus familiaris*. Though the former is the current Linnean term in most favor, the latter reflects the belief that the dog is but a sub-species of wolf, having evolved therefrom.

In the U.S. alone, there are an estimated 75 million owned, pet dogs, both purebred and mixed breed. These animals are but a fraction of the stray, free-ranging, and owned dogs worldwide. The ubiquity, and, to a great degree, the success of this species at living with and among humans makes them a compelling subject.

Parts of the Volume

The current volume is divided into three parts, each highlighting one of the different vantages relevant to providing a full understanding of the dog. **Part I** includes chapters providing orientation on the subject, such as the perceptual abilities of dogs and the effect of interbreeding. Surprisingly, while olfaction is the dog's primary sensory modality, few academic papers on or about dogs begin with an assessment of this skill and describe its relation to the behavior of the species. This volume remedies that, opening with **Gadbois and Reeve's** detailed summary of the science of canine olfaction, from physiology to proposed processing systems, and including various methods for assessing and training dogs. They discuss the relevance of the species' olfactory ability in the context of experimental work and applied fields, such as in disease detection. Next, **Serpell and Duffy** describe the contributions of a dog's breed on its behavior, using the well-validated owner questionnaire developed by Serpell and others (C-BARQ) to hypothesize about the bases for various behaviors, both desired and undesired. In this way, they explore the genetic, functional, and inbreeding-related origins of behavior. Finally, canid ethological work with dogs is reviewed and advocated for by **Bekoff**, who uses results from play and scent-marking studies to highlight the value of observational studies of dogs, reminding us of the opportunities available, and the methods necessary, to study this animal in detail.

Part II reviews observational and experimental results from studies of physical and social cognition, such as learning and social referencing. To begin, **Huber, Range, and Virányi** describe results from their and others' labs on what is broadly called "social learning" in dogs: essentially, learning how to do an act by seeing it done. They consider the different kinds of learning, including social facilitation and various levels of imitation. By critically assessing the differences between kinds of imitation, they provide an acute lens on the current state of the field.

Perhaps the most widely researched field of dog cognition is "social cognition." Even without the white sclera of the human eye which, it has been posited, allows humans to more distinctly see the gaze casting out from the iris, dogs share with us the willingness, even interest, in making eye contact. **Prato-Previde and Marshall-Pescini** argue that the social cognitive ability of dogs arises in part from this change. They review the wide-ranging studies of how dogs use their own, conspecifics', and humans' gaze—from interspecies social referencing, to dogs' use of human communicative cues—as well as exploring the meaning and communicative value of this "looking." Elaborating on that work, **Rossi, Smedema,**

Parada, and Allen make the case that social cognitive skills of dogs reflect in part a kind of coevolution between humans and early domestic dogs. They describe preliminary work with an eye-tracking device which can determine gaze in a free-moving dog that is participating in a gaze-following task. By assessing the way the dog visually scans human gestures, they contribute to a better understanding of what the dog is experiencing in various experimental settings. This research, as with all work on social cognition, is also an exploration of the commonalities of the dog–human social group.

Finally, **Fiset, Nadeau-Marchand, and Hall** investigate the physical cognitive abilities of hand-reared, captive (but not pet or tame) wolves as a way to reflect upon what of *Canis lupus* is left in *Canis familiaris*. In particular, they report results of tests of development of object permanence and sensorimotor skills in young wolves, and compare these to the available results from domestic dogs. The differences as well as the similarities put a useful lens on what is changed by the process of domestication and artificial selection.

Part III nicely ties up and reflects upon the work in the fields of dog cognition and behavior to date, in chapters reviewing the various conceptual and methodological approaches in the field, testing anthropomorphisms made of dogs, and developing practical application for behavioral and cognitive results to be used in animal welfare. **Fugazza and Miklósi** begin by critically assessing the methods used to study dogs in much of the research described above. Since researchers of the field come from a wide array of backgrounds, methods do not always converge, and the corpus of knowledge on the species is being built, they argue, somewhat haphazardly. They encourage a consensus based in ethology and on sharing data.

Among those estimated 75 million owned dogs in the U.S., there are presumably a few that are not anthropomorphized by their owners. But not many. Anthropomorphisms are relevant to the drawing of a complete picture of the species dog because they reflect an *anthropocentric* attitude that not only are dogs smaller, furrier versions of humans, but also that they are valuable or interesting (as scientific subjects or even as animals) only insofar as they resemble us. **Horowitz and Hecht** instead argue for replacing this perspective with a more dog-centered research program. They describe work from their lab testing the context of behavior that prompts common anthropomorphisms of dogs. Like Fugazza and Miklósi, they reflect on the dog research to date and identify anthropocentric as well as more dog-centric elements of published research paradigms.

Next, **Udell, Lord, Feuerbacher, and Wynne** follow Fiset et al. in looking at hand-raised, captive wolf behavior, but in their case they use it to argue for a revised understanding of who the dog is. In particular, they describe their work which reassesses the cognitive and developmental differences between wolves and dogs. Also reflecting on the approach of the dog research to date, they observe that most research in the field is with owned dogs in first-world countries, which, they argue, is not just the numerical minority of dogs currently alive, but also may not be a representative sample of the extant species members.

The volume ends with a chapter considering how our growing, if incipient understanding of the dog can be used practically to affect the lives of all dogs.

An appraisal of the good measures of health, emotional experience, as well as cognitive abilities of the dogs can be used to propose a wide-ranging dog welfare framework, as begun by **Rooney and Bradshaw**. Their chapter is a comprehensive and important integration of many of the topics of this volume into the foundations of a practicum for vets, handlers, kennel workers, and dog owners alike.

Domestic Dog Cognition and Behavior: The Scientific Study of Canis familiaris highlights the state of the field in this new, provocative line of research. Chapters considering past methods and work and initiating novel lines of inquiry draw a fuller picture of the behavior of the domestic dog than has ever been done. These pages also represent a move toward considering and studying domestic dogs for their own sake, not only insofar as they reflect back on human beings.

Alexandra Horowitz

References

Asher, L., Diesel, G., Summers, J. F., McGreevy, P. D., & Collins, L. M. (2009). Inherited Defects in Pedigree Dogs. Part 1: Disorders Related to Breed Standards. *The Veterinary Journal, 182*, 402–11.
Bekoff, M., & Byers, J. (Eds.). (1998). *Animal play: Evolutionary, comparative, and ecological perspectives.* Cambridge: Cambridge University Press.
Belyaev, D. K. (1979). Destabilizing selection as a factor in domestication. *Journal of Heredity, 70*, 301–308.
Bloom, P. (2004). Can a dog learn a word? *Science, 304*, 1605–1606.
Boyko, A., et al. (2009). Complex population structure in African village dogs and its implication for inferring dog domestication history. In *Proceedings of the National Academy of Science, 106*, 13903–13908.
Call, J., Bräuer, J., Kaminski, J., & Tomasello, M. (2003). Domestic dogs (*Canis familiaris*) are sensitive to the attentional state of humans. *Journal of Comparative Psychology, 117*, 257–263.
Clutton-Brock, J. (1999). *A natural history of domesticated mammals.* Cambridge: Cambridge University Press.
Coppinger, R., & Coppinger, L. (2001). *Dogs: A startling new understanding of canine origin, behavior and evolution.* New York: Scribner.
Fagen, R. (1981). *Animal play behavior.* Oxford: Oxford University Press.
Fuller, J. L., & Fox, M. W. (1969). The behavior of dogs. In E. S. E. Hafez (Ed.),*The behaviour of domestic animals* (pp. 438–481). Baltimore: Williams & Wilkins Co.
Hale, E. B. (1969). Domestication and the evolution of behaviour. In E. S. E. Hafez (Ed.), *The behaviour of domestic animals* (pp. 22–42). Baltimore: Williams & Wilkins Co.
Horowitz, A. C. (2002). *The behaviors of theories of mind, and A case study of dogs at play.* Ph.D. Dissertation, University of California, San Diego.
Horowitz, A. (2009). Attention to attention in domestic dog (*Canis familiaris*) dyadic play. *Animal Cognition, 12*, 107–118.
Pavlov, I. P. (1927). *Conditional reflexes: An investigation of the physiological activity of the cerebral cortex.* Oxford, England: Oxford University Press.
Schwab, C., & Huber, L. (2006). Obey or not obey? Dogs (*Canis familiaris*) behave differently in response to attentional states of their owners. *Journal of Comparative Psychology, 120*, 169–175.

Serpell, J. (Ed.). (1995). *The domestic dog: Its evolution, behaviour, and interaction with people.* Cambridge: Cambridge University Press.

Shettleworth, S. J. (1998). *Cognition, evolution, and behavior.* New York: Oxford University Press.

Townshend, E. (2009). *Darwin's dogs: How Darwin's pets helped form a world-changing theory of evolution.* London: Frances Lincoln Limited.

Trut, L. N. (1999). Early canid domestication: The farm-fox experiment. *American Scientist, 87,* 160–169.

Vilà, C., Savolainen, P., Maldonado, J. E., Amorim, I. R., Rice, J. E., & Honeycutt, R. L., et al. (1997). Multiple and ancient origins of the domestic dog. *Science, 276,* 1687–1689.

Contents

Part I Orientation: Perceptual and Breed Effects on Behavior and Early Ethological Research

1 Canine Olfaction: Scent, Sign, and Situation 3
  Simon Gadbois and Catherine Reeve

2 Dog Breeds and Their Behavior 31
  James A. Serpell and Deborah L. Duffy

3 The Significance of Ethological Studies: Playing and Peeing 59
  Marc Bekoff

Part II Behavior and Cognition: Observational and Experimental Results

4 Dog Imitation and Its Possible Origins 79
  Ludwig Huber, Friederike Range and Zsófia Virányi

5 Social Looking in the Domestic Dog 101
  Emanuela Prato-Previde and Sarah Marshall-Pescini

6 Visual Attention in Dogs and the Evolution
  of Non-Verbal Communication 133
  Alejandra Rossi, Daniel Smedema, Francisco J. Parada
  and Colin Allen

7 Cognitive Development in Gray Wolves: Development
  of Object Permanence and Sensorimotor Intelligence
  with Respect to Domestic Dogs 155
  Sylvain Fiset, Pierre Nadeau-Marchand and Nathaniel J. Hall

Part III The Future of Dog Research: Critical Reassessment of Methods and Practice, and Practical Applications

8 **Measuring the Behaviour of Dogs: An Ethological Approach** 177
   Claudia Fugazza and Ádam Miklósi

9 **Looking at Dogs: Moving from Anthropocentrism to Canid *Umwelt*** 201
   Alexandra Horowitz and Julie Hecht

10 **A Dog's-Eye View of Canine Cognition** 221
   Monique A. R. Udell, Kathryn Lord, Erica N. Feuerbacher and Clive D. L. Wynne

11 **Canine Welfare Science: An Antidote to Sentiment and Myth** 241
   Nicola Rooney and John Bradshaw

Part I
Orientation: Perceptual and Breed Effects on Behavior and Early Ethological Research

Chapter 1
Canine Olfaction: Scent, Sign, and Situation

Simon Gadbois and Catherine Reeve

Abstract Canine olfaction is a rich field of study for the behavioural sciences and neurosciences, and it is rich in interdisciplinary connections. This chapter will explore the neurocognitive and neuroconative bases of olfaction (the neurophysiological foundations of cognition and motivation), and discuss the behavioural, psychological, and semiotic dimensions of scent processing. It will cover the basic psychophysics of olfaction and the methodologies allowing us to explore this sensory modality, as well as the complex cognitive and motivational dimensions of scent. This chapter will open with an overview of the different disciplines involved in the study of canine olfaction. Some basic anatomy and neuroscience will be reviewed, mostly with direct reference to behaviour and associated psychological processes (e.g., cognitive, motivational, and affective systems). For the behavioural aspect of olfaction, a discussion of the contrasting, yet complementary methods of ethology and experimental psychology will be examined. The importance of both field and laboratory research will be highlighted. Olfaction "in context" will also be discussed in reference to zoosemiotics and in order to understand the canine olfactory psychoethology in its most meaningful and functional dimension: processing "signs" (including symptoms as with dogs trained for biomedical applications such as symptom detection). We will conclude with a short commentary on the human-canine sensory symbiosis with sniffer dogs.

1.1 The Sciences of Canine Olfaction

Canids, like most mammals (and many other vertebrates, such as reptiles), live in an olfactory world. Their *Umwelt,* or "sensory world", is impressively different from ours (see also Horowitz and Hecht, this volume). Observing our dogs exploring

S. Gadbois (✉) · C. Reeve
Department of Psychology and Neuroscience, Neuroscience Institute, Dalhousie University, Halifax, NS, Canada
e-mail: sgadbois@dal.ca

their social landscape by relentlessly sniffing scent marks (mostly invisible, unless you live where yellow snow is a possibility) left by others, we understand that we are not in the same sensory-perceptual world. We are just starting to understand the amount of information that animals process from chemical messages. Like us with our emails and social site postings, dogs and other canids have their own world of "peemails" and "Nosebook" to explore, create and manipulate (Harrington and Asa 2003; Bekoff 2001; Allen et al. 1999; Wells and Bekoff 1981).

The study of olfaction has traditionally focused on mammals and insects. In mammals, rats and mice have been the primary model systems, at least in experimental psychology and neuroscience. Recent focus on dogs, and working dogs more specifically, seems to have sparked an interest in the scientific study of olfactory processing in canines. Different theoretical, conceptual, and methodological perspectives have contributed to the science of canine olfaction over the past century or so, either directly or indirectly. Here, we will advocate a generalist, synthetic, and broad-reaching perspective on canine olfaction. We believe in a full integration of experimental psychology (mostly psychophysics and animal learning theory), behavioural biology (mostly ethology because of its strong focus on proximate questions), neuroscience (behavioural, cognitive, affective, and social) and zoosemiotics. Many new applications surface every year (from bed bug detection to telephone poll rot detection)—most either unknown by the scientific community, or known in their own parallel (non-academic) world of "research and development" (R&D) and applied types of research with low inter-disciplinary diffusion. An important part of the scientific contribution to canine olfaction has been fringe and marginal for decades. Canine olfaction can be discussed in relation to the natural environment of the animal or in the context of laboratory conditions. Applied canine olfaction is also a growing area of investigation and often relates to "quasi-experimental" approaches and the industrial, R&D model of applied research. We will discuss some of these applications later.

The field of 'zoosemiotics' deserves a brief introduction. Sebeok (1968, 1977) conceptualized the field around the idea of 'semiotics', or the 'theory of signs'. This perspective applies well to "semiochemicals": chemicals used as signs. At first glance, the field seems to duplicate the study of animal communication (e.g., Bradbury and Vehrencamp 2011). But interestingly, communication, according to zoosemioticians, is only one of three sign processes, or semiosis. Communication describes the "classical" perspective in the field of ethology: exchange of information between a sender and a receiver. Zoosemiotics also makes room for "representation" (when a sender is producing a sign without the presence of a receiver, arguably for an intended receiver or clearly identified receiver, or if you will, a "to whom it may concern" message) and "signification" (when a receiver is present and processing a sign, without the emitter or sender being present). A clear case of representation would be when a canine sender/emitter is urinating, potentially scent marking its territory, but without any other dog in the vicinity. Signification would be when a urine mark or defection is found, and processed by the receiver, without the sender/emitter present, and without the assumption that the receiver was the intended target (see also Bekoff, this volume, on urination patterns). The elegant nuance here is that intentionality in the processing of pheromones or allomones (pheromones crossing the species barrier) is not assumed in either signification or representation. In other words, a "sign" (including an olfactory one, or semiochemical) can provide information without being necessarily produced in the context of communication (in which case, the term "signal" is used).

This brings us to the title of this chapter, "Canine Olfaction: Scent, Sign, and Situation". It goes without saying that the stimuli discussed here are odorants or scents. They are signs as defined by zoosemiotics, and they are always in context. Without having to take a radical "ecological" or behavioural ecology perspective on the issue, ethology, with its focus on direct observation of observable behaviours, and its interest in social and developmental issues, as well as neurophysiological and fine-grained analysis of motor patterns (Fentress and Gadbois 2001), provides tools and an "in-context" framework that complements zoosemiotics, not to mention the highly formalist approaches of psychophysics and animal learning research.

1.2 A Neurocognitive and Neuroconative Perspective on Olfaction

This section will address how mental processes (cognition) and motivations (conation) interact to produce and modulate olfactory behaviours. The neuroscience of mammalian olfaction is a vast area of research, but the work on canines is limited, mostly because of the potential invasiveness of the research. Much of what will be mentioned here emanates from rodent and human research: most vertebrates share the fundamental structure of the olfactory system, as well as its mechanisms, and the homogeneity within the mammalian brain is truly impressive (Panksepp 1998; Panksepp and Biven 2012). The olfactory system is fundamentally linked to the limbic system or paleo-mammalian brain (MacLean 1990). Our perspective in this chapter is neurocognitive. We will discuss two areas of theoretical interest in our lab: the neurocognitive issues behind olfactory processing and learning in relation to training scent processing canines and the issue of learning and motivation from a "soft" pharmacological perspective that one of us (SG) calls the "dopamine hypothesis" (Gadbois 2010).

Because of our neurocognitive focus we will redirect the reader interested in the neuroanatomical and neurophysiological foundations of olfaction to excellent reviews (e.g., Buck 2000; Menini 2009; Shepherd 1994; Wilson and Stevenson 2006; Zelano and Sobel 2005). For the purpose of this chapter, we will focus mostly on the olfactory cortex per se (Haberly 1998; Price 2003), that is, the cortical and peri-cortical part of the olfactory system (the neo-mammalian and paleo-mammalian parts of the olfactory system, respectively, per MacLean's (1990) terminology). We will not get into the distinction between the primary (or main) olfactory system and secondary (or accessory) olfactory system, well known from countless reviews on the mammalian vomeronasal organ (VNO) and processing of pheromones and allomones, but we will discuss one important neurocognitive distinction in olfactory processing: the distinction between smelling (implicit processing, involuntary) and sniffing (explicit processing, exploratory) as explained by Sobel et al. (1998). Although Sobel et al. studied human brains, it is reasonable to think that some aspects of this distinction between implicit and explicit neurocognitive processing could take place in canids.

1.2.1 Three Main Neuroanatomical Components of Olfactory Cognition and Conation

We will now describe briefly three main components of the olfactory system that are involved in more "cognitive" processing of olfactory information, as well as being involved in motivational mechanisms underlying olfaction.

1.2.1.1 Pyriform Cortex

The pyriform cortex is also called the 'prepyriform cortex' or 'primary olfactory cortex': information from the olfactory bulb (an integral component of the limbic system and the first sub-system involved in processing olfactory information from the primary and secondary olfactory systems) is directly wired to the pyriform cortex, which then feeds information to the frontal lobes and the orbitofrontal cortex (via the thalamus). The pyriform cortex is believed to play an important role in olfactory detection and discrimination (Price 2003; Wilson and Sullivan 2011), especially the posterior pyriform cortex. (The anterior pyriform cortex is more involved in the basic analysis of the chemical structure of the odorant. See Sect. 1.3.1.1 below).

1.2.1.2 Entorhinal Cortex

The entorhinal cortex feeds directly into the hippocampus, frontal cortex, and orbitofrontal cortex. It has an important role in memory, especially spatial memory, and thus may play an important role in navigation and possibly tracking and trailing in dogs. It is often defined as the main interface between the hippocampus and the neocortex (frontal and orbitofrontal cortices). The structure is also linked to the amygdala and seems to be involved in autonomic nervous system responses to odours. Emotional memories driven or triggered by smell may involve the entorhinal cortex in significant ways. It is fundamentally part of the hippocampal complex and therefore involved in spatial memory and orientation. See Sect. 1.3.1.2 below.

1.2.1.3 Orbitofrontal Cortex

Often labelled as the 'secondary olfactory cortex', this part of the cortex is known to be important in decision making, as well as some cognitive processing of reward, especially expectation and anticipation of rewards (Kringelbach and Berridge 2009). It is therefore involved in the explicit processing of odours (sniffing, as opposed to smelling).

1.2.2 Motor and Motivational Factors in Olfaction

The words 'motor', 'motion', 'motivation', and 'emotion' are etymologically related, from *movere*, "to move" in Latin. They all have a neuropharmacological connection as well: dopamine. This important excitatory brain neurotransmitter is of great importance in the motor system at the cortical level (e.g., frontal lobes) and subcortical level (the limbic system, the basal ganglia, and associated structures such as the ventral tegmental area, the substantia nigra, the nucleus accumbens, etc.). It is also implicated in activity in the olfactory system, including the olfactory tubercle in the olfactory cortex (where it plays a role in the overall "reward" system of the brain; the pyriform cortex is rich in dopamine and dopamine receptors), and finally the periglomerular cells in the olfactory bulb. Cognitively, dopamine is associated with a broad range of cognitive functions, including attentional processes, which are conceptually linkable to anticipatory processes at the conative level (in fact, those constructs may be less conceptually and practically distinguishable than currently believed).

Dopamine is involved motivational processes as well. As addiction to dopamine agonists (amphetamines, cocaine) may suggest, the role of dopamine in modulating the motivational system of addicts is remarkable. Interestingly, dopamine agonists are known to increase motivation and anticipation, as well as olfactomotor behaviours and olfactory activity in the olfactory system. In fact, dopamine activity in the lateral hypothalamus is associated with "stimulus-bound processes" (Panksepp 1998), such as exploratory olfactomotor behaviour—in other words, sniffing. Other behaviours may be associated with this as well, such as mouthing, licking, whisking (Deschênes et al. 2011) and more involved motor behaviours such as searching, exploring, and manipulating behaviours. Incentive salience is a characteristic of reward-predicting stimuli that define a system called the "WANTING system" by Berridge and associates (Berridge 2001, 2004). This system is discussed by Panksepp as being the SEEKING system, and similar theories exist elsewhere: Gray's "Behavioural Activation System" (Gray 1987) and Depue's "Behavioural Facilitation System" (Depue 2000). Berridge's theory is more integrated in our opinion as it explains very well the balance between an arousal state of anticipation for the reward (wanting or seeking the reward: what ethologists labelled appetitive behaviours) and the opposing system that takes over when the reward is acquired and being consumed. Here Berridge and Panksepp agree on some of the details, including the idea that those two systems are incompatible in the sense that if one is activated, the other one is not (in normal conditions, and as Berridge would explain, addiction would be an exception), and that what really motivate animals to do things, including learning, is anticipating the reward, not consuming it: it is wanting it, not liking it. Berridge explains well his integration of those two systems and their impact on behaviour and learning (Berridge 2001; Berridge and Robinson 1998; Berridge et al. 2009). Some neo-behaviourist theories of conditioning postulated such quasi-cognitive or cognitive factors in learning.

Neurocognitively and neuroconatively, "anticipations" and "expectations" seem to be modulated by the dopaminergic system—although it is not the only neurotransmitter system involved (see Table 1.1 showing how the LIKING system taps into endorphins, in fact suppressing the WANTING system, therefore suppressing behaviour). As a reminder, dopamine is the central neurotransmitter in the WANTING/SEEKING system. It is important in olfaction and motor behaviour in general, and therefore plays a role in olfactomotor behaviours, including the basic behaviour of sniffing that is enjoying its own scientific literature in recent years (in humans, rodents, and canines; see Mainland and Sobel 2006; Sobel et al. 1998; Kepecs et al. 2005; Panksepp 1998). Sniffing is an exploratory behaviour that has many important roles in olfaction: it actively participates in the input of the olfactory stimulus, it can be modulated to account for different odorant concentrations, and it can modulate the pattern of neural activity (e.g., brain waves).

In relation to canines, in 1992, Arons and Shoemaker demonstrated that some dog breeds have higher baseline levels of dopamine than others. Border collies and huskies have high dopamine levels; livestock guarding dogs' levels are lower. It is difficult not to think about the role of dopamine in some human disorders when thinking of dopamine and dog breeds that seem to follow the pattern. Think of border collies. The terms "hyperactive", "obsessive", "compulsive", etc., are frequently used to describe individuals of that breed. This basically translates into the intriguing possibility that individual differences and breed differences in baseline dopamine levels may have a direct impact on cognition, motivation, learning, and overall olfactory behaviour and performance. It is interesting to note that our most successful laboratory and field work dogs are Border collies. A selection bias may be at play here since we recruit dogs volunteered by their owners eager to find an occupation for their overactive pets who appear in need of stimulation, but if we look at the retention of individuals (the ones that make the cut for advanced laboratory or field training), Border collies dominate the roster. As a general rule, they are good and hard workers: motivated, persistent, good sniffers, and their attentional focus can be channelled (in most cases) very well.

Our "dopamine hypothesis" essentially highlights the possibility that "software"-level characteristics (neurochemical and neural-level mechanisms and processes) are more important than the often-touted "hardware" characteristics. For example, we never had any luck with Bloodhounds and other hunting "scent dogs" mostly because of motivational issues, resilience, ability to work long hours or consistently, and overall performance and energy levels. By contrast, the "work ethics" of high-dopamine breeds, like the Belgian Malinois, Jack Russell, and Parsons, is remarkable.

1 Canine Olfaction: Scent, Sign, and Situation

Table 1.1 Theoretical relationships between anticipatory and reward systems of the brain

	Appetitive behaviours	Consummatory behaviours
Traditional system(s)	• SEEKING system (Panksepp) • Behavioural activation system (Gray) • Behavioural facilitation system (Depue)	"Reward system"
Berridge's perspective	WANTING system	LIKING system
Associated brain areas	Hypothalamus, basal ganglia and associated structures	Limbic system: Amygdala, Hippocampus, Septum, etc.
Associated behaviours	Exploratory behaviours: foraging, stimulus-bound sniffing, mouthing, licking	"Feeling good"
Associated neurotransmitters	Dopamine	Endorphins

Although we have not yet had the opportunity to work with these breeds, some field biologists working with wildlife conservation canines believe Jack Russells are the ultimate detector breed: Engeman et al. (1998) calls them "the unique detector dogs". We know local dog trainers and handlers in the Canadian Maritime provinces working in bed bug detection that would share that belief.

1.3 Between Nose, Brain, and Mind: Cognitive Processes

1.3.1 Neurocognitive Sub-Systems

The visual system has two pathways (or streams) of processing information from the outside world (Schneider 1969). One is the WHAT system for object recognition, and the other is the WHERE system for spatial vision and localization. From an evolutionary perspective it is believed that those systems evolved in order to make sense of the immediate threats and potential foraging opportunities afforded to the animal. In this section, we argue that the olfactory system can be conceptualized the same way. It may be too early to determine the neuroanatomical boundaries and localization of these subsystems (assuming it is even relevant), but at least the processes involved can be identified. To those two main systems, we will add one that may be of crucial importance to olfaction: HOW MUCH. Table 1.2 summarizes the perspective we propose.

1.3.1.1 The WHAT System

The psychophysics literature makes a clear case for the distinction between detection, discrimination, and identification. Those three processes are part of the "what" system. We will briefly describe the processes involved. Note that in terms of higher

Table 1.2 Olfactory neurocognitive systems and corresponding neuroanatomical centres

WHAT	WHERE	HOW MUCH
Detection	Searching	Scaling
Discrimination	Trailing	
Identification (e.g., matching)	Tracking	
Pyriform cortex	*Entorhinal cortex*	*Olfactory bulb; cortical?*

level (cognitive) processing, we have already identified the pyrifom cortex as an important role player in the WHAT system.

Detection defines the identification of one stimulus (e.g., grapefruit oil) or stimulus category (e.g., citrus essential oil) among background noise or interference. Note that we recognize the importance of early stimulus generalization when inferring categorical detection.

Discrimination defines the identification of one stimulus (e.g., grapefruit oil, referred to as the S+ or positive stimulus) or stimulus category (citrus oils) as contrasted to another often similar stimulus (e.g., orange oil, the S− or negative stimulus) or another category (e.g., floral oils).

Identification is a process by which a more explicit knowledge of the stimulus is made. In humans, for example, "naming" the stimulus would be a demonstration of this level of discriminatory process. Matching-to-sample tasks attempt to get at that level of investigation; that is, the hope is that matching a sample with a target among many other choices is an indication that the animal explicitly "identifies" the target as the "same as" the sample. By definition, identification is preceded by detection and discrimination. We will use the laboratory technique of simultaneous matching-to-sample as an example. Imagine a set of four exemplars: lavender, grapefruit, sandalwood, and bergamot oils. Each of these odorants can be used as the sample to be matched with one or more instances of the same odours. For example, if presented with a lavender oil sample, a dog may be required to pick the matching sample in an array (matrix, line-up, or any other arrangement) of two, three, or x number of choices that can include any of the oils in the initial set and/or distractors. Another example is same-or-different judgments: dogs are trained to investigate two odours and simply indicate if the samples are the same or different (by pressing or poking a paddle, for example). This specific type of learning, although at first glance simple and elegant, has been found to be very difficult if not impossible to acquire by dogs in our lab. Colleagues in developmental psychology have pointed out to us that even children have a hard time with matching-to-sample (non-matching-to-sample tasks being often acquired more readily) and same-or-different judgment tasks (Diamond et al. 1999, Diamond 2006; Overman 1990; Premack 1983).

1.3.1.2 The WHERE System

The localization of olfactory stimuli is crucial in the context of finding food and mates, just to mention the most obvious. We have identified the entorhinal cortex as being an integral part in olfactory processing in the context of spatial memory and likely localization of odours. It should be no surprise that canids, as predators, have been efficient at using their olfactory sense to survive and reproduce. What is less obvious is how this works beyond the WHAT system. Assuming the canine knows the target scent (can detect it from background interference, can discriminate it from similar odours that may be less relevant, and can identify it in more complex situations), the issue of finding it when no other sensory modality can help (especially vision and audition) is less obvious. This is where laboratory conditions fail to give a full sense of the complexity of the processes necessary to "find" target odours.

The "sniffer dog" literature often distinguishes between trailing and tracking dogs. There is in fact a significant confusion between these two processes, and in some ways the distinction may be somewhat artificial or irrelevant to brain and behavioural organization. But since it is an accepted conceptualization and often defines specific training methods, protocols, and even dogs, we will include the nuance in the WHERE system category. Not unlike the processes in the WHAT system, the WHERE system addresses an incremental level of complexity in terms of processing the stimulus. In this case, the stimulus is entirely "in context": in a dynamic environment, meaning that the animal needs to be in foraging mode and move around. Our experience in lab conditions suggests that a motor involvement in active searching involving rooting and burying to find an odour source may be facilitating detection and identification, despite the added olfactory noise coming from the substrate. Our hypothesis is consistent with data presented by Hall et al. (2013). We are currently investigating this intriguing hypothesis further. This factor may also explain the "field effect", that is, the often radical and counter-intuitive loss of performance in dogs that experience field conditions after laboratory training. Motor integration between basic locomotor functions and olfactomotor functions may be crucial for the system to work efficiently. Therefore, searching, trailing, and tracking are uncommon areas of research but promising behaviours to study. This is particularly evident with some of our dogs in the scent processing program at the Canid Behaviour Research Lab that are trained in the lab and later transferred to the field. For half of our dogs, at minimum, it is almost impossible to bring them back to work in lab conditions. They seem to have lost all motivation for the low stimulation (and contamination) of the laboratory environment (it is possible that the absence of cues associated with reward would be the cause). Alternatively, dogs may simply not be stimulated enough cognitively—something that we address often with Border collies that seem to need being constantly challenged. Our discussion of neuroconative processes addresses the potential reasons behind this phenomenon documented by other teams (e.g., Smith et al. 2003) that we have labelled "field effect". It is one type of "motivational collapse" (also our term) that is often reported by dog handlers and trainers familiar with working dogs.

Searching The first step in localization, before the stimulus is acquired, is to search. Searching requires the animal to have an identified target, and in applied settings, it may require the dog to memorize biologically irrelevant stimuli (e.g., looking for drugs or explosives). The important dimension of this step is that the stimulus has not yet been detected. The early stages of foraging behaviour are essentially "searching" behaviours.

Trailing Trailing is often defined as searching, at least in the early stages, but also may suggest that the stimulus is acquired, but not yet localized. In other words, cues are detected that announce the presence of the target, but the exact localization or path taken by the target is not yet identified (and may never be). This process requires significant amount of "air scenting" or sampling the air, as opposed to "ground scenting" or investigating the ground.

Tracking Tracking is much more specific and there is a consensus on the definition of the term. In tracking, the target is acquired, and the path taken by the moving target is also identified and followed (with different levels of spatial accuracy).

1.3.1.3 The HOW MUCH System

There are at least two situations, natural and artificial, that may require the dog to assess the quantity of molecules present in the environment. Volatiles will be distributed according to a specific gradient influenced by contextual conditions (temperature, humidity, barometric pressure, and most importantly, air movements such as drafts and wind), not to mention the distance between the dog and the target (and obviously all this in relation to the actual saliency of the stimulus). Much of this category is akin to the "scaling" process known in psychophysics. Training and experimental conditions may require a dog to identify a threshold and, for instance, give a positive response if the stimulus is above threshold and give a negative response or no response if the stimulus is below that threshold. An example would be a dog trained to identify *Varroa destructor* (parasitic mites) and *Nosema apis* (fungus) in beehives. Both are potentially important factors in colony collapse disorder (CCD). Most beehives in North America are infected with some level of *Varroa* and *Nosema*, but the applied issue would be to train a dog to identify hives infected beyond a specific threshold, highlighting the need for an immediate intervention.

Interestingly, it is unclear whether the HOW MUCH system would actually be able to discriminate within the actual volumetric quantity of a given stimulus (of biological significance or having a primary incentive value—e.g., food). A study by Horowitz et al. (2013) suggests that pet dogs may not differentiate between low and high quantities of food based on olfactory cues alone. It is possible that a larger differential between small and large amounts would have resulted in more significant results (in terms of physical or chemical volatility), or perhaps the differential incentive value of the stimuli was minimal enough to keep them indifferent in their choice of food source.

The HOW MUCH system likely serves a function in processing gradients—in other words, helping the dog to determine the direction of a source of volatiles.

This specific question has been discussed in the context of directionality of tracks. The basic question is as simple as "Do dogs know if they backtrack or forward-track a target?", and as a corollary, "Can they make a mistake?". It seems logical that, in order to survive, wolves would have had to "know" where their prey was going, as backtracking would be counterproductive and maladaptive. A debate has been ongoing regarding this issue with dogs (Thesen et al. 1993; Steen and Wilson 1990; Wells and Hepper 2003; Hepper and Wells 2005). Interestingly, Wells and Hepper (2003) found that dogs were not good at "detecting" direction. Only 36.3 % of the dogs studied could do this consistently. Steen and Wilson (1990) suggested that the training (read "learning" for ethological, non-artificial contexts) may be of importance in determining if dogs track in the right direction. Thesen et al. (1993) identified three stages in tracking behaviour: a "searching" phase (before the stimulus or target is acquired, as described above), a "deciding" phase (when the dog determines the directionality of the moving target), and a "tracking" phase. They also found the dogs to be more accurate and consistent than those in the study by Wells and Hepper (2003). Note that this literature and perspective on the HOW MUCH system may link it directly to the WHERE system. It may even suggest that it is a sub-system of the WHERE system or simply needs to be fully merged with it.

Most of the traditional fundamental research on canine olfaction and applied research with sniffer dogs has focused on the WHAT system (e.g., odour discriminations) and the HOW MUCH system focusing on detection thresholds, in other words, more traditional psychophysics experiments (see Helton 2009a, b and Lit 2009 for reviews; see also the work of the Auburn University College of Veterinary Medicine group, e.g., Furton and Myers 2001). Research on tracking per se is at its infancy, mostly because of the methodological constraints imposed by moving subjects during searching, trailing, and tracking but also because of the challenges presented by field work.

1.4 Methodology: Psychophysics, Olfactory Learning, and Cognition

Section 1.3 presented our three-system perspective on the sensory and cognitive processes that work in synergy to process information. Now we move on to the methods to investigate olfactory capacity. Studying olfaction in animals has been a challenge in that what are salient odours for most non-human mammals (with exceptions, e.g., cetaceans) are for humans "invisible" and often undetectable stimuli. Experimental psychology has provided effective tools to study sensory processes in animals using mostly operant methods (Blough 1966; Blough and Blough 1977). Quantitative tools in human psychophysics have also contributed to the application of Signal Detection Theory (SDT) to olfactory stimuli, either in detection tasks (one stimulus in a noisy environment) or discrimination and identification tasks (discriminating between two stimuli). While we are not covering it here, the basics of SDT for canine olfactory processing are discussed in Helton (2009a, b). McNicol (2005) is a short yet useful resource to cover the basics of SDT and MacMillan and Creelman (2005) is a comprehensive resource to cover advanced applications of SDT. The latter includes its use in designs such as two-alternative forced choice designs (2AFC) and multiple alternative forced choice (mAFC), same-different, matching-to-sample, and oddity design (triangular method) (Lit 2009). Although written with humans and mostly visual stimuli in mind, olfactory stimuli can be used with these approaches. It is also worth mentioning that although SDT is usually used as a parametric tool, a non-parametric version of SDT also exists (Pastore et al. 2003).

1.4.1 Habituation-Dishabituation

Slotnick and Schellinck (2002) also review methodologies used with rodents, including an interesting non-operant technique called "habituation-dishabituation". The method is often used in our lab before training dogs on a specific scent when we are in the early stages of a project. As habituation (and dishabituation) are non-associative, "simple" forms of learning, no training is required. We use this technique to test the ability of dogs to naturally detect two given odours. For example, in 2009, our laboratory started a project with Parks Canada that required sniffer dogs to find and potentially track Eastern Ribbon Snakes (*Thamnophis sauritus sauritus*), a species-at-risk in Nova Scotia. One of our worries was that Common Garter Snakes (*Thamnophis sirtalis*) are very common in the same habitat and areas where the dogs were going to work. Both species are of the same genus and to a human nose, smell quite the same. The procedure typically includes five trials. The first four are the habituation phase, when the dogs are exposed to the target scent for five minutes. During that time, the duration of sniffing (sniffing time) is recorded. The dogs are given a break of fifteen minutes between each exposure. Typically, by trial four, the sniffing time has been reduced dramatically. On trial five, the new scent (Common Garter Snake) is introduced. This is the dishabituation phase. If the dogs perceive the smell as different, it is assumed that the sniffing time will increase dramatically from trial four because of the novelty of the smell (Gheusi et al. 1997; Vaché et al. 2001). It is expected that the sniffing time would approach the sniffing time of the first trial. In our case (Gadbois et al. in prep), all dogs increased their sniffing time significantly, more than doubling the sniffing time for trial one (see the Fig. 1.1). This can be interpreted as a strong novelty effect, suggesting dogs can naturally discriminate the two smells. A control condition—a cotton ball without the smell—is always added within each trial. This immediate, within-trial control allows the experimenter to determine if the dog is sensing the target odours.

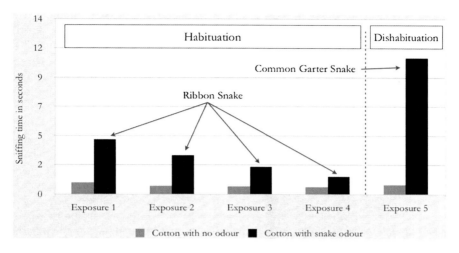

Fig. 1.1 Data showing Zyla's (expert sniffer dog) habituation/dishabituation test comparing Eastern Ribbon Snake scent to Common Garter Snake scent (Gadbois et al. in prep)

1.4.2 Errorless Discrimination Training

Our lab has also worked on adapting Errorless Discrimination Training (EDT) (Terrace 1963a, 1963b, 1964, 1966) to olfactory discriminations with some success. Using the same example as above (training the Ribbon Snakes as the S+ or target scent versus Common Garter Snakes as S− or non-target scent) we use an adaptation of EDT. Terrace developed the procedure for colour (wavelength) discrimination in pigeons. As far as we know, EDT has never been applied to scent detection and discrimination. It differs from traditional discrimination training in a number of ways:

- There is no peak shift: i.e., gradual biases in discriminations away from the target stimulus do not occur.
- There is a large reduction of errors compared to traditional discrimination learning (~ 45 times fewer errors) (Terrace 1963a, b).
- EDT is based on excitatory conditioning only. In other words, mistakes (errors) are not required for learning. Thus, there are no negative emotions during training (e.g., frustration, helplessness, stress, anxiety, etc.), so the training does not become aversive.

In traditional discrimination training methods (TDT henceforth), a target odour is presented (S+) and a non-target odour is presented (S−). Presentations of S+ and S− can be simultaneous or sequential. Responses to S+ are always rewarded, at least initially, and responses to S− are never rewarded. EDT's procedure is simpler, yet somewhat counterintuitive for most people familiar with TDT: initially, only S+ is presented, and a response specific to S+ (e.g., "sit") is rewarded. Soon after the beginning of the training, S− is presented alongside S+ but in very low concentrations. The concentration of S− is gradually and slowly increased until it is at the same level as S+. This process is called "fading-in". Theoretically, dogs learn to ignore S−, and to only responded to S+. As mentioned above, errors made by animals with this type of training are radically reduced compared to TDT. The main drawback of this technique is that transfer of learning (during re-training) may be longer and more difficult for the dog. EDT dogs are more likely to become a "one-smell-dog" than TDT dogs, but they are very accurate for that one smell and the learned behaviour is very resistant to extinction.

The problem with EDT is to find a way to deliver the stimuli with a fading-in procedure in place. Olfactometers could be used, but we decided to explore a simpler and cheaper method. Our training used the Ribbon Snake odour (S+: a swab of a live animal). Part of the training takes place with the scent of other snake species or distractor odours (S−: food, dog smell, Common Garter Snake odour). A small rectangular aquarium is used to present the odours. Each sample (S+ and S−) sits at the bottom of the aquarium. Two containers (baby food jars or stainless steel spice containers) are placed in the aquarium, above water level, one containing the S+ (Ribbon Snake smell), the other the S− (e.g., Garter Snake smell). The location (left or right) of the S+ is randomly determined between trials. The Ribbon Snake sample is placed directly in the well, but the S− smell is contained in an ice cube that gradually melts, increasing ever so gradually the saliency of the smell. Since we do not control the presentation of the smell in discrete presentations, the dog is taken in and out of the testing room. In our case, ice cubes would take between 4 and 5 h to melt completely. Dogs are then invited to visit the experimental room and are rewarded (e.g. with play, food, or praise) for any attention paid to the S+ and simply ignored for any time spent sniffing the S−. The saliency of the S− (and the S+) can be increased towards the end of the exposures or accelerated overall to reduce the length of the session by using a hot plate or other heating device. The aquarium has a cover and multiple small holes allowing the scent to diffuse out of it. Time spent smelling the target side of the box is measured (and as mentioned above, counterbalanced between trials to avoid lateral preferences and learning) and rewarded. Within 3 to 5 h (or depending on the volume of the frozen sample), the dog typically sniffs only the S+ and ignores the S−. This type of learning can be transferred and generalized to procedures requiring the shaping of a more explicit response (e.g., nose poking and holding for a few seconds at the target stimulus). Our experience with this technique so far has been promising, although we often run into the problem of having a dog perfectly able to distinguish between the S+ and S−, but unable to then quickly learn a desired operant response to signal the presence of a target (unpublished data).

1.4.3 Line-Ups: Memory Load Issues in Scent Processing

The line-up is another very common method in olfactory discrimination and identification. It is traditionally used by the police forces and the military and has become common in many modern applications for training as well as experimental and biomedical trials (e.g., cancer detection). For example, Schoon and Haak (2002) wrote what is often considered the standard for forensic work with canines. Their use of line-ups in the context of a matching-to-sample problem has many potential applications. In their case, they describe the training of dogs to match objects found at a crime scene with a potential perpetrator. At first glance, the concept seems to make sense as dogs act here as witnesses. Following discussions with a colleague at Dalhousie University, Dr. John Christie, a cognitive psychologist, we realized that the dogs may be "expert witnesses", but they are not comparable to expert witnesses participating in a photo or person line-up as they were not at the crime scene. In other words, we are not testing their memory of a past event, but we are instead testing their sensory-perceptual matching of the odour of a person, and the odour of an object with which they were in contact. So the question then becomes: Why make this task a memory task?

In the early 1970s, a former mentor of one of us (SG), Werner Honig, introduced into animal learning, from human cognitive psychology, the concept of working memory. Honig was by training a traditional operant conditioning researcher who edited seminal books in the field (Honig 1966; Honig and Staddon 1977). Growing increasingly dissatisfied with a "pure", traditional approach to instrumental conditioning, Honig became receptive to the memory literature in humans and applied many of the theoretical and conceptual foundations of human cognition to animal learning (he contributed a number of edited books on the topic, e.g., Honig and James 1971; Hulse et al. 1978; Honig and Fetterman 1992, and many articles addressing working memory in pigeons: Honig 1978, 1981, 1984). Although his work was never applied to canines and was largely restricted to pigeons (a traditional model in classical behaviourism), the concept of working memory in animals is now largely accepted in animal cognition textbooks. The basic idea is that information presented to animals may stay in memory for a short period of time, for the time necessary to complete a task. Working memory, as a specific type of short term memory, is prone to interference, and tends to fade rapidly when not in use.

Let us go back to the line-up: Remember that a line-up is a special case of matching-to-sample—the sample being the "cue" given to the dog, i.e., what to find. The general procedure in a formal test of "expert canine testimony" would be to present the dog with the smell of a suspect, or sample (e.g., sweat sample on a cotton ball) and ask the dog to walk a line-up of containers containing objects, one of which could be the potential target. Conversely, the object found at the crime scene could be the sample, and a line-up of sweat samples from different people (including the suspect) could be available for investigation. Note that the most standard procedure includes six containers with samples to sniff: one target and five foils or distractors. The position of the target is typically randomly determined by throwing a die. Note also that dogs are typically trained to not respond to blank line-ups, so they should know that a "no target" condition is possible, thus reducing false alarms. In addition, they should be trained to identify two instances of the same target in the same line-up to encourage them to complete the sampling of the line-up even if the target is in an early position.

Table 1.3 Sequential position of target and performance of two expert detection dogs in a line-up procedure

	Position 1	Position 2	Position 3	Position 4	Position 5	Position 6
Dog 1	97.7 %	93.7 %	90.6 %	82.7 %	28.5 %	10.5 %
Dog 2	97.7 %	90.6 %	93.7 %	79.3 %	35.7 %	13.1 %
# trials	45	32	32	29	28	38

The working memory issue is easy to miss. Taking into consideration that the dog is not an actual witness of a crime trying to remember information about a crime scene, it is puzzling that we would increase interference with such a procedure. Added interference comes from the information in working memory after the inspection of each station containing a sample to inspect. Since this is done sequentially in one scan of all six containers, you can imagine that by the time the dog inspects the fifth and sixth samples, it may not remember the characteristics of the cue or initial sample.

We decided to put this hypothesis to the test with a mini-experiment with two dogs that were considered experts at line-ups and were of equal overall performance (80–100 %, averaging 90 %). Dogs were trained on diluted essential oils and in this case, we identified a specific target (lavender oil) to find in the line-up. We measured the accuracy of the dogs based on the position of the target in the line-up. The position of the target was randomly determined by throwing a die over 200 trials. As expected, if the target was early or close in the line-up (positions 1 and 2), accuracy was high (>90 %). But if the target was late or far in the line-up (positions 5 and 6), accuracy dropped significantly (<40 %) (Table 1.3). To go back to the forensic example with canine "expert witnesses", imagine now that the target associated with a perpetrator has been randomly assigned to position 6. The dogs have less than a 15 % chance of making an accurate match. Note also that we are not working with degraded stimuli. In fact, even if significantly diluted, essential oils are so strong and salient that we stopped using them a few years ago in favour of tea and other infusions in the early training phases of detection-, discrimination, or matching dogs.

After realizing this pattern was common when using line-ups, and since we are not interested in mnemonic performance but rather psychophysical accuracy at the sensory and perceptual level, we started using simpler tasks with a reduced number of potential choices. Simplicity is our best ally in those situations. If possible, using simple designs and procedures (e.g., go/no-go, a single-scent task, or 2AFC or 3AFC if more than one operant response is necessary) is preferable. In other words, as discussed in Lit (2009), reducing the cognitive demand of the task will accelerate learning and likely increase performance. By analogy, in the case of the line-up, making sure that the matching-to-sample is simultaneous (i.e., the options are made available immediately, as opposed to a delayed matching-to-sample), and a small number of options is offered, would help performance. In other words, the spatial and temporal contiguities should be such that the task does not tax memory processing resources, but only sensory-perceptual processing resources.

1.4.4 Remote Scenting: Attentional Load Issues in Scent Processing

Sometimes the presence of the dog "on site", especially in applied settings, is not desirable or even safe. Field conditions are sometimes too hazardous for the dogs to be present (land mine detection dogs can get injured or killed in the field), sample collection of rare species sign with wildlife conservation canines may be infrequent and occur over large expanses of terrain or in remote and difficult to access areas (collecting scats and identifying potential latrine sites for Eastern Cougars), or the dog is simply not able to easily work on premises (e.g., in clinics and hospitals for diagnostic detection or in interactions with patients). In those examples, practical reasons would normally exclude dogs from being part of the detection, searching, tracking, etc. Remote-scenting protocols were developed with these cases in mind, and an important historical first case was the development of Remote Explosive Scent Tracing (REST) with land mine detection in mind (Fjellanger et al. 2002; McLean et al. 2003; see Helton 2009a, b for a short review). One other clear advantage of this method addresses the issue of attention and an attempt with remote scenting to reduce attention demands by having the dogs work in controlled, consistent, and familiar indoor conditions. Microclimatic and micrometeorological conditions (temperature, humidity, air movements) can be controlled and delivery methods can be designed to optimize scent perception.

1.4.5 Ethological Approaches and Future Lines of Investigation

The study of canine olfactory psychophysics and learning has certainly benefited from the tradition of behaviourism and behaviour analysis, including applied behaviour analysis. Indeed, applied research with sniffer dogs seems to be the main impetus for funding and research opportunities in olfactory cognition, especially if an experimental approach is favoured. Unfortunately, there is an immense gap in the study of olfaction in canines: The true ethological field approach when looking into mid- to long-range tracking and trailing processes. Potential approaches to this may include the use of optical tracking methods (as used in path integration in canines by Séguinot et al. 1998) and GPS/GIS technology, when the technology becomes less expensive and more accessible.

A clear understanding and deep knowledge of animal learning and cognition is without a doubt useful to the development of experimental programs seeking to understand the fundamental and applied dimensions of canine olfaction.

1.5 Medical Detection and Assistance Canines: Cancer, Diabetes, and Epilepsy

The challenges of applied research are numerous. Although not our primary applied research area, we will present a quick review of a fascinating emergent area of biomedical research involving the training of dogs for detection of and assistance for medical problems.

The relationship between owner and pet can be a very fulfilling and rewarding experience. Recent advances in the field of health research suggest that dogs may soon be more than just our pets and that the nature of our relationship with them could change drastically.

1.5.1 Cancer Detection

The first evidence of a dog's ability to detect disease came from the well-known report by Williams and Pembroke (1989) in the Lancet, in which a woman sought medical attention after her dog persistently sniffed a mole on her leg. Upon clinical examination, the spot was discovered to be malignant melanoma. What was it the dog smelled that interested it so much? Advances in technology suggest that diseases such as cancer likely have a "signature scent", characterized by the volatile organic compounds (VOCs) being released (Szulejko et al. 2010).

In the 1970s, Linus Pauling et al. (1971) found 200 VOCs in exhaled human breath. Since then, more than 3400 VOCs have been documented in human breath (Phillips et al. 1999). VOCs released in the breath, urine, and tissues may provide a window into biological processes, with certain biological markers indicating specific medical conditions (Bijland et al. 2013; Buszewski et al. 2012; Miekish et al. 2004; Szulejko et al. 2010).

In 1999, Phillips et al. obtained breath samples from 50 healthy individuals and analyzed the biological components of the samples using an analytical technique called Gas Chromatography—Mass Spectrometry. The analysis revealed a total of over 3400 different VOCs. Importantly though, each individual breath sample was found to have an average of 204.2 VOCs, and only 27 VOCs were found to be present in every sample. This demonstrates a huge level of variability between individuals' breath samples. Such variation is likely due to differences in diet, drugs, medication, metabolism, and health status, to name a few factors (Phillips et al. 1999).

Can dogs detect differences in the VOCs being emitted by their owners? Can they even smell the VOCs? Evidence would suggest that this is the case. The burgeoning science of canine olfaction has elucidated just how sensitive the nose of a dog is. For example, using an olfactometer, Waggoner et al. (1998) showed that dogs were able to detect a target odour present in one part per billion in the presence of a distracting odour at a concentration of twenty parts per million, and

Pearsall and Verbruggen (1982) reported that dogs can smell some odours at one part per trillion.

The ability of dogs to sniff out melanoma was empirically tested by Pickel et al. (2004). Using two well-trained dogs, one of which was experienced in cancer detection, Pickel et al. first confirmed the dogs' ability to detect melanoma tissue using a variety of search tasks. Researchers then placed between 8 and 30 adhesive bandages on human participants, one of which covered the site of the cancerous tissue. The first dog successfully identified the correct bandage on 6 out of 7 patients, while the second dog was successful in sniffing out melanoma on 3 out of 4 patients.

Horvath et al. (2008) showed that a naive Riesenschnauzer (neither the dog's previous level of training, or experience in olfactory detection was mentioned) was able to detect ovarian cancer from cancerous ovary tissues with 100 % sensitivity and 97.5 % specificity. Moreover, control tissues in this study included abdominal fat, muscle, small bowel tissue, healthy postmenopausal ovarian tissue, and some tissues from an area just adjacent to the tumor. Given the impressive results, this study serves as excellent evidence that cancerous cells have a distinct odor that is reportedly detectable by dogs.

Detection of cancer in urine and fecal samples has yielded more conflicting results. Willis et al. (2004) reported that after a seven month training period, dogs with no prior experience could detect bladder cancer in urine samples at 41 % accuracy (as compared to 14 % expected by chance). Using urine samples to test dogs' ability to detect breast or prostate cancer, Gordon et al. (2008) had ten dogs trained by professional dog trainers. The reported successful detection rate for breast cancer was 22 % and only 18 % for prostate cancer.

More promisingly, Cornu et al. (2011) were able to train a Belgian Malinois with no prior experience to detect prostate cancer from urine samples with a sensitivity and specificity of 91 % in a period of 24 months. Furthermore, Sonoda et al. (2011) tested a trained cancer-detection Labrador retriever's ability to detect colorectal cancer in both breath and watery stool samples from patients with different stages of colorectal cancer. Control samples were obtained from patients with other colorectal conditions such as chronic inflammatory disease and various forms of colitis. The dog was able to detect colorectal cancer in the breath samples with 91 % sensitivity and 99 % specificity, and in the watery stool samples with 97 % sensitivity and 99 % specificity.

Still, based on the available literature, it would appear that testing the detection of cancer by dogs is more consistently successful with the use of breath samples. McCulloch et al. (2006) used a three-phase training program that spanned only a couple of weeks to train naive sniffing dogs to detect lung and breast cancer from breath samples. Testing revealed the dogs' ability to detect breast cancer with a specificity of 98 and 88 % sensitivity, and lung cancer with 99 % specificity and sensitivity.

Ehmann et al. (2012) had four family dogs trained by professional dog trainers to detect lung cancer from breath samples (no mention of length of training). In this study, sample controls were from patients with a non-malignant lung disease.

Here, dogs were reported to successfully detect lung cancer at a sensitivity of 72 % and specificity of 94 %.

Empirical studies of the ability of dogs to detect cancer are still in their infancy. Inconsistent findings are likely the result of differing training programs and sample collection techniques (Moser and McCulloch 2010), as well as breed-specific behavioural profiles as suggested by our dopamine hypothesis (Sect. 1.2). However, given the reported ability of dogs to successfully detect cancer despite potential confounds and biologically comparable control stimuli (Ehmann et al. 2012; Horvath et al. 2008), these studies provide an extremely promising and intriguing area of study that warrants much further investigation.

1.5.2 Diabetes Detection

Anecdotal evidence suggests that dogs (and cats) may be able to prevent health complications in individuals with insulin-dependent diabetes by signaling impending hypoglycemic events in their owners (Chen et al. 2000; Wells et al. 2008, 2011). In a series of case studies, anecdotal evidence from individuals with diabetes suggests that their dogs were aware of fluctuations in their blood sugar levels before they experienced symptoms. Furthermore, some dogs woke owners during the night, and one even signaled through a closed bedroom door (Chen et al. 2000).

Although there are companies claiming to train hypoglycemia detection dogs (e.g., CARES, Canine Assistance Rehabilitation Education and Services), there are currently no empirical studies confirming this ability in dogs, and there is only speculation as to what the dog is detecting before a hypoglycemic event. Researchers hypothesize dogs may be using olfactory cues such as a change in the chemical composition of their owners' sweat (sweating is a common symptom of hypoglycemia), or that signaling dogs are acutely aware of the behavioural changes accompanying hypoglycemia in their owners (Wells et al. 2008).

In an attempt to elucidate the mechanism with which dogs may detect hypoglycemia, our Canid Behaviour Research Team at Dalhousie University has teamed-up with colleagues at the IWK Health Centre in Halifax, Drs. Elizabeth Cummings and Elizabeth McLaughlin. We began in early 2013 a series of projects to examine potential biochemical routes of detection. Our dogs are selected based on motivation levels and their performance in detecting low-saliency stimuli. Those dogs that are selected are tested on their ability to detect glycemic changes in breath, sweat, and saliva samples from individuals with Type 1 diabetes. The dogs are presented with a forced choice task that requires them to match hypoglycemic samples with hypoglycemic samples, in the presence of normoglycemic samples from the same individual. Successful matching would indicate the dog's ability to discriminate between glycemic levels from samples in vitro, in the absence of the actual patient. Although the project is still very young, preliminary results are inconclusive. In the future, we would like to test the hypothesis that what dogs may be detecting in their owner is actually a myriad of physiological and behavioural changes, therefore detection would only be possible in vivo. If this is found to be the case, we would also like to test the idea that hypoglycemia-detection dogs are responding to a generalized stress response. Anecdotal evidence may suggest that trained hypoglycemia detection dogs respond to a variety of biologically stressful events (e.g., asthma attack) in their owners and others and this may be an indication that physiological stress markers (e.g., increase in adrenalin, cortisol levels, etc.) may be detected, and not glycemic VOCs per se (e.g., personal communication, Sarah Holbert of CARES, March 2013).

1.5.3 Seizure Alert Dogs

Until recently, only anecdotal reports of dogs signalling oncoming seizures in their owners existed. However, recent empirical evidence has shown that dogs can indeed detect seizures and can be trained to do so reliably (Kirton et al. 2008, and for reviews see Brown and Goldstein 2011; Dalziel et al. 2003).

Strong et al. (1999) successfully trained six dogs to anticipate and signal an impending seizure in a family member. Following a training period of six months, the dogs learned to associate seizures with pleasurable events and consistently signalled 15–45 min before a seizure. An unexpected result was that owners reported a reduction in the frequency of seizures. Therefore, in 2002, Strong et al. examined this directly by following epileptic patients 24 weeks after acquiring a trained seizure detection dog. As reported by patients in the 1999 study, a reduction in the frequency of seizures in almost all patients was observed (9 out of 10 patients, mean reduction of 43 %). Given that seizures are often preceded by anxiety (Betts 1981) and that owners of seizure detection dogs have reported increases in well-being (Kersting et al. 2009), it is possible that owning a trained seizure detection dog provides feelings of comfort and safety (as discussed above), thereby reducing anxiety and as a result, seizure frequency.

As with hypoglycemia detection dogs, it is not known what signals from the owner alert the dog before a seizure. However, in this case, researchers appear confident that trained dogs are recognizing and responding to minute changes in the behaviour of their owner (Brown and Strong 2001), but detection of physiological changes cannot be ruled out without further investigation (Wells 2007).

1.5.4 Where to Go From Here?

Canine detection of disease and the use of dogs as assistance dogs is an extremely intriguing field. Evidence suggests that dogs can be trained to detect different forms of cancer using olfactory cues from multiple biological channels. Testing the VOCs in human biological samples provides an interesting alternative to current screening methods for cancer. Depending on the cancer being tested for, current screening techniques can be expensive, inaccurate, increase exposure to radiation, and can result in unnecessary biopsies (Jett 2005; Gotzche and Nielsen 2006). The ability of dogs to detect impending hypoglycemic events in diabetic owners is a phenomenon that merits further study. Based on the successful training of seizure detection dogs, there is reason to believe that the validity of hypoglycemia-detection dogs may be empirically confirmed in the near future. Taken together, the literature presented here suggests that in the future we will see dogs not only as human's best friend, but as our partners in health care, providing detection of, assistance for, and treatment of disease.

1.6 Human-Canine Sensory Symbiosis and Appeal for a Renewed (Situated) Science of Canine Olfaction

Much of our applied work at the Canid Behaviour Research Lab at Dalhousie University is based on wild canid research and the use of sniffer dogs as "wildlife conservation canines", helping us find our target species (e.g., coyotes, various species-at-risk, or invasive species) in unobtrusive and non-invasive ways. Although one of us (SG) has been using dogs in this capacity since the early 1990s, it was not until a student (Flannery and Gadbois, unpublished manuscript) decided to write a literature review on the topic that we realized the potential of this association between humans as field researchers and dogs as research assistants. As Hewes (1994) discusses, the symbiosis between humans and wolves or early dogs may have been a question of survival, and the complementarity of our sensory ecologies—visual humans, olfactory wolves—may have been the start of a remarkable (mutualistic) symbiosis. This never became more obvious to me than when a few years ago, in the scenic and majestic scenery of the Cape Breton Highlands in Nova Scotia, we were looking for a pack of coyotes and a suspected moose carcass site. Our sniffer dog Zyla was air scenting to localize the coyotes, pulling in one direction, and ravens were converging in a slightly different direction (towards, we realized later, the moose carcass). I could not stop thinking about how this all made sense. Early humans would have relied on scavengers and predators to locate food, and would have quickly realized that the keen sense of smell of wolves was an asset.

As suggested and highlighted in our discussion of the biomedical (and companionship, when assistance complements the detection work) applications of canine olfaction, we are only at the beginning of the realization of the amazing potential this partnership can offer.

Acknowledgements Sincere thanks to Alexandra Horowitz for inviting us to participate in this exciting project. Thank you to all the dogs and their owners that participated in our research over the years, the many hundreds of volunteers working in the lab, Honours and graduate students (for a full list, see http://gadbois.org/simon/team.html). Simon Gadbois wants to take the opportunity to thank the many mentors who inspired him over the years, "in order of appearance": Louis Gadbois, Ward O'Neill, Marvin Krank, Werner Honig, Vincent LoLordo, John Fentress, Peter McLeod, William Moger, and Fred Harrington. You have no idea how much you all contributed to shape and focus that mind of mine and force it to always want to synthesize and keep an open mind, and like coyotes, be happy to be a generalist.

References

Allen, J. J., Bekoff, M., & Crabtree, R. L. (1999). An observational study of coyote (*Canis latrans*) scent-marking and territoriality in Yellowstone National Park. *Ethology, 105*, 289–302.
Arons, C. D., & Shoemaker, W. J. (1992). The distribution of catecholamines and beta-endorphin in the brains of three behaviorally distinct breeds of dogs and their F1 hybrids. *Brain Research, 594*, 31–39.
Bekoff, M. (2001). Observations of scent-marking and discriminating self from others by a domestic dog (*Canis familiaris*): Tales of displaced yellow snow. *Behavioural Processes, 55*(2), 75–79.
Berridge, K. C. (2001). Reward learning: Reinforcement, incentives and expectations. In D. L. Medin (Ed.), *Psychology of learning and motivation* (Vol. 40, pp. 223–278). San Diego: Academic Press.
Berridge, K. C. (2004). Motivation concepts in behavioral neuroscience. *Physiology & Behavior 81*(2), 179–209.
Berridge, K. C., & Robinson, T. E. (1998). What is the role of dopamine in reward: Hedonics, learning, or incentive salience? *Brain Research Reviews, 28*(3), 308–367.
Berridge, K. C., Robinson, T. E., & Aldridge, J. W. (2009). Dissecting components of reward: 'Liking', 'wanting', and learning. *Current Opinion in Pharmacology, 9*, 65–73.
Betts, T. (1981). Epilepsy: Questions and answers. *Nursing Mirror, 153*, 6–9.
Bijland, L. R., Bomers, M. K., & Smulders, Y. M. (2013). Smelling the diagnosis. A review on the use of scent in diagnosing disease. *The Netherlands Journal of Medicine, 71*(6), 300–307.
Blough, D., & Blough, P. (1977). Animal psychophysics. In W. K. Honig & J. E. R. Staddon (Eds.), *Handbook of operant behaviour* (pp. 514–539). Englewood Cliffs, NJ: Prentice-Hall Inc.
Blough, D. S. (1966). The study of animal sensory processes by operant methods. In W. K. Honig (Ed.), *Operant behavior: Areas of research and application* (pp. 345–379). New York, NY: Meredith Publishing Company.
Bradbury, J. W., & Vehrencamp, S. L. (2011). *Principles of animal communication* (2nd ed.). Sunderland, MA: Sinauer Associates Inc.
Brown, S. W., & Goldstein, L. H. (2011). Can seizure-alert dogs predict seizures? *Epilepsy Research, 97*, 236–242.
Brown, S. W., & Strong, V. (2001). The use of seizure-alert dogs. *Seizure, 10*, 39–41.
Buck, L. B. (2000). Smell and taste: The chemical senses. In E. R. Kandel, J. H. Shwartz, & T. M. Jessell (Eds.), *Principles of neural science* (4th ed., pp. 625–647). New York, NY: McGraw-Hill Companies.
Buszewski, B., Ligor, T., Jezierski, T., Wenda-Piesik, A., Walczak, M., & Rudnicka, J. (2012). Identification of volatile lung cancer markers by gas chromatography-mass spectrometry: Comparison with discrimination by canines. *Analytical and Bioanalytical Chemistry, 404*, 141–146.
Chen, M., Daly, M., & Williams, G. (2000). Non-invasive detection of hypoglycaemia using a novel, fully biocompatible and patient friendly alarm system. *British Medical Journal, 321*, 1565–1566.

Cornu, J. N., Cancel-Tassin, G., Ondet, V., Girardet, C., & Cussenot, O. (2011). Olfactory detection of prostate cancer by dogs sniffing urine: A step forward in early diagnosis. *European Urology, 59*, 197–201.

Dalziel, D. J., Uthman, B. M., McGorray, S. P., & Reep, R. L. (2003). Seizure-alert dogs: A review and preliminary study. *Seizure, 12*, 115–120.

Depue, R. (2000). *Neurobehavioral systems, personality and psychopathology.* New York, NY: Springer.

Deschênes, M., Moore, J., & Kleinfeld, D. (2011). Sniffing and whisking in rodents. *Current Opinion in Neurobiology, 22*, 1–8.

Diamond, A. (2006). Bootstrapping conceptual deduction using physical connection: Rethinking frontal cortex. *Trends in Cognitive Sciences, 10*, 212–218.

Diamond, A., Churchland, A., Cruess, L., & Kirkham, N. (1999). Early developments in the ability to understand the relation between stimulus and reward. *Developmental Psychology, 35*, 1507–1517.

Ehmann, R., Boedeker, E., Friedrich, U., Sagert, J., Dippon, J., Friedel, G. et al. (2012). Canine scent detection in the diagnosis of lung cancer: Revisiting a puzzling phenomenon. *European Respiratory Journal, 39*, 669–676.

Engeman, R. M., Vice, D .S., Rodriguez, D. V., Gruver, K. S., Santos, W. S., & Pitzler, M. E. (1998). Effectiveness of the detector dogs used for deterring the dispersal of Brown Tree Snakes. *Pacific Conservation Biology, 4*, 256–260.

Fentress, J. C., & Gadbois, S. (2001). The development of action sequences. In E. M. Blass (Ed.), *Handbooks of behavioral neurobiology: Developmental psychobiology, developmental neurobiology and behavioral ecology: Mechanisms and early principles* (Vol. 13, pp. 393–431). New York: Kluwer Academic Publishers.

Fjellanger, R., Andersen, E. K., & McLean, I. G. (2002). A training program for filter-search mine detection dogs. *International Journal of Comparative Psychology, 15*, 277–286.

Flannery, M., & Gadbois, S. (2013). *The use of scent detection dogs in wildlife conservation.* Manuscript in preparation.

Furton K. G., & Myers L. J. (2001). The scientific foundations and efficacy of the use of canines as chemical detectors for explosives. *Talanta, 54*(3), 487–500.

Gadbois, S. (2010). Canine behavioural neuroscience: From canine science in shackles to new opportunities. In Proceedings of the 2nd Canine Science Forum, Vienna, Austria.

Gadbois, S., Demontfaucon, M., Mousse, D., & Flannery, M. (in prep). *Ribbon Snake Conservation Canines in Kejimkujik National Park.*

Gheusi, G., Goodall, G., & Dantzer, R. (1997). Individually distinctive odours represent individual conspecifics in rats. *Animal Behaviour, 53*, 935–944.

Gordon, R. T., Schatz, C. B., Myers, L. J., Kosty, M., Gonczy, C., Kroener, J. et al. (2008). The use of canines in the detection of human cancers. *The Journal of Alternative and Complementary Medicine, 14*, 61–67.

Gotzche, P. C., & Nielsen, M. (2006). Screening for breast cancer with mammography. *Cochrane Database of Systematic Reviews, 4*, CD001877.

Gray, J. A. (1987). *The psychology of fear and stress.* New York, NY: Cambridge University Press.

Hall, N. J., Smith, D. W., Wynne, C. D. L. (2013). Training domestic dogs (Canis lupus familiaris) on a novel discrete trials odor-detection task. *Learning and Motivation, 44*(4), 218–228.

Haberly, L. B. (1998). Olfactory cortex. In G .M. Shepherd (Ed.), *The synaptic organization of the brain* (4th ed.), (pp. 377–416). New York, NY: Oxford University Press.

Helton, W. S. (2009a). Attention in dogs: Sustained attention in mine detection as case study. In W. S. Helton (Ed.), *Canine ergonomics. The science of working dogs* (pp. 83–97). Boca Raton, FL: Taylor and Francis Group.

Helton, W. S. (2009b). Overview of scent detection work. In W. S. Helton (Ed.), *Canine ergonomics. The science of working dogs* (pp. 83–97). Boca Raton, FL: Taylor and Francis Group.

Hepper, P. G., & Wells, D. L. (2005). How many footsteps do dogs need to determine the direction of an odour trail? *Chemical Senses, 30*, 291–298.

Harrington, F. H., & Asa, C. S. (2003). Wolf communication. In D. Mech & L. Boitani (Eds.), *Wolves. Behaviour, ecology, and conservation.* (pp. 66–103). Chicago, IL: University of Chicago Press.

Hewes, G. W. (1994). Evolution of human semiosis and the reading of animal tracks. In W. Nöth (Ed.), *Origins of semiosis. Sign evolution in nature and culture* (pp. 139–149). Berlin, Germany: Walter de Gruyter & Co.

Honig, W. K., & James, P. H. R. (1971). *Animal memory.* New York, NY: Academic Press.

Honig, W. K. (1978). Studies of working memory in the pigeon. In S. H. Hulse, H. Fowler, & W. K. Honig (Eds.), *Cognitive processes in animal behavior* (pp. 211–247). Hillsdale, NJ: Lawrence Erlbaum Associates.

Honig, W. K. (1966). *Operant behavior: Areas of research and application.* New York: Appleton-Century-Crofts.

Honig, W. K. (1981). Working memory and the temporal map. In N. E. Spear & R. R. Miller (Eds.), *Information processing in animals: Memory mechanisms* (pp. 167–197). Hillsdale, NJ: Lawrence Erlbaum Associates

Honig, W. K. (1984). Contributions of animal memory to the study of animal learning. In H. L. Roitblat, T. G. Bever, & H. S. Terrace (Eds.), *Animal cognition* (pp. 29–44). Hillsdale, NJ: Lawrence Erlbaum Associates.

Honig, W. K. & Fetterman, J. G. (1992). *Cognitive aspects of stimulus control.* Hillsdale, NJ: Lawrence Erlbaum Associates.

Honig, W. K. & Staddon, J. E. R. (1977). *Handbook of operant behavior.* Englewood Cliffs, NJ: Prentice-Hall.

Horowitz, A., Hecht, J., Dedrick, A. (2013). Smelling more or less: Investigating the olfactory experience of the domestic dog. Learning and motivation, 44, 207–217.

Horvath, G., Järverud, G. K., Järverud, S., & Horváth, I. (2008). Human ovarian carcinomas detected by specific odor. *Integrative Cancer Therapy, 7*(2), 76–80.

Hulse, S. H., Fowler, H., & Honig, W. K. (1978). *Cognitive processes in animal behavior.* Hillsdale, NJ: Lawrence Erlbaum Associates.

Jett, J. R. (2005). Limitations of screening for lung cancer with low-dose spiral computer tomography. *Clinical Cancer Research, 11,* 4988s–4992s.

Kepecs, A., Uchida, N., & Mainen, Z. F. (2005). The sniff as a unit of olfactory processing. *Chemical Senses, 31*, 167–179.

Kersting, E., Belényi, B., Topál, J., & Miklósi, A. (2009). Judging the effect of epilepsy-seizure alert dogs on human well-being by a self-administered questionnaire. *Journal of Veterinary Behavior, 4*(2), 84.

Kirton, A., Winter, A., Wirrel, E., & Snead, O. C. (2008). Seizure response dogs: Evaluation of a formal training program. *Epilepsy & Behaviour, 13*, 499–504.

Kringelbach, M. L., & Berridge, K. C. (2009). Towards a functional neuroanatomy of pleasure and happiness. *Trends in Cognitive Sciences, 13*, 479–487.

Lit, L. (2009). Evaluating learning tasks commonly applied in detection dog training. In W. S. Helton (Ed.), *Canine ergonomics. The science of working dogs* (pp. 99–114). Boca Raton, FL: Taylor and Francis Group.

MacLean, P. D. (1990). *The triune brain in evolution: Role in paleocerebral functions.* New York: Plenum Press.

Macmillan, N. A., & Creelman, C. D. (2005). *Detection theory. A user's guide* (2nd ed.). Mahwah, NJ: Lawrence Erlbaum Associates, Inc.

Mainland, J., & Sobel, N. (2006). The sniff is part of the olfactory percept. *Chemical Senses, 31*, 181–196.

McCulloch, M., Jezierski, T., Broffman, M., Hubbard, A., Turner, K., & Janecki, T. (2006). Diagnostic accuracy of canine scent detection in early- and late-stage lung and breast cancers. *Integrative Cancer Therapies, 5*(1), 30–39.

McLean, I. G., Bach, H., Fjellanger, R., & Akerblom, C. (2003). Bringing the minefield to the detector: Updating the REST concept. *Proceedings of EUDEM2-SCOT, 1*, 156–161.

McNicol, D. (2005). *A primer of signal detection theory*. Mahwah, NJ: Lawrence Erlbaum Associates, Inc.

Menini, A. (2009). *The neurobiology of olfaction*. Boca Raton, FL: CRC Press.

Miekish, W., Schubert, J. K., & Noeldge-Schomburg, G. F. E. (2004). Diagnostic potential of breath analysis—focus on volatile organic compounds. *Clinica Chimica Acta, 347*, 25–39.

Moser, E., & McCulloch, M. (2010). Canine scent detection of human cancers: A review of methods and accuracy. *Journal of Veterinary Behaviour, 5*, 145–152.

Overman, W. H. (1990). Performance on traditional matching to sample, non-matching to sample, and object discrimination tasks by 12–32-month-old children. In A. Diamond (Ed.), *The development and neural bases of higher cognitive functions, annals of the New York academy of sciences* (Vol. 608, pp. 365–393). New York, NY: New York Academy of Sciences.

Panksepp, J., & Biven, L. (2012). *The archaeology of mind: Neuroevolutionary origins of human emotions*. New York, NY: W.W. Norton.

Panksepp, J. (1998). *Affective neuroscience. The foundations of human and animal emotions*. New York, NY: Oxford University Press.

Pastore, R. E., Crawley, E. J., Berens, M. S., & Skelley, M. A. (2003). "Nonparametric" A' and other modern misconceptions about signal detection theory. *Psychonomic Bulletin & Review, 10*(3), 556–569.

Pauling, L., Robinson, A. B., Teranishi, R., & Cary, P. (1971). Quantitative analysis of urine vapor and breath by gas–liquid partition chromatography. *Proceedings of the National Academy of Science, 68*, 2374–2376.

Pearsall, M. D., & Verbruggen, H. (1982). *Scent. Training to track, search, and rescue*. Loveland, CO: Alpine Publications.

Phillips, M., Herrera, J., Krishnan, S., Zain, M., Greenberg, J., & Cataneo, R. N. (1999). Variation in volatile organic compounds in the breath of normal humans. *Journal of Chromatography B, 729*, 75–88.

Pickel, D., Manucy, G. P., & Walker, D. B. (2004). Evidence for canine olfactory detection of melanoma. *Applied Animal Behaviour Science, 89*, 107–116.

Premack, D. (1983). The codes of man and beasts. *Behavioral and Brain Sciences, 6*(1), 125–137.

Price, J. L. (2003). The olfactory system. In: G. Paxinos (Ed.), *The human nervous system* (2nd ed)., (pp. 1198–1212). San Diego, CA: Elsevier Academic Press.

Schneider, G. E. (1969). Two visual systems. *Science, 163*(3870), 895–902.

Schoon, G. A., & Haak, R. (2002). *K9 suspect discrimination: Training and practicing scent identification line-ups*. Calgary, Alberta: Detselig Enterprises.

Sebeok, T. A. (1968). *Animal Communication: Techniques of study and results of research*. Bloomington, IN: Indiana University Press

Sebeok, T. A. (1977). *How animals communicate*. Bloomington, IN: Indiana University Press.

Séguinot, V., Cattet, J., & Benhamou, S. (1998). Path integration in dogs. *Animal Behaviour, 55*, 787–797.

Shepherd, G. M. (1994). *Neurobiology* (3rd ed.). New York, NY: Oxford University Press.

Slotnick, B., & Schellinck, H. (2002). Methods in olfactory research with rodents. In S. A. Simon & M. Nicolelis (Eds.), *Frontiers and methods in chemosenses* (pp. 21–61). Boca Raton, FL: CRC Press.

Smith, D. A., Ralls, K., Hurt, A., Adams, B., Parker, M., Davenport, B., et al. (2003). Detection and accuracy rates of dogs trained to find scats of San Joaquin kit foxes (*Vulpes macrotis mutica*). *Animal Conservation, 6*, 339–346.

Sobel, N., Prabhakaran, V., Desmond, J. E., Glover, G. H., Goode, R. L., Sulliva, E. V., et al. (1998). Sniffing and smelling: Separate subsystems in the human olfactory cortex. *Nature, 392*, 282–286.

Sonoda, H., et al. (2011). Colorectal cancer screening with odour material by canine scent detection. *Gut, 60*, 814–819.

Steen, J. B., & Wilson, E. (1990). How do dogs determine the direction of tracks? *Acta Physiologica Scandinavica, 139*(4), 531–534

Strong, V., Brown, S., & Walker, R. (1999). Seizure-alert dogs - fact or fiction? *Seizure, 8*, 62–65.

Strong, V., Brown, S., Huyton, M., & Coyle, H. (2002). Effect of trained Seizure Alert Dogs® on frequency of tonic-clonic seizures. *Seizure, 11*, 402–405.

Szulejko, J. R., McCulloch, M., Jackson, J., McKee, D. L., Walker, J. C., & Touradj, S. (2010). Evidence for cancer biomarkers in exhaled breath. *IEEE Sensors Journal, 10*(1), 185–210

Terrace, H. S. (1963a). Discrimination learning with and without errors. *Journal of Experimental Analysis of Behavior, 6*, 1–27.

Terrace, H. S. (1963b). Errorless transfer of a discrimination across two continua. *Journal of Experimental Analysis of Behavior, 6*, 223–232.

Terrace, H. S. (1964). Wavelength generalization after discrimination learning with and without errors. *Science, 144*, 78–80.

Terrace, H. S. (1966). Stimulus control. In W. K. Honig (Ed.), *Operant behavior: Areas of research and application* (pp. 271–344). New York: Appleton-Century-Croft.

Thesen, A., Steen, J. B., & Doving, K. B. (1993). Behaviour of dogs during olfactory tracking. *Journal of Experimental Biology, 180*, 247–251.

Vaché, M., Ferron, J., & Gouat, P. (2001). The ability of Red Squirrels (*Tamiasciurus hudsonicus*) to discriminate conspecific olfactory signatures. *Canadian Journal of Zoology, 79*, 1296–1300.

Waggoner, L. P., Jones, M., Williams, M., Johnston, J. M., Edge, C., & Petrousky, J. A. (1998). Effects of extraneous odors on canine detection. *SPIE Proceedings, 2575*, 355–362.

Wells, D. (2007). Domestic dogs and human health: An overview. *British Journal of Health Psychology, 12*, 145–156.

Wells, D. L., & Hepper, P. G. (2003). Directional tracking in the domestic dog, *Canis familiaris*, *84*(4), 297–305.

Wells, D. L., Lawson, S. W., & Siriwardena, A. N. (2008). Canine responses to hypoglycemia in patients with Type 1 Diabetes. *The Journal of Alternative and Complementary Medicine, 14*(10), 1235–1241.

Wells, D. L., Lawson, S. W., & Siriwardena, A. N. (2011). Feline responses to hypoglycemia in people with Type 1 Diabetes. *The Journal of Alternative and Complementary Medicine, 17*(2), 99–100.

Wells, M. C., & Bekoff, M. (1981). An observational study of scent-marking in coyotes, *Canis latrans*. *Animal Behaviour, 29*(2), 332–250.

Williams, H., & Pembroke, A. (1989). Sniffer dogs in the melanoma clinic? *Lancet, 333*(8640), 734.

Willis, C. M., Church, S. M., Guest, C. M., Cook, W. A., McCarthy, N., Bransbury, A. J., et al. (2004). Olfactory detection of human bladder cancer by dogs: Proof of principle study. *British Medical Journal, 329*, 712–714.

Wilson, D. A., & Sullivan, R. M. (2011). Cortical Processing of Odor Objects. *Neuron 72*, 506–519.

Wilson, D. A. & Stevenson, R. J. (2006). *Learning to smell: Olfactory perception from neurobiology to behavior*. Baltimore: John Hopkins University Press.

Zelano, C., & Sobel, N. (2005). Humans as an Animal Model for Systems-Level Organization of Olfaction. *Neuron, 48*, 431–454.

Chapter 2
Dog Breeds and Their Behavior

James A. Serpell and Deborah L. Duffy

Abstract Domestic dogs display an extraordinary level of phenotypic diversity in morphology and behavior. Furthermore, due to breeding practices introduced during the nineteenth century, these phenotypic traits have become relatively 'fixed' within breeds, allowing biologists to obtain unique insights regarding the genetic bases of behavioral diversity, and the effects of domestication and artificial selection on temperament. Here we explore differences in behavior among the 30 most popular dog breeds registered with the American Kennel Club based on owner responses to a standardized and validated behavioral questionnaire (C-BARQ©). The findings indicate that some breed-associated temperament traits (e.g. fear/anxiety) may be linked to specific gene mutations, while others may represent more general behavioral legacies of 'ancient' ancestry, physical deformity, and/or human selection for specific functional abilities. They also suggest that previous efforts to relate dog breed popularity to behavior may have failed due to the confounding effects of body size.

2.1 Introduction

Despite much speculation, and an ongoing supply of somewhat contradictory molecular and archaeological discoveries, it is still not known precisely where or when the dog (*Canis familiaris*) was first domesticated. Based on available evidence it seems likely that domestication had occurred by around 15,000 years ago, but it remains unclear where it happened and whether it was a single, isolated event or the result of multiple domestications in different parts of Europe and Asia (Larson et al. 2012; Thalman et al. 2013). Regardless of location and timing,

---

J. A. Serpell (✉) · D. L. Duffy
Center for the Interaction of Animals and Society, School of Veterinary Medicine, University of Pennsylvania, Philadelphia, PA 19104, USA
e-mail: serpell@vet.upenn.edu

however, it is likely that the process of domestication occurred in stages. Since the grey wolf (*Canis lupus*)—the putative ancestor—is typically more fearful or neophobic and potentially more aggressive towards humans than most dogs, the first stage in the process probably involved relatively intense selection for 'tameness' or docility (Coppinger and Schneider 1995). In a related canid, the domesticated silver fox (*Vulpes vulpes*), deliberate, experimental selection for tameness resulted within a few generations in dramatic reductions in human-directed fearfulness and aggression and increases in prosocial behavior, as well as a wide variety of correlated changes in physiology and morphology (Trut et al. 2009). It seems reasonable to postulate that early semi-domesticated wolves/dogs went through a similar process, and came out the other side of it looking and behaving quite different from the ancestral species (Coppinger and Schneider 1995). This new animal, the domestic dog, then experienced thousands of years of unprecedented diversification.

One of the unique things about dogs that distinguishes them from all other domestic animals is that they are, above all, products of human selection for behavior. The majority of domestic species—cows, sheep, pigs, chickens, and so on—are largely the consequence of selection for production-related traits, such as growth rates, feed conversion, muscle and/or fat mass, egg production, fur or hair quality, etc.[1] In contrast, dogs have traditionally been valued for their ability to perform an extraordinary variety of working and social roles, including that of security guards, hunting aides, beasts of burden, weapons of war, entertainers, fighters, shepherds, guides, garbage disposers, and pets, to name just a few. Thus, when we look at dogs as a whole, we see a species that has undergone a remarkably rapid adaptive radiation into a capricious ecological niche defined by the diverse instrumental and social demands of human beings. This history of adaptive radiation is to some extent still preserved in the genomes of what we now call 'breeds' of dog (Boyko et al. 2010), although the term itself is problematical and tends to mean different things to different people.

2.2 What is a Breed?

The domestic dog currently comprises a bewildering variety of different breeds—more than 400 according to some accounts—that differ dramatically in physical appearance and behavior. Judging from archaeological discoveries, early artistic representations and various written accounts, recognizable types of dogs—sight hounds, mastiffs, scent hounds, spaniels, terriers, lapdogs, and so on—have existed since ancient times (Clutton-Brock 1995). Most of these early dog types

---

[1] Although dogs have also been employed, from time to time, as food items or as a source of fiber, such uses were relatively limited and localized, and most of the types of dog developed for such purposes are now extinct (Serpell 1995).

represented *natural* breeds or *landraces* that evolved as a consequence of geographic isolation, random genetic processes such as 'founder effect' and 'drift', and local adaptation to both natural and artificial (human) selection. In general, the distinctive physical and behavioral characteristics of these natural breeds reflected their various functions within the human cultures in which they evolved. Most early accounts of dog breeds categorized them according to the various jobs they performed (Sampson and Binns 2006; Young and Bannasch 2006) and this system of classification according to 'function' is still retained by modern kennel clubs. The American Kennel Club's current breed group classification into sporting, hound, working, terrier, toy, non-sporting, herding, and miscellaneous groups provides an obvious example.

Paleolithic and Neolithic humans may have had aesthetic preferences regarding the appearance of their dogs, but it is safe to assume that they were primarily concerned with the whole package of traits. In other words, they were seeking an animal that displayed the appropriate behavior—for herding sheep, protecting property, chasing hares, or whatever the task—while also possessing the right physical attributes, be it size, speed, visual acuity, coat length and quality, and so on, to enable it to perform these tasks well. In the sense that individual dogs were probably favored with extra food or access to mates when they did their jobs effectively, and were abandoned, traded or culled when they did not, these early 'breeds' were certainly products of human selection, but the process was largely unconscious rather than goal-directed, and little effort would have been made to prevent these animals from mating with whomever they chose. This somewhat haphazard approach eventually gave rise to the so-called 'foundation' breeds from which contemporary dogs are all ultimately derived.

Modern 'purebred' dogs are an entirely different story. In current dog breeding circles, the term "breed" refers to a population of closely related animals of similar appearance that is bred and maintained from a known foundation stock through genetic isolation and deliberate selection. For any modern dog to be successfully registered as purebred, both its parents and grandparents must also have been registered members of the same breed, which means that essentially all modern dog breeds are closed breeding populations (Ostrander 2007). The idea of 'fixing' the characteristics of dog varieties by genetic isolation and inbreeding is less than 200 years old, having originated from the hobby breeding of prize-winning poultry and livestock in England during the middle of the nineteenth century (Ritvo 1987). In some cases, it is claimed that modern purebred dogs are direct descendants of ancient or foundational stock but usually the genetic evidence for continuity is shaky at best (Larson et al. 2012). In reality, the lines of descent between modern and ancestral breeds have been thoroughly obscured by the effects of arbitrary selection for unusual or extreme aspects of physical appearance combined with deliberate hybridization between existing breed types to produce new, true-breeding strains that combine the attributes of the parental lines.

Despite this uncertainty, the remarkable phenotypic variation among modern dog breeds and their effective 'fixation' though genetic isolation presents biologists with a unique opportunity to explore both the genetic bases of canine temperament and the impact of domestication and artificial selection on the evolution of behavior (Ostrander and Galibert 2006). Breed differences in behavior and temperament are also relevant to prospective dog owners who may wish to acquire an animal that is likely to be compatible with their own personalities and lifestyles.

2.3 Measuring Breed Differences in Behavior

When discussing breed differences in behavior, it is helpful to distinguish between *breed-specific* behavior patterns, such as the retriever's love of water, the pointer's 'point', or the Border collie's tendency to show 'eye', and more general breed differences in personality or temperament. The former are often viewed as unique or defining characteristics of particular breeds while the latter are considered aspects of each breed's overall character or behavioral 'style', and are often alluded to in the written standards for the breed. For example, the AKC temperament standard for the Belgian malinois states that this breed is, "confident, exhibiting neither shyness nor aggressiveness in new situations," is "reserved with strangers but is affectionate with his own people," is "naturally protective of his owner's person and property without being overly aggressive," and "possesses a strong desire to work and is quick and responsive to commands from his owner."

A variety of techniques have been used to measure these more general kinds of breed differences in behavior in dogs, including standardized behavioral tests (Scott and Fuller 1965; Svartberg and Forkman 2002; Svartberg 2006), expert opinions (Bradshaw and Goodwin 1998; Hart and Hart 1985; Hart and Miller 1985; Takeuchi and Mori 2006), and questionnaire surveys of dog owners and handlers (Duffy et al. 2008; Ley et al. 2009; Serpell and Hsu 2005; Turcsán et al. 2011). There are benefits and disadvantages to each of these approaches.

In theory, standardized behavioral tests ought to provide the most objective behavioral evaluations since they are based on direct observations of actual behavior. On the other hand, such tests are relatively laborious and time-consuming to conduct and, unless they are performed repeatedly on each dog over an extended period, the results are likely to be strongly influenced by the animal's emotional and motivational state at the time of testing (Serpell and Hsu 2001). In their pioneering early work, Scott and Fuller (1965) used a combination of behavioral observations and 'performance tests' to investigate the genetic basis for behavioral differences among five different dog breeds: basenjis, beagles, cocker spaniels, Shetland sheepdogs and wire-haired fox terriers. Puppies from each of these breeds were reared in identical conditions in order to reduce environmental influences on their behavioral development. Their performance was then evaluated at various ages on a series of standardized tests designed to measure such traits as overall emotional reactivity, response to handling and leash restraint, problem solving ability, trainability, aggressiveness, and tendency to bark. Analysis of variance by individual, litter and breed indicated strong and statistically significant effects of breed on the expression of many of these traits. Scott and Fuller then carried out hybridization experiments using some of these breeds, such as basenjis and cocker spaniels, to produce F_1 and F_2 hybrids and backcrosses to each of the parent strains. Given the basic rules of Mendelian inheritance, the ways in which the different traits segregated out in these hybrid generations then provided clues to their genetic origins. For example, they found that the hybrid puppies' responses to approach and handling by a stranger (a trait the authors labeled 'tameness') were consistent with the actions of a single dominant gene for wildness in basenjis and a corresponding recessive gene for tameness in cocker spaniels. Other traits, such as problem-solving ability, were far more complex, however, and did not reveal any clear patterns of genetic inheritance (Scott and Fuller 1965).

Svartberg (2006) studied behavioral differences in 31 breeds of dogs subjected to a standardized test known as the Dog Mentality Assessment (DMA) that comprises 10 subtests and measures four distinct canine 'personality' traits: *playfulness, curiosity/fearlessness, sociability*, and *aggressiveness*. He found statistically significant breed differences in all of these traits, *although*, surprisingly, none of these differences were related to the breeds' original functional roles based on breed groupings (e.g. herding dogs, working dogs, terriers, and gun dogs). More recently, across-breed, genome-wide association (GWAS) studies have had some success identifying chromosome regions and possible candidate genes associated with canine personality traits such as *boldness* (Jones et al. 2008; Vaysse et al. 2011).

Hart and Hart (1985) pioneered the use of expert opinion as a technique for characterizing breed differences in behavior in dogs. In this method, canine 'experts' (e.g. obedience judges and veterinarians) are asked to rank a random subset of seven common breeds on 13 separate behavioral characteristics judged to be important to a majority of dog owners: e.g. watchdog barking, snapping at children, obedience training, destructiveness, excitability, etc. The respondents' rankings are then converted into deciles, each containing five or six breeds, with the highest decile representing the most extreme expression of the behavior. Analysis of variance subsequently determined that each of the traits could be used to discriminate between breeds, although some did so more reliably than others. Other researchers have since applied the same technique to examine breed differences in other countries, and have tended to find similar rankings for the same breeds (Bradshaw and Goodwin 1998; Notari and Goodwin 2007; Takeuchi and Mori 2006). It remains unclear, however, whether agreements among experts, either within or between countries, reflect true differences in breed behavior or shared opinions based on breed stereotypes (Duffy et al. 2008).

An alternative to canvassing the views of experts is to ask dog owners to provide personality or behavioral assessments of their dogs, and then use these assessments to investigate differences among breeds. While potentially more subjective than direct observations of behavior, such assessments are arguably less susceptible to cultural stereotypes than expert opinions, and, if large sample sizes are used, the effects of individual subjective biases can be greatly reduced (Jones and Gosling 2005). Such surveys can be divided into those that focus on relatively broad, overarching personality dimensions such as *boldness, sociability*, or *extraversion*, and those that have investigated more specific phenotypic traits such as *trainability* or particular types of aggression.

Ley et al. (2009) used the Monash Canine Personality Questionnaire (MCPQ) to explore breed differences in a sample of 455 Australian dogs. The MCPQ comprises a series of 41 descriptive adjectives that loaded on five personality subscales when subjected to factor analysis (*extroversion, motivation, training focus, amicability, and neuroticism*). Although not able to compare individual breeds due to small sample sizes, they investigated personality differences across the seven breed groups recognized by the Australian National Kennel Club. Relatively few significant differences were identified: working dogs and terriers were rated as significantly more extroverted, and toy breeds less extroverted. Working dogs and gundogs were rated significant higher for *training focus*, while toys and hounds were rated as lower for this subscale. The authors also noted that dogs' scores on the *neuroticism* subscale (a measure of fear and anxiety) correlated negatively with weight and height, while scores on *amicability* (a measure of sociability) correlated positively with weight and height.

In another study, Turcsán et al. (2011) invited the owners of 5733 dogs belonging to 98 breeds to rate their dogs on four broad personality traits: *Trainability, boldness, calmness,* and *dog sociability*. The results were then used to compare breeds belonging to the conventional breed groups recognized by the AKC (functional classification) and those identified as being more closely related according to genetic analyses (Parker et al. 2004). Significant breed differences in the four personality factors were observed. The results also indicated that dogs belonging to herding breeds tended to be significantly more trainable than hounds, working dogs, toys, and non-sporting breeds, and sporting breeds were more trainable than non-sporting ones. Terriers also scored higher for *boldness* than hounds or herding dogs. Neither the *calmness* nor *dog sociability* traits were able to discriminate reliably between breed groups. With respect to the five clusters of breeds identified as genetically related, the so-called 'ancient breeds' cluster was found to be less trainable than the herding/sighthound cluster, and dog breeds in the mastiff/terrier cluster scored higher for *boldness* than those in either the ancient breed, herding/sighthound, or hunting clusters. Again, *calmness* and *dog sociability* failed to discriminate between the clusters. The authors interpreted these results as indicating that some traits such as *trainability* and *boldness* are partly determined by genetic factors and by different histories of human selection for these functional traits. The distributions of the other two personality traits seemed to bear no relation to either functional or genetic breed classifications.

Three previous studies have used the *Canine Behavioral Assessment and Research Questionnaire* (C-BARQ) to investigate breed differences in behavior. (See below for further elaboration of this method.) Serpell and Hsu (2005) invited the owners of 1563 dogs belonging to 11 common breeds to assess them on the *trainability* factor of the C-BARQ. Highly significant breed differences in *trainability* were detected. In two breeds with distinct field and show-bred lines, show-bred dogs obtained significantly lower *trainability* scores. In general, sporting dog and working dog breeds (English springer spaniel, golden retriever, Labrador retriever, poodle, rottweiler, and Shetland sheepdog) tended to obtain high scores for this factor while hounds (basset hound and dachshund), terriers (West Highland white terriers, and Yorkshire terriers), and the Siberian husky obtained low scores. The authors argued that these results were consistent with the differential effects of human selection for social cognitive skills in particular breeds, and lines within breeds, that historically tended to work in close partnership with people.

In a second study, Duffy et al. (2008) surveyed two separate samples of owners (1,521 breed club members and 3,791 pet owners) of 33 breeds of dog on the four aggression factors of the C-BARQ, and again found highly significant breed differences in behavior. Breeds that were common to both samples also ranked similarly on three of the four aggression factors. Small breeds, such as dachshunds, Chihuahuas and Jack Russell terriers, tended to obtain high scores on all or most aggression factors, while other breeds only displayed higher than average scores in specific contexts (e.g. *dog-directed aggression* in akitas and pit bulls). The authors concluded that aggression in small breed dogs was motivated primarily by fear, due to their greater vulnerability, and that owners of such breeds were also more tolerant of aggression due to the relatively limited risks of severe bites. In contrast, other breeds showed evidence of differential human selection for aggressive responses in specific contexts. The study also detected higher rates of *owner-directed aggression* in show-bred lines of English springer spaniels compared with field-bred dogs, and the reverse pattern for Labrador retrievers, a result the authors attributed to different popular sire effects in the two breeds.

More recently, McGreevy et al. (2013) explored the relationship between average C-BARQ factor and item scores and skull shape (cephalic index), body weight, and body size in a sample of 8,301 dogs belonging to 49 different breeds. A highly significant inverse relationship was detected between breed-specific body height and a large number of problematic behaviors including a range of fear/anxiety-related behaviors, *owner-directed aggression, attachment and attention-seeking*, and house-soiling when left alone. Body weight also correlated inversely with *excitability* and *hyperactivity*. These findings suggested that, across breeds, behavior tends to become more problematic as size decreases. The study also found that brachycephalic breeds tended to obtain lower scores for predatory *chasing*.

2.4 Current Study

As well as varying greatly in size, shape, and behavior, dog breeds also differ in popularity, with some breeds (e.g. Labrador retriever) maintaining rather consistent levels of popularity over time while others (e.g. Irish setter) have been subject to relatively sudden and rapid fluctuations in popularity (Herzog 2006). One possible explanation for this variation in dog breed popularity is that some breeds possess temperament traits that render them functionally better at serving as pets than others. In a recent study, Ghirlanda et al. (2013) used C-BARQ comparisons to test this hypothesis on a selection of 80 breeds of known popularity, but failed to detect any association with breed behavioral characteristics. In the current study we use breed-specific C-BARQ data to re-examine this possible relationship by focusing on the behavioral traits of the 30 most popular breeds currently registered by the American Kennel Club (AKC).

2.4.1 Methods

2.4.1.1 The Sample

Table 2.1 provides the characteristics of the sample of breeds included in the present analysis. It consists of the 30 most popular breeds registered by the American Kennel Club (AKC), including both standard and miniature or toy-size variants of two breeds: the dachshund and poodle. It should be noted that, while there are no significant differences in sex ratio between the breeds sampled, breeds do differ significantly in terms of age and percent neutered. The values of N for each breed reflect the numbers available in the C-BARQ database at the time of sampling. The online version of the C-BARQ (http://www.cbarq.org) has been freely available to dog owners since it was created in 2006. Although initially publicized via notices sent to Philadelphia area veterinary clinics and the top 20 USA breed clubs (based on AKC registrations), the availability of the survey subsequently spread by word of mouth. The current sample of owner reports is therefore self-selected. While this may be considered a potential source of bias in the data, there is no *a priori* reason to believe that such biases would affect different breeds unequally.

2.4.1.2 About the C-BARQ

The 'gold standard' of behavioral measurement is the direct, unmediated observation and recording of all instances of an animal's behavior over time (Martin and Bateson 1993). However, because most dogs in developed countries live inside people's homes where it is impractical to observe them for extended periods, it is sometimes necessary to develop different kinds of measurement techniques in order to study their behavior. The *Canine Behavioral Assessment and Research Questionnaire* (C-BARQ) is one such technique that relies on measurement by proxy (Hsu and Serpell 2003). In other words, instead of observing and measuring the animal's behavior directly, the C-BARQ records indirect behavioral information provided by the dog's owner, guardian, or handler. This approach relies on two important assumptions: first, that the dog's owner (or handler) knows more about its typical behavior than anybody else does, and second, that this knowledge of the dog's typical behavior can be extracted from the person in a form that is quantitative, reliable, and reasonably accurate. The first of these assumptions seems intuitively reasonable given that the owner lives with the dog most of the

Table 2.1 Demographic characteristics of the sample of dogs used in the study

Breed	N	% Female	% Neutered	Age (+/− SD)
Australian Shepherd	406	48.3	74.9	4.28 +/− 3.13
Beagle	188	45.7	78.7	4.80 +/− 3.33
Boston Terrier	62	37.1	85.5	3.87 +/− 2.22
Boxer	206	49.0	74.8	4.39 +/− 2.93
Bulldog	49	57.1	61.2	3.84 +/− 3.04
Cavalier King Charles Spaniel	75	52.0	77.3	3.70 +/− 2.61
Chihuahua	273	45.8	61.2	4.03 +/− 3.23
Cocker Spaniel (American)	213	47.4	78.4	4.48 +/− 3.39
Dachshund	127	42.5	78.0	5.18 +/− 3.88
Dachshund (Miniature)	78	53.8	84.6	4.47 +/− 3.61
Doberman Pinscher	314	50.6	61.5	4.37 +/− 2.94
English Springer Spaniel	138	46.4	61.6	4.73 +/− 3.29
French Bulldog	34	47.1	70.6	3.55 +/− 2.76
German Shepherd	781	51.5	66.6	4.09 +/− 3.12
German Shorthaired Pointer	70	42.9	71.4	4.61 +/− 3.31
Golden Retriever	605	49.1	71.9	4.80 +/− 3.50
Great Dane	138	52.2	77.5	3.84 +/− 2.84
Havanese	121	48.8	58.7	2.94 +/− 2.56
Labrador Retriever	1120	49.7	76.2	4.06 +/− 3.19
Maltese	109	45.0	83.5	4.43 +/− 3.52
Mastiff (English)	196	34.2	54.1	2.53 +/− 2.04
Miniature Schnauzer	119	47.9	79.8	4.91 +/− 3.30
Pembroke Welsh Corgi	85	47.1	84.7	4.69 +/− 3.41
Pomeranian	140	41.4	57.1	4.13 +/− 3.52
Poodle (Standard)	314	45.2	67.8	4.46 +/− 3.28
Poodle (Toy)	70	50.0	61.4	5.47 +/− 4.05
Pug	105	47.6	79.0	4.33 +/− 2.90
Rottweiler	416	47.8	57.7	4.01 +/− 2.80
Shetland Sheepdog	179	48.0	71.5	5.16 +/− 3.37
Shih Tzu	133	45.1	74.4	5.16 +/− 3.76
Siberian Husky	150	47.3	73.3	4.61 +/− 3.41
Yorkshire Terrier	110	43.6	74.5	5.02 +/− 4.32

time and is likely to have observed its reactions and behavior across a wide range every day circumstances. The second assumption is more speculative and requires empirical verification.

The C-BARQ currently consists of 100 questionnaire items that ask respondents to use a series of five point ordinal rating scales (from 0 to 4) to indicate their dogs' typical responses to a variety of everyday situations and stimuli during the recent past. Depending on the type of behaviour being measured, the scales rate either severity (aggression, fear/anxiety, excitability) or frequency (all other categories of behavior) (Duffy and Serpell 2012). Participants are instructed to answer all questions. However, if they are unable to answer a question because they have never observed the dog in the specified situation, they have the option to select "not observed/not applicable" and the item is then treated as a missing value. Using factor analysis, 68 of the original items were condensed into 11 behavioral factors. Two new factors (*energy* and *dog rivalry*) were added subsequently, and one of the original factors (*dog-directed aggression/fear*) was divided into two (*dog-directed aggression* and *dog-directed fear*) to create a total of 14 factors. This factor structure has been found to be remarkably consistent irrespective of breed, sex, or geographic location (Duffy et al. 2008; Hsu and Serpell 2003; Hsu and Sun 2010; Nagasawa et al. 2011; van den Berg et al. 2006, 2010). Twenty-two miscellaneous items are also included in the C-BARQ as stand-alone behavioral measures. High scores are less favourable for all items and factors with the exception of *trainability*, for which high scores are more desirable. For the purposes of analysis, each dog's factor score is calculated as the average of its scores for the questionnaire items pertaining to that factor. Breed averages are based on the sum of the individual dog factor scores divided by the value of N for each breed (See Table 2.1).

To date, the various C-BARQ factors and items have been shown to have adequate internal reliabilities (Cronbach's alpha ≥ 0.7), and acceptable test-retest and inter-rater reliabilities (Duffy et al. 2008; Duffy and Serpell 2012; Jakuba et al. 2013). Initially, 7 of the original 11 subscales were validated using a panel of 200 dogs previously diagnosed with specific behavior problems (Hsu and Serpell 2003). More recently, other studies have provided criterion validation of the C-BARQ by demonstrating associations between the various factor and item scores and, for example, training outcomes in working dogs (Duffy and Serpell 2012), the performance of dogs in various standardized behavioral tests (Barnard et al. 2012; De Meester et al. 2008; Svartberg 2005), and neurophysiological markers of canine anxiety and compulsive disorders (Vermeire et al. 2011, 2012).

2.4.2 Results

2.4.2.1 Aggression

Initial factor analysis of C-BARQ questionnaire data extracted three factors that measured different manifestations of aggression in dogs: *Stranger-directed aggression* (10 items), *owner-directed aggression* (8 items) and *dog-directed aggression* (4 items) (Hsu and Serpell 2003). As their names suggest, the first two of these factors measure aggression directed toward either unfamiliar or familiar people, respectively, while the third refers to contexts in which the dog directs aggressive threats or actions toward unknown or unfamiliar dogs. A fourth factor, *dog rivalry* (4 items), was later added to the C-BARQ to cover aggression directed toward other familiar dogs living in the same household (Duffy et al. 2008). Average scores on all of the C-BARQ aggression factors tend to be skewed toward

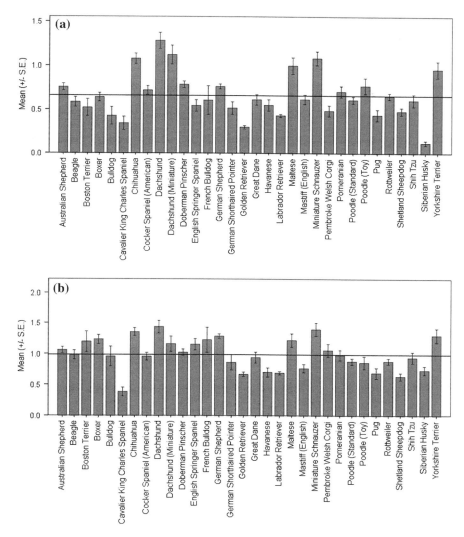

Fig. 2.1 Breed average C-BARQ scores for: **a** *stranger-directed aggression*, **b** *dog-directed aggression*, **c** *owner-directed aggression*, and **d** *dog rivalry* in the 30 most popular AKC breeds (including miniature and toy variants of the dachshund and poodle). The horizontal line represents the average score for this population of dogs

zero (especially in the case of *owner-directed aggression*), hence the expanded Y axes in the charts (Fig. 2.1a–d). This probably reflects a history of relatively intense selection against disruptive levels of aggression in dogs, particularly when directed toward human members of the same household. Despite the limited variation in the data, breed differences for all four factors are statistically highly significant (Kruskal-Wallis P values < 0.0001).

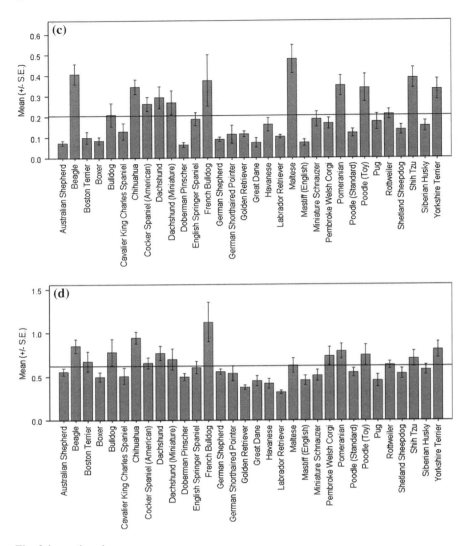

Fig. 2.1 continued

For *stranger-directed aggression*, several small or toy breeds stand out as having scores above the population average (e.g. Chihuahua, standard and miniature dachshund, Maltese, miniature schnauzer, toy poodle, and Yorkshire terrier). Two guard dog breeds, the Doberman and the German Shepherd, also obtain somewhat high scores on this factor, as does the Australian shepherd. In contrast, the bulldog, cavalier King Charles spaniel, golden retriever, Labrador retriever, pug, and Siberian husky all obtain below-average scores. A rather similar array of

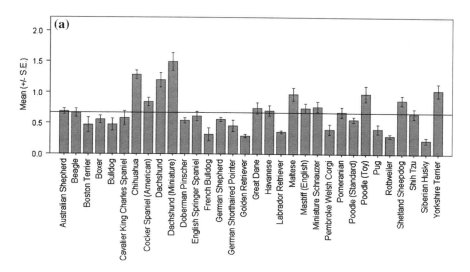

Fig. 2.2 Breed average C-BARQ scores for: **a** *stranger-directed fear*, **b** *dog-directed fear*, **c** *nonsocial fear*, **d** *separation related problems*, and **e** *touch sensitivity* in the 30 most popular AKC breeds

high and low breeds characterizes the *dog-directed aggression* factor, although in this case the boxer, English springer spaniel and French bulldog also join the ranks of the more aggressive breeds.

The *owner-directed aggression* and *dog rivalry* factors also show considerable overlap in terms of breed distribution. Again, small or toy breeds tend to score highest for these two types of aggression, particularly the beagle, Chihuahua, American cocker spaniel, both dachshunds, French bulldog, Maltese, Pomeranian, toy poodle, shi tzu and Yorkshire terrier, whereas all of the large or medium-sized breeds tend to score low on these factors.

2.4.2.2 Anxiety and Fear

Four distinct C-BARQ factors measure various dimensions of anxiety or fear. They include *stranger-directed fear* (4 items), *(unfamiliar) dog-directed fear* (4 items), *nonsocial fear* (e.g. loud noises, novel objects, etc.—6 items) and *separation-related behavior* (anxiety—8 items). A fifth factor, *touch sensitivity* (4 items), also appears to be an expression of fear in relation to being touched, handled, or groomed. Breed differences on all of these factors were highly significant (Kruskal-Wallis P value <0.0001) and, as with the aggression factors, scores for the various fear factors tend to be skewed toward zero; again presumably an effect of generations of selection in favor dogs that are less neophobic than their wild ancestor.

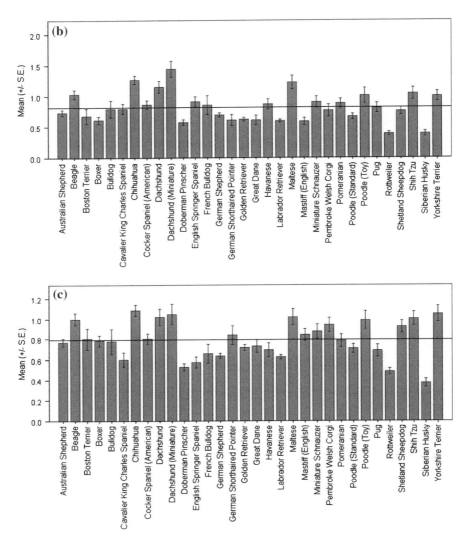

Fig. 2.2 continued

The most striking observation to be made from a cursory examination of the charts for these behaviors is the consistency with which the same group of breeds scores higher than the population average across the different contexts (Fig. 2.2a–e). Almost without exception they belong to either small or toy breeds, and, even within breeds, the dwarf or miniature versions (e.g. miniature dachshund and toy poodle) are significantly more fearful or anxious than their larger counterparts. Six of these breeds (Chihuahua, dachshund, miniature dachshund, Maltese, toy poodle, and Yorkshire terrier) score high in all five contexts, and two more (beagle and shih tzu) score high in four out of five.

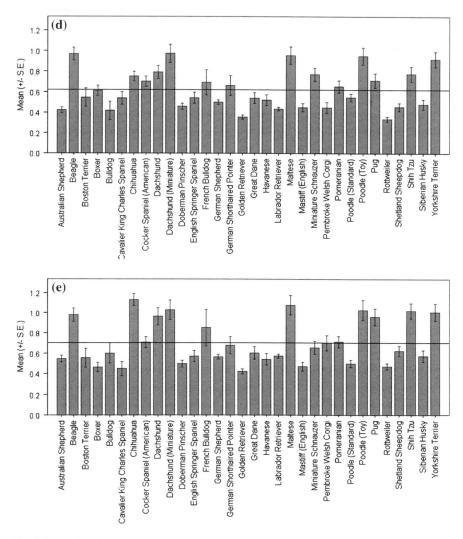

Fig. 2.2 continued

In contrast, many of the brachycephalic breeds (Boston terriers, bulldogs, French bulldogs, and pugs) tend to obtain low scores for fear, as do the golden retriever, Labrador retriever, rottweiler, and Siberian husky.

2.4.2.3 Attachment and Attention-Seeking

The C-BARQ *attachment and attention-seeking* factor comprises six questionnaire items. Dogs that score high on this factor tend to stay close to their owners, solicit

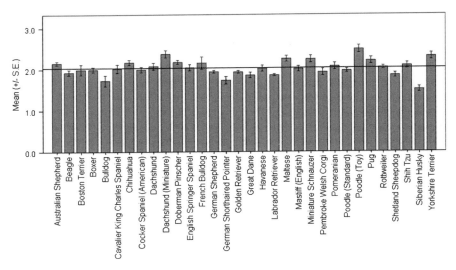

Fig. 2.3 Breed average C-BARQ scores for *attachment/attention-seeking* in the 30 most popular AKC breeds (including miniature and toy variants of the dachshund and poodle)

more affection and attention, and display agitation when the owner gives attention to third parties, such as other people or dogs (Hsu and Serpell 2003). Although breed differences for this factor are still highly significant statistically (Kruskal-Wallis P Value <0.0001), it is apparent from the chart (Fig. 2.3) that only a small number of breeds diverge markedly from the population average compared with most of the other C-BARQ factors, and only one—the Siberian husky—stands out. This breed is often described in the literature as being somewhat aloof, and its low score on this factor confirms this anecdotal observation. Interestingly, the breeds that tend to score higher for *attachment and attention-seeking* tend to be the same small or toy breeds that score high on the various aggression and fear factors (e.g. Chihuahua, miniature dachshund, Maltese, toy poodle and Yorkshire terrier). This might suggest that the *attachment/attention-seeking* behavior of these breeds is partly motivated by fear or anxiety.

2.4.2.4 Predatory Chasing

The C-BARQ *chasing* factor refers to the tendency of some dogs to display predatory chasing of cats, squirrels, birds, and/or other small animals, when given the opportunity. It comprises 4 questionnaire items. Again, there are large and highly significant breed differences in the expression of this trait (Kruskal-Wallis P <0.0001) with some breeds exhibiting high average scores (German short-haired pointer, miniature schnauzer, and Siberian husky) and others low (bulldog, Chihuahua, English mastiff, pug, and shih tzu) (see Fig. 2.4). Although the three

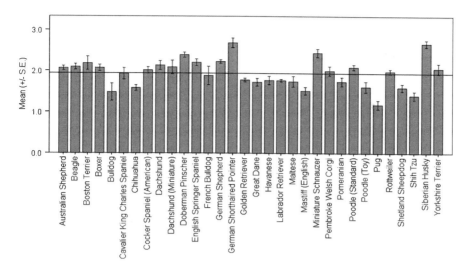

Fig. 2.4 Breed average C-BARQ scores for *chasing* in the 30 most popular AKC breeds (including miniature and toy variants of the dachshund and poodle)

highest scoring breeds have relatively little in common in other respects, all of them are known historically for their role in hunting or vermin control and their strong predatory drive. In the case of the lowest scoring breeds, anatomical and physical constraints may limit these dogs' ability to chase potential prey species, even if they possess the inclination to do so.

2.4.2.5 Excitability

Dogs that obtain a high score on the C-BARQ *excitability* factor (6 items) tend to display strong reactions to potentially exciting or arousing events, such as going for walks or car trips, doorbells ringing, the arrival of visitors, or the owner arriving home after a period of absence. Such dogs also have difficulty calming down after such events (Hsu and Serpell 2003). As with *attachment and attention-seeking*, there is a surprising degree of uniformity among the selected breeds respecting their average scores on this factor, and no particular breed or breeds stands out, although there are highly significant breed differences (Kruskal-Wallis $P < 0.0001$). The three lowest scoring breeds (bulldog, English mastiff, and Siberian husky) are well known for being relatively 'laid back', and many of the high scoring breeds belong to the same group of small or toy breeds that also tend to score high for aggression, fearfulness, and *attachment and attention-seeking* (e.g. both dachshunds, Maltese, toy poodle, and Yorkshire terrier). In addition, the Pomeranian, miniature schnauzer, and many of the brachycephalic breeds (Boston terrier, boxer, French bulldog, and pug) tend to obtain higher than average scores for *excitability* (Fig. 2.5).

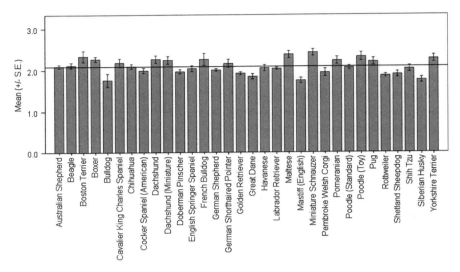

Fig. 2.5 Breed average C-BARQ scores for *excitability* in the 30 most popular AKC breeds (including miniature and toy variants of the dachshund and poodle)

2.4.2.6 Energy

The C-BARQ *energy* factor consists of just two questionnaire items: "playful, puppyish, boisterous" and "active, energetic, always on the go." Although the majority of the selected breeds show average energy levels close to the population mean for this factor, several breeds stand out (Kruskal-Wallis P <0.0001). In particular, the Australian shepherd, boxer, Doberman pinscher, German shorthaired pointer and Maltese obtain relatively high average scores for *energy*, while the bulldog, Great Dane and English mastiff obtain relatively low scores (Fig. 2.6). With the exception of the Maltese, most of the former are originally working breeds so an abundance of energy may have been selected for. Among the latter, the bulldog's various medical issues may account for its lack of energy, and the two giant breeds may also be lethargic for morphological reasons.

2.4.2.7 Trainability

Eight questionnaire items comprise the C-BARQ *trainability* factor. Dogs that score high on this factor are attentive to their owners and willing to obey basic commands, are not easily distracted, respond positively to correction, tend to be fast learners, and readily fetch or retrieve thrown objects/toys (Hsu and Serpell 2003). As with all C-BARQ traits, there are highly significant differences across the 30 most popular breeds in the expression of this factor (Kruskal-Wallis P <0.0001), with some breeds (e.g. Australian shepherd, Doberman pinscher, English springer spaniel, golden and Labrador retrievers, standard poodle,

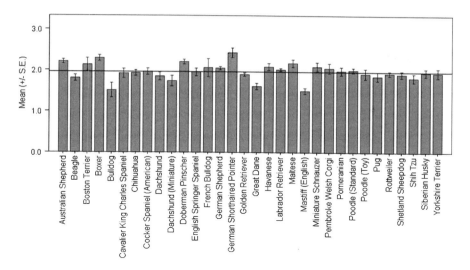

Fig. 2.6 Breed average C-BARQ scores for *energy* in the 30 most popular AKC breeds (including miniature and toy variants of the dachshund and poodle)

rottweiler and Shetland sheepdog) tending to obtain high average scores for *trainability*, while other breeds (e.g. beagle, dachshund, pug, and Yorkshire terrier) obtain relatively low scores (Fig. 2.7). In general, most of the breeds classified as belonging to the AKC's hound, toy, terrier and non-sporting groups fall below the population average for this trait, whereas breeds belonging to the sporting and herding groups tend to be above the average. Breeds in the heterogeneous "working dog" group are inconsistent, with some obtaining a high average score (rottweiler) and others a low one (Siberian husky).

2.4.2.8 Miscellaneous Problems: Persistent Barking, House Soiling, Escaping/Roaming

In addition to the main factors, the C-BARQ also comprises a number of individual questionnaire items that were retained as sources of information about relatively specific patterns of behavior. Some of these items also reveal pronounced differences in behavior across breeds. For example, the item, *"barks persistently when alarmed or excited"* reveals striking differences across breeds (Fig. 2.8, Kruskal-Wallis $P < 0.0001$). All the small or toy breeds previously associated with aggression, fearfulness, *attachment and attention seeking*, and *excitability* exhibit higher than average levels of *persistent barking*, and in this they are joined by the miniature schnauzer and Shetland sheepdog, which displays the highest level of this behavior of any of the sampled breeds. Conversely, most of the larger and brachycephalic breeds display relatively low levels of barking, and the Siberian husky rarely barks.

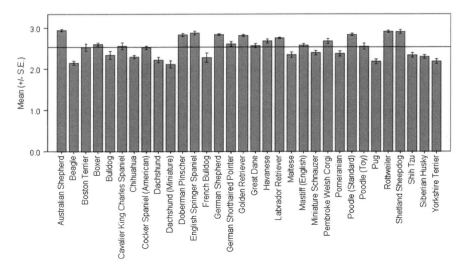

Fig. 2.7 Breed average C-BARQ scores for *trainability* in the 30 most popular AKC breeds (including miniature and toy variants of the dachshund and poodle)

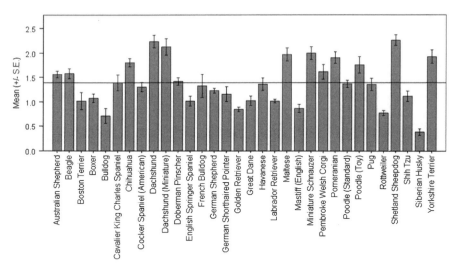

Fig. 2.8 Breed average C-BARQ scores for *persistent barking* in the 30 most popular AKC breeds (including miniature and toy variants of the dachshund and poodle)

Another C-BARQ 'miscellaneous' item, *"urinates when left alone at night or during the daytime,"* reveals an even more striking disparity between large and small breed dogs (Kruskal-Wallis P <0.0001). With the exception of the beagle and the bulldog, all of the breeds with higher then average scores for this behavior are in the toy or miniature size range. Conversely, apart from the cavalier King

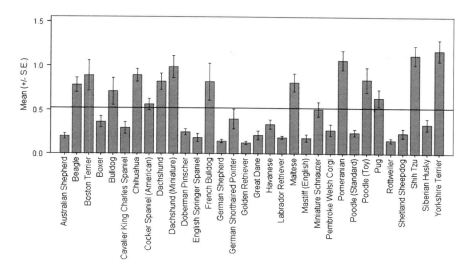

Fig. 2.9 Breed average C-BARQ scores for *urination when left alone* in the 30 most popular AKC breeds (including miniature and toy variants of the dachshund and poodle)

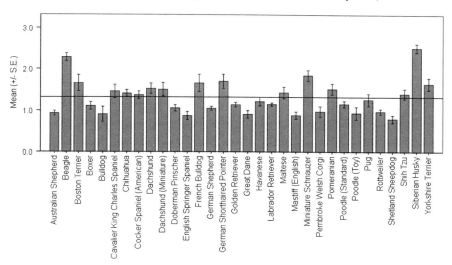

Fig. 2.10 Breed average C-BARQ scores for *escaping/roaming* in the 30 most popular AKC breeds (including miniature and toy variants of the dachshund and poodle)

Charles spaniel, Havanese and Shetland sheepdog, all of the low scoring breeds are medium to large breed dogs (Fig. 2.9).

Finally, the item, *"escapes or would escape from home or yard, given the chance,"* points to a distinctive flight risk for beagles and Siberian huskies compared with most of the other sampled breeds, as well as some risk for Boston Terriers, French bulldogs, German short-haired pointers, miniature schnauzers, and Yorkshire terriers (Fig. 2.10, Kruskal-Wallis $P < 0.0001$).

2.5 Discussion

In general respects, the observed breed differences in behavior in the present study resembled some of those obtained in previous analyses. For example, Hart and Hart (1985), Ley et al. (2009) and Turcsán et al. (2011) also found that herding and sporting dog breeds tended to obtain higher scores for *trainability* (although measured in different ways), and Ley et al. (2009) also detected positive associations between small body size and *neuroticism* (i.e. anxiety-related behavior) and less sociable personality. McGreevy et al. (2013) detected similar correlations, but this study was based on a comparable dataset to the one we used, so some overlap is inevitable. While replication of results by independent studies suggests that these apparent breed differences in behavior reflect some sort of underlying biological reality, it remains unclear the extent to which they are caused by specific genes; differential histories of human selection for functional traits; differences in the early environment, socialization and training of dogs of different breeds; and/or systematic biases in how the owners of different breeds evaluate them in behavioral surveys. Furthermore, it must be emphasized that individual variation in C-BARQ scores within breeds are often as great or greater than the differences between breeds, and this limits our ability to talk about breed-specific or breed-typical personality traits based on these kinds of measures.

Genes have certainly been demonstrated to play important roles in the expression of behavior in dogs (Scott and Fuller 1965; Mackenzie et al. 1986; Serpell 1987). Attempts to estimate the heritability of canine temperament and performance traits have typically obtained highest values for traits reflecting anxiety/fearfulness, sociability, boldness, and various forms of aggression (Goddard and Beilharz 1982; Liinamo et al. 2007; Pérez-Guisado et al. 2006; Saetre et al. 2006; Scott and Bielefelt 1976; Willis 1995; Wilson and Sundgren 1998). In the present study, a possible direct genetic affect may help to account for the consistently higher levels of fearfulness and reactivity in the small or toy breeds of dog. For example, molecular geneticists have been able to identify alleles associated with the *Insulin-like Growth Factor-1* (*IGF1*) gene that are present in all toy or miniature dog breeds, but absent from wolves and other wild canids, and very rare among large breed dogs (Sutter et al. 2007; Gray et al. 2010). Earlier studies also found significantly lower levels of circulating *IGF1* in the serum of miniature poodles compared with standard poodles, and in small breed dogs compared with larger ones (Eigenmann et al. 1984; Guler et al. 1989). More interestingly, in a study of German shorthaired pointers deliberately selected for nervousness/fearfulness,[2] an inverse linear correlation was detected between the severity of the dogs' fearful behavior and their serum *IGF1* levels (Uhde et al. 1992). This might suggest that, in addition to affecting growth and stature, the *IGF1* gene has plieotropic effects on temperament that tend to make small dogs more fearful,

---

[2] This population of nervous pointers was created by researchers during the 1960s to serve as an animal model of human anxiety disorders.

more reactive and excitable, more likely to display defensive aggression and anxious attachment, and more likely to urinate when left alone. A genetic correlation between size and temperament might make sense from an evolutionary perspective since smaller individuals are likely to be more vulnerable than larger ones. However, it is also possible that the observed correlation between small size and fearful/anxious temperament is due to an environmental factor. Small breed dogs may learn to be more risk averse due to their greater vulnerability to harm or injury, and/or the owners and breeders of such dogs may be more tolerant of their behavioral issues because of their smaller size, and/or they may provide them with less early exposure to unfamiliar environments and social interactions during the sensitive period for socialization, thereby rendering them more fearful and anxious. Choosing between these various alternatives is beyond the scope of the present study but future investigations of this apparent 'small breed effect' might benefit from comparing small or toy breeds that show the effect consistently (e.g. Chihuahua, dachshund, toy poodle) with those that don't (e.g. cavalier King Charles spaniel, Havanese).

Less specific genetic factors may also help to explain the relatively eccentric C-BARQ factor and item scores for the Siberian husky. This breed has been classified as an 'ancient breed' based on its apparent degree of genetic relatedness to the wolf (Parker et al. 2004). Its high scores for *chasing* and *escaping/roaming*, and low scores for *trainability*, *attachment/attention-seeking* and *persistent barking* certainly tend to render it more wolflike in behavior than any of the other popular breeds in our sample, although these traits might also reflect evolutionary convergence rather than genetic relatedness. In addition, it should be emphasized that this breed also displays exceptionally low levels of both social and *nonsocial* fear: a characteristic that would generally be considered atypical of wolves.

Several of the observed breed differences in the study are most plausibly accounted for by reference to the original functional (working) roles of the breeds involved. For example, the relatively low scores for the fear-related factors in Doberman pinschers, German shepherds, and rottweilers, and the high scores of two of these breeds for *stranger-directed aggression* make sense in the context of their widespread use as guard dogs. Similarly, the beagle's apparent tendency to escape and roam is probably a legacy of its hound-like propensity to hunt by 'following its nose', while the high scores obtained for *chasing* in German shorthaired pointers and miniature schnauzers may also reflect past selection for attention to, and pursuit of, potential prey. The breed differences identified for the *trainability* factor also fit within a functional framework. Because they traditionally performed relatively complex tasks in tandem with human partners, sporting and herding breeds have presumably been selected for social-cognitive skills that enhance their working performance. Such skills would likely include many of the elements of the *trainability* factor such as attention and focus, responsiveness to signals and directions provided by humans, and quickness to learn. In contrast, toy breeds are not required to work, and hounds and terriers typically work more or less independently of human direction.

Some breeds may also suffer from straightforward morphological and anatomical constraints on their behavior. The low scores for *chasing* and *energy* of the some of the brachycephalic and giant breeds are likely to be at least partly a consequence of their lack of stamina due either to their exceptionally large body size or congenital deformation of the respiratory tract and axial skeleton.

With respect to the issue of dog breed popularity, the current findings may also help to explain why Ghirlanda et al. (2013) were unable to detect any consistent relationship between popularity and behavioral characteristics in their study. Miniature and toy breed dogs have grown markedly in popularity throughout the last decade (Euromonitor International 2013), despite the evidence provided here that they are likely to display a range of more severe behavioral problems than larger breeds. Since the effects of canine behavioral problems, such as aggression, excitability, or house soiling, are likely to scale with body size, it follows that dog owners are going to be more tolerant of the same behavior problems produced by a small breed dog compared with a larger one. In which case, any underlying association between breed popularity and behavior will tend to be confounded by the effects of body size.

Finally, these results provide support for the use of the C-BARQ as a behavioral assay for measuring behavioral phenotypes in dogs. According to some authorities, the field of behavioral genetics has been held back in recent years by the lack of reliable behavioral phenotyping techniques (Hall and Wynne 2012; Spady and Ostrander 2008). Hopefully, future genetic association studies will help to determine whether the C-BARQ or other behavioral measurement techniques will be able to fulfill this important role.

References

Barnard, S., Siracusa, C., Reisner, I., Valsecchi, P., & Serpell, J. A. (2012). Validity of model devices used to assess canine temperament in behavioral tests. *Applied Animal Behaviour Science, 138*, 79–87.

Boyko, A. R., Quignon, P., Li, L., et al. (2010). A simple genetic architecture underlies morphological variation in dogs. *PLoS Biology, 8*, e1000451.

Bradshaw, J. W. S., & Goodwin, D. (1998). Determination of behavioural traits of pure-bred dogs using factor analysis and cluster analysis; a comparison of studies in the USA and UK. *Research in Veterinary Science, 66*, 73–76.

Clutton-Brock, J. (1995). Origins of the dog: Domestication and early history. In J. Serpell (Ed.), *The domestic dog, its evolution, behaviour and interactions with people* (pp. 8–20). Cambridge: Cambridge University Press.

Coppinger, R., & Schneider, R. (1995). Evolution of working dogs. In J Serpell (Ed.), *Domestic dog, its evolution, behaviour and interactions with people* (pp. 22–47). Cambridge: Cambridge University Press

De Meester, R., De Bacquer, D., Peremens, K., Vermeire, S., Planta, D. J., Coopman, F., et al. (2008). A preliminary study on the use of the socially acceptable behaviour test as a test for shyness/confidence in the temperament of dogs. *Journal of Veterinary Behavior, 3*, 161–170.

Duffy, D. L., Hsu, Y., & Serpell, J. A. (2008). Breed differences in canine aggression. *Applied Animal Behaviour Science, 114*, 441–460.

Duffy, D. L., & Serpell, J. A. (2012). Predictive validity of a method for evaluating temperament in young guide and service dogs. *Applied Animal Behaviour Science, 138*, 99–109.

Eigenmann, J. E., Patterson, D. F., Zapf, J., & Froesch, E. R. (1984). Insulin-like growth factor I in the dog: A study in different dog breeds and in dogs with growth hormone elevation. *Acta Endocrinology, 105*, 294–301.

Euromonitor International. (2013). Passport global market information database, Retrieved September 11 2013.

Ghirlanda, S., Acerbi, A., Herzog, H., & Serpell, J. A. (2013). Fashion vs. function in cultural evolution: The case of dog breed popularity. *PLoS ONE, 8*(9), e74770.

Goddard, M. E., & Beilharz, R. G. (19820. Genetic and environmental factors affecting the suitability of dogs as guide dogs for the blind. *Theoretical and Applied Genetics, 62*, 97–102.

Gray, M. M., Sutter, N. B., Ostrander, E. A., & Wayne, R. K. (2010). The IGF1 small dog haplotype is derived from middle eastern gray wolves. *BMC Biology, 8*, 16.

Guler, H-P., Binz, K., Eigenmann, E., Jaggi, S., Zimmermann, D., Zapf, J., & Froesch, E. R. (1989). Small stature and insulin-like growth factors: Prolonged treatment of mini-poodles with recombinant human insulin-like growth factor I. *Endocrinologica (Copen.), 121*, 456–464.

Hall, N. J., & Wynne, C. D. (2012). The canid genome: Behavioral geneticists' best friend? *Genes Brain and Behavior, 11*, 889–902.

Hart, B., & Hart, L., (1985). Selecting pet dogs on the basis of cluster-analysis of breed behavior profiles and gender. *JAVMA, 186*, 1181–1185.

Hart, B., & Miller, M. F. (1985). Behavioral profiles of dog breeds. *JAVMA, 186*, 1175–1180.

Herzog, H. (2006). Forty-two thousand and one Dalmatians: Fads, social contagion, and dog breed popularity. *Society and Animals, 14*, 383–397.

Hsu, Y., & Serpell, J. A. (2003). Development and validation of a questionnaire for measuring behavior and temperament traits in pet dogs. *JAVMA, 223*, 1293–1300.

Hsu, Y., & Sun, L. (2010). Factors associated with aggressive responses in pet dogs. *Applied Animal Behaviour Science, 123*, 108–123.

Jakuba, T., Polcova, Z., Fedakova, D., Kottferova, J., Marekova, J., Fejsakova, M., et al. (2013). Differences in evaluation of a dog's temperament by individual members of the same household. *Society and Animals, 21*, 582–589.

Jones, A. C., & Gosling, S. D. (2005). Temperament and personality in dogs (*Canis familiaris*): A review and evaluation of past research. *Applied Animal Behaviour Science, 95*, 1–53.

Jones, P., Chase, K., Martin, A., Davern, P. Ostrander, E. A., & Lark, K. G. (2008). Single-nucleotide-polymorphism-based association mapping of dog stereotypes. *Genetics, 179*, 1033–1044.

Larson, G., Karlsson, E. K., Perri, A., et al. (2012). Rethinking dog domestication by integrating genetics, archeology, and biogeography. *Proceedings of the National Academy of Sciences USA, 109*, 8878–8883.

Ley, J. M., Bennett, P. M., & Coleman, G. J. (2009). A refinement and validation of the Monash Canine Personality Questionnaire (MCPQ). *Applied Animal Behaviour Science, 116*, 220–227.

Liinamo, A-E., van den Berg, L., Leegwater, P. A. J., et al. (2007). Genetic variation in aggression-related traits in golden retriever dogs. *Applied Animal Behaviour Science, 104*, 95–106.

Martin, P., & Bateson, P. P. G. B. (1993). *Measuring behaviour*. Cambridge: Cambridge University Press.

MacKenzie, S. A., Oltenacu, E. A. B., & Houpt, K. A. (1986). Canine behavioral genetics—a review. *Applied Animal Behaviour Science, 15*, 365–393.

McGreevy, P., Georgevsky, D., Carrasco, J., Valanzuela, M., Duffy, D. L., & Serpell, J. A. (2013). Dog behavior co-varies with height, bodyweight and skull shape. *PLoS ONE, 8*(12), e80529.

Nagasawa, M., Tsujimura, A., Tateishi, K., Mogi, K., Ohta, M., Serpell, J. A., & Kikusui, T. (2011). Assessment of the factorial structure of the C-BARQ in Japan. *Journal of Veterinary Medical Science , 73,* 869–875.

Notari, L., & Goodwin, D. (2007). A survey of behavioural characteristics of pure-bred dogs in Italy. *Applied Animal Behaviour Science, 103,* 118–130.

Ostrander, E. A. (2007). Genetics and the shape of dogs. *American Scientific, 95,* 406–413.

Ostrander, E. A., & Galibert, F. (2006). Forward. In E. A. Ostrander, U. Giger & K. Lindblad-Toh (Eds.), *The dog and its genome* (pp. xiii-xix). Cold Spring Harbor, NY: Cold Spring Harbor Laboratory Press.

Parker, H. G., Kim, L. V., Sutter, N. B. et al. (2004). Genetic structure of the purebred domestic dog. *Science, 304,* 1160–1164.

Pérez-Guisado, J., Lopez-Rodríguez, R., & Munoz-Serrano, A. (2006). Heritability of dominant-aggressive behaviour in English Cocker Spaniels. *Applied Animal Behaviour Science, 100,* 219–227.

Ritvo, H. (1987). *The animal estate: English and other creatures in the Victoria age.* Cambridge, MA: Harvard University Press.

Saetre, P., Strandberg, E., Sundgren, P-E., Pettersson, U., Jazin, E., & Bergström. (2006). The genetic contribution to canine personality. *Genes, Brain and Behavior, 5,* 240–248.

Sampson, J., & Binns, M. M. (2006). The Kennel club and the early history of dog shows and breed clubs. In E. A. Ostrander, U. Giger & K. Lindblad-Toh (Eds.), *The dog and Iis genome* (pp. 19–30). Cold Spring Harbor, NY: Cold Spring Harbor Laboratory Press.

Scott, J. P., & Bielfelt, S. W. (1976). Analysis of the puppy testing program. In C. J. Pfaffenberger, J. P. Scott, J. L. Fuller, B. E. Ginsburg & S. W. Bielfelt (Eds.), *Guide dogs for the blind: Their selection, development and training* (pp. 39–75). Amsterdam: Elsevier.

Scott, J. P., & Fuller, J. L. (1965). *Genetics and the social behavior of the dog.* Chicago: Chicago University Press.

Serpell, J. A. (1987). The influence of inheritance and environment of canine behaviour: Myth and fact. *Journal of Small Animal Practice, 22,* 949–766.

Serpell, J. A. (1995). The hair of the dog. In J. A. Serpell (Ed.), *The domestic dog: Its evolution, behaviour and interactions with people* (pp. 258–264). Cambridge: Cambridge University Press.

Serpell, J. A., & Hsu Y. (2001). Development and validation of a novel method for evaluating behavior and temperament in guide dogs. *Applied Animal Behaviour Science, 72,* 347–364.

Serpell, J. A., & Hsu, Y. (2005). Effects of breed, sex, and neuter status on trainability in dogs. *Anthrozoös, 18,* 196–207.

Spady, T. C., & Ostrander, E. A. (2008). Canine behavioral genetics: Pointing out the phenotypes and herding up the genes. *The American Journal of Human Genetics, 82,* 10–18.

Sutter, N. B., Bustamante, C. D., Chase, K., et al. (2007). A single IGF1 allele is a major determinant of small size in dogs. *Science, 316,* 112–115.

Svartberg, K. (2005). A comparison of behaviour in test and in everyday life: Evidence of three consistent boldness-related personality traits in dogs. *Applied Animal Behaviour Science, 91,* 103–128.

Svartberg, K. (2006). Breed-typical behavior in dogs—historical remnants or recent constructs? *Applied Animal Behaviour Science, 96,* 293–313.

Svartberg, K., & Forkman, B. (2002). Personality traits in the domestic dog (*Canis familiaris*). *Applied Animal Behaviour Science, 79,* 133–155.

Takeuchi, Y., & Mori, Y. (2006). A comparison of the behavioral profiles of purebred dogs in Japan to profiles of those in the United States and the United Kingdom. *The Journal of Veterinary Medical Science, 68,* 789–796.

Thalman, O., Shapiro, B., Cui, P., et al. (2013). Complete mitochondrial genomes of ancient canids suggest a European origin of domestic dogs. *Science, 342,* 871–874.

Trut, L., Oskina, I., & Kharlamova, A. (2009). Animal evolution during domestication: The domesticated fox as a model. *Bioessays, 31,* 349–360.

Turcsán, B., Kubinyi, E., & Miklósi, A. (2011). Trainability and boldness traits differ between dog breed clusters based on conventional breed categories and genetic relatedness. *Applied Animal Behaviour Science, 132*, 61–70.

Uhde, T. W., Malloy, L. C., & Slate, S. O. (1992). Fearful behavior, body size, and serum *IGF-I* levels in nervous and normal pointer dogs. *Pharmacology Biochemistry and Behavior, 43*, 263–269.

van den Berg, L., Schilder, M. B. H., de Vries, H., Leegwater, P. A. J., & van Oost, B. A. (2006). Phenotyping of aggressive behavior in golden retriever dogs with a questionnaire. *Behavior Genetics, 36*, 882–902.

van den Berg, S. M., Heuven, H. C. M., Van den Berg, L., Duffy, D. L., & Serpell, J. A. (2010). Evaluation of the C-BARQ as a measure of stranger-directed aggression in three common dog breeds. *Applied Animal Behaviour Science, 124*,136–141.

Vaysse, A., Ratnakumar, A., & Derrien, T. et al. (2011). Identification of genomic regions associated with phenotypic variation between dog breeds using selection mapping. *PLOS Genetics, 7*, e1002316.

Vermeire, S. T., Audenaert, K. R., & De Meester, R. H. et al. (2011). Neuro-imaging the serotonin 2A receptor as a valid biomarker for canine behavioral disorders. *Research in Veterinary Science, 91*, 465–472.

Vermeire, S. T., Audenaert, K. R., & De Meester, R. H. et al. (2012). Serotonin 2A receptor, serotonin transporter and dopamine transporter alterations in dogs with compulsive behavior as a promising model for human obsessive-compulsive disorder. *Psychiatric Research, 201*, 78–87.

Willis, M. B. (1995). Genetics of dog behaviour with particular reference to working dogs. In Serpell J (ed), *The domestic dog, its evolution, behaviour and interactions with people* (pp. 51–64). Cambridge: Cambridge University Press.

Wilsson, E., & Sundgren, P-E. (1998). Behaviour test for eight-week old puppies—heritabilities of tested behaviour traits and its correspondence to later behaviour. *Applied Animal Behaviour Science, 58*, 151–162.

Young, A., & Bannasch, D. (2006). Morphological variation in the dog. In E. A. Ostrander, U. Giger, K. Lindblad-Toh (Eds.), *The dog and its genome* (pp. 47–65). Cold Spring Harbor, NY: Cold Spring Harbor Laboratory Press.

Chapter 3
The Significance of Ethological Studies: Playing and Peeing

Marc Bekoff

Abstract The ease of observing and reliably identifying dogs makes them prime candidates for ethological and observational studies of a wide variety of behaviors including social play, social dominance, social organization, and urination patterns. In this chapter I discuss research on social play behavior and urination/scent-marking patterns. Through long-term observational studies, we have catalogued the behaviors of play, including play requests, communication of intentions, and arbitration and negotiation of "fair play." Using this behavioral category as a model, we can discuss questions of the evolution of morality and social justice. Similarly, by detailed study of scent-marking behavior, we can deduce the evolutionary history of different patterns of elimination. Finally, a systematic ethological approach is contrasted with the casual-observational approach of popular literature on dogs.

3.1 Introduction

Domesticated dogs are fascinating mammals. We created them in our own image, often favoring traits that compromise their health and longevity, and they also vary greatly in size, shape, mass, color, coat, personality, and behavior (see also Duffy and Serpell, this volume). And, because they are easy to observe and to identify reliably in various environments, dogs are wonderful candidates for ethological and observational studies that are concerned with a wide variety of behaviors including social play, social dominance, social organization, and urination patterns and olfactory communication.

M. Bekoff (✉)
Ecology and Evolutionary Biology, University of Colorado, Boulder,
CO 80309-0334, USA
e-mail: marc.bekoff@gmail.com

In this essay I focus on research on social play ("play") behavior and also write briefly on urination/scent-marking patterns to show how much we have learned from detailed ethological studies of our 'best friends'. Not only have we learned much about the nitty–gritty of details of what dogs do when they play (for example, how they ask another dog to play, how they announce their intentions, and how they apologize and forgive transgressions against the rules of 'fair play'), and why different patterns of elimination may have evolved; we have also generated some theories about the evolution of social behavior and moral sentiments, and begun to answer 'big' questions about such areas as moral behavior, fairness, peace, and social justice in animals. Fair play is tightly linked to the evolution of social tolerance, social reciprocity, individual fitness, and peaceful relationships among group-living animals. The study of play in dogs involves non-invasive research that can readily be conducted in dog parks and in various non-captive settings, generates results that can be used to enrich the lives of individuals, and provides information that can be used to more fully understand and integrate dogs and other canids into society. Careful observations from 'citizen scientists' can also help us along. We also can learn a lot about human play from studies of various canids.

The popular literature abounds with books about dogs perpetuating myths about behavior that are based on casual observations rather then on detailed systematic studies still prevail. This is most unfortunate because dogs' cognitive and emotional capacities do not have to be embellished to make them more interesting or alluring (Bekoff 2007, 2013a, b). Nor ought the study of dogs be only to learn more about how they compare to other animals, such as nonhuman primates. In fact, because the wide diversity of extant canids share a common heritage, they lend themselves nicely to studies in which dogs are compared to wild relatives such as wolves and coyotes. No longer pushed aside as not worth studying because they are 'mere artifacts', dogs are wonderful subjects for a wide variety of non-invasive studies including the use of functional Magnetic Resonance Imaging (fMRI) (Berns 2013).

3.2 Observing Animals: There are no Substitutes

There are no substitutes for careful observation and description: this stage of study is critical for the generation of experiments, models, and theory. The wide-ranging and comparative importance of ethological investigations was highlighted in 1973 when three ethologists, Konrad Lorenz, Niko Tinbergen, and Karl von Frisch (for his work on "bee language") jointly won the Nobel Prize in Physiology or Medicine "for their discoveries concerning organization and elicitation of individual and social behaviour patterns" (http://www.nobelprize.org/nobel_prizes/medicine/laureates/1973/). Each keenly observed animals, devised novel and often incredibly simple experiments, and offered useful and enduring theories concerning the evolution of behavior. Lorenz also stressed that behavior was not only something an animal "did", but also something an animal "had": a structure (similar to a bodily organ) or behavioral phenotype on which evolution could operate.

In my studies I take a strongly evolutionary and ecological approach using Niko Tinbergen's (1951, 1963) integrative ideas about the questions with which ethological studies should be concerned: namely, evolution, adaptation, causation, and ontogeny (development and the emergence of individual differences). Concerning the tendency of some scientists to overlook the importance of ethological and detailed observational studies, Tinbergen (1963) noted, "… we might forget that naïve, unsophisticated, or intuitively guided observation may open our eyes to new problems. Contempt for simple observation is a lethal trait in any science, and certainly in a science as young as ours." Tinbergen (1963) also claimed, "[b]ecause subjective phenomena cannot be observed objectively in animals, it is idle to either claim or to deny their existence." So, in his view, for example, while we cannot really know that animals find play to be enjoyable, they just might. Tinbergen did not claim that animals do not have emotional lives. To this end, Gordon Burghardt (1997) suggested adding "subjective experience" to Tinbergen's scheme. (For further discussions about how Tinbergen's ideas can be applied to behavioral studies in general see Kappeler et al. 2013 and Barrett et al. 2013, and to cognitive ethological inquiries in particular, see Jamieson and Bekoff 1993).

With this framework in mind, we can ask, concerning play: why did play evolve; how does it promote survival value and reproductive fitness and allow individuals to come to terms with social situation in which they find themselves; what causes play; how does play develop; and what is the experience of animals in play—the emotional side of play. As time has past studies of play have become much more detailed and theoretically driven and a chronology of this progress can be found in Bekoff (1974a), Symons (1978), Fagen (1981), Bekoff and Byers (1981, 1998), Burghardt (2005), and Pellis and Pellis (2010). This essay is not meant to be a review of the field. However, it does show that detailed observations of social play, a behavior tossed aside because it was a waste of time to study, can inform the development of "big theories" in a number of different areas.

3.3 What Is this Thing Called Play?

What is play? The deceptively simple question has troubled researchers for many years. A well-received definition of social play developed with John Byers (Bekoff and Byers 1981; for further discussion see Burghardt 2005) is "social play is an activity directed toward another individual in which actions from other contexts are used in modified forms and in altered sequences." Our definition centers on what animals do when they play—the structure of play—rather than on possible functions of play. Nonetheless our definition of play could in some instances be problematic in that it would seem to apply, for example, "to stereotypical behaviors such as the repetitive pacing or excessive self-grooming sometimes evinced by caged animals" (Allen and Bekoff 1997). It is difficult to see how to state a non-arbitrary restriction on the range of behaviors that may constitute play (Colin Allen, personal communication). Gordon Burghardt's (1997) later characterization of play as having five criteria attempts to do this. He notes that "Playful activities can be characterized as being (1) incompletely functional in the context expressed; (2) voluntary, pleasurable, or self-rewarding; (3) different structurally or temporarily from related serious behavior systems; (4) expressed repeatedly at least during at least some part of an animal's life span; and (5) initiated in benign situations" (2005, p. 382).

During play, actions may also be changed in their form and intensity and combined in a wide variety of unpredictable sequences (Bekoff and Byers 1981; Burghardt 2005). In juvenile chimpanzees the unpredictability of play increases compared to infants, and indeed the play sessions are more complex and variable in pattern use (Cordoni and Palagi 2011). In polecats, coyotes, and American black bears biting in play fighting is inhibited when compared to biting in real fighting (Bekoff 2004b). Clawing in bears is also inhibited and less intense (Henry and Herrero 1974), an example of 'self-handicapping' that is observed in many diverse species. Play among bears also is non-vocal, and biting and clawing are directed to more parts of their playmates' body during play than during aggression.

Play sequences may also be more variable and less predictable than those performed in 'true' predation or aggression, for example, because individuals are mixing actions from a number of different contexts. Because there are more actions for individuals to choose from it is not surprising that sequences are significantly more variable (Bekoff and Byers 1981; Hill and Bekoff 1977), which is to say that it is more difficult to predict which actions will follow one another during play than, for example, during real predation or aggression during which sequences of motor patterns are more highly structured. When dogs, coyotes, or wolves play one might see sequences of "biting, chasing, wrestling, body slamming, wrestling, mouthing, chasing, lunging, biting, and wrestling, whereas during aggression it would be more likely to see threatening, chasing, lunging, attacking, biting, wrestling, and then one individual submitting to the other" (Bekoff 2005, p. 125). Young canids do not show gender differences in play (Bekoff 1974b; Biben 1983).

In the wild, animals do not spend a lot of time engaging in social play. However, this does not mean that it is not important to play. Much animal play also is spontaneous in that it is common to see two animals sniffing the ground or walking about and then begin to play when they cross one another's path or bump into one another. The amount of time and energy that young mammals devote to various types of play is usually less than ten percent of their total time and energy budgets (Bekoff and Byers 1998). For example, in captive adult wolves the amount of play is about nine percent of the total time budget (Cordoni 2009). In most species play occurs mainly during infant and juvenile life. Adults do engage in social play but usually less so than the young of their species (for a notable exception, see Palagi 2006).

3.3.1 A Rearview and Prospective View of Social Play

Early in my career, many colleagues, especially fieldworkers, told me forthrightly that it was a waste of time to study play behavior. Some people also told me that "real" ethologists do not study dogs because they are artifacts—merely "creations of man"—and we cannot really learn much about the behavior of wild animals by studying them. In the 1970s, it seemed that only veterinarians and those people interested in practical applications of behavioral data studied dogs. Indeed, at two meetings in 2013, this historical mistake was revisited and soundly rejected, and the present volume shows just how important studies of dogs truly are.

Concerning play, some people thought of it as a wastebasket into which behavior patterns that were difficult to understand should be tossed, or that understanding play was not important for researchers interested in behavioral development (Lazar and Beckhorn 1974), whereas others, including my Ph.D. mentor, Michael W. Fox, realized that play was essential to normal social, cognitive, and physical development and that people just had not taken the time to study it in detail (for discussions about possible functions of play see Bekoff and Byers 1981; Burghardt 2005; Palagi et al. 2004; Palagi 2011; Pellis and Pellis 2010; Spinka et al. 2001). One reason why studying play has been difficult is because it is a hodge-podge or kaleidoscope of lots of different activities from various social contexts including predatory, mating, and agonistic behavior (Bekoff 1972), and it takes a lot of time to learn about the details of this behavior. For example, it can take many hours to conduct frame-by-frame analyses of as little as 10 min of play captured on video, but these sorts of analyses are essential to gaining an understanding of this behavioral phenotype.

In order to get the ball rolling on detailed comparative systematic studies of play, I organized a symposium that centered on play (Bekoff 1974a). I also began detailed studies focusing on what animals do in play that have now lasted more than four decades. My studies of play are based on careful observation and analyses of videotape. I watch tapes of play one frame at a time to catalogue the animals' behavior and determine how they exchange information about their motivations, intentions, and desires to play.

Following ethological traditions, my first step was to develop a lengthy and detailed ethogram—a list of actions (Bekoff 1972, 1974b). A completeness analysis was performed that showed that we had noted the fifty or so actions that were most used and the probability of adding a new, as yet unobserved action, was extremely low. Data were collected using direct observation and filming, some of which at had no obvious connection to the then scanty extant theories about the evolution and development of play. However, over time, the *zeitgeist* changed and many data found homes as new hypotheses and theories materialized.

One example of the relevance of play to other cognitive and evolutionary questions is the obvious fact that play is a voluntary and cooperative behavior. As such, it is linked to the development of social skills and animals learning 'right' from 'wrong': a process that is important in the development of fairness and moral sentiments (Bekoff 2004a; Bekoff and Pierce 2009; Cordoni and Palagi 2011; Dugatkin and Bekoff 2003; Pierce and Bekoff 2012), as well as social justice (Brosnan 2012; Pierce and Bekoff 2012). By studying play we may be able to learn about what may be going on in an individual's mind, what they feel, what they are thinking about, what they want, and what they are likely to do during a social encounter. By using play and other special contexts (such as greetings, courtship, or ritualized fighting) to communicate about relationships, animals can convey intentions and emotions and negotiate and re-negotiate the terms of a relationship while minimizing the risk of injury or misunderstanding (Ward and Smuts 2007).

My early research focused on the importance of "bows" in the initiation of play, conveying the message "I want to play with you" (Bekoff 1974b, 1977a). In dogs, coyotes, and wolves, bowing takes the form of crouching on their forelimbs, raising their hind ends in the air, and often barking and wagging their tails. At the time, I did not yet see how they were related to how canids punctuate play sequences and tell others, in essence, "I am going to bite you hard but it is still play" or "I am sorry I just bit you so hard, please forgive me" (Bekoff 1995). What I and others have found is that bows rarely occur outside of the context of social play, and occur throughout play sequences (although they are most commonly are performed at the beginning or towards the middle of playful encounters). And, of course, bows have to be seen by other dogs. In her research on play in dogs, Horowitz (2009) discovered that play signals were "sent nearly exclusively to forward-facing conspecifics while attention-getting behaviors were used most often when a playmate was facing away, and before signaling an interest to play." The play bow is a highly ritualized and stereotyped movement resembling Modal Action Patterns (Barlow 1977). Bows are highly stereotyped, distinctive, and recognizable—but not identical, as Fixed Action Patterns are (Bekoff 1977a). There also are auditory (play sounds such as play panting), olfactory (play odors), and tactile (touch) play signals (Bekoff and Byers 1981; Burghardt 2005; Fagen 1981; Horowitz 2009; Pellis and Pellis 2010).

In my own research I did not look at play bouts as having been 'won' or 'lost' mainly because they were not in any obvious way related to an individual's position in the social hierarchy, in the leadership of their group, or of their social status with the individual with whom they were playing. Burghardt (2010) also noted that there were not individual winners and losers in play. Comparative research has shown that play may be important in the development of motor and physical training, cognitive/motor training, or in the development of other social skills. However, it is difficult to generalize about possible functions of play across species. For example, play fighting (also called 'rough-and-tumble' play) does not appear to be important in the development of motor training for fighting skills in laboratory rats (Pellis and Pellis 2010).

Play also may serve a number of functions simultaneously, for example, socialization, exercise, practice, cognitive development, or training for the unexpected (Spinka et al. 2001), the last theory being based on the kaleidoscopic (unpredictable) nature of play sequences. Play may also have an "anxiolytic effect" by reducing anxiety during tense situations (e.g. pre-feeding time) and preventing escalation to an aggressive encounter. For example, chimpanzees, bonobos, and juvenile gorillas show an increase in social play during pre-feeding periods compare to other times (Palagi et al. 2004, 2006, 2007). No matter what the functions of play may be many researchers believe that play is provides important nourishment for brain growth and helps to rewire the brain, increasing the connections between neurons in the cerebral cortex. (For more information on comparative studies about brains and play see Lewis and Barton 2006; Pellis and Pellis 2010; Graham 2010, and Graham and Burghardt 2010).

I did not see at the time how individual patterns of play could be related to the development of social bonds and individual dispersal patterns (Bekoff 1977b) or to evolutionary questions about individual reproductive fitness (Bekoff 2004a; Bekoff and Pierce 2009). And, while I focused on visual signals I did not pay attention to how dogs sought attention from others using vocal signals. In the late 1990s I had the pleasure of helping to train Alexandra Horowitz as she began her work on visual attention and play (Horowitz 2002, 2009). I am thrilled that I and others persisted in studying this behavior because now it is clear that detailed studies of play can inform the development of "big theories" concerning the evolution of social behavior, fairness, cooperation, moral behavior, cognitive capacities (including whether animals have a "theory of mind"), and individual survival and reproductive fitness (Bekoff and Pierce 2009; Pierce and Bekoff 2012).

3.3.2 Fair Play

For instance, based on extensive research, we have discovered the potential relevance of play to the development of morality. Dogs practice what we call 'fair' play, whose four 'rules' are "Ask first, be honest, follow the rules, and admit when you're wrong" (Bekoff and Pierce 2009). When the rules of play are violated and when fairness breaks down, so does play (Pierce and Bekoff 2009, 2012)—and not just among dogs. For example, in juvenile chimpanzees, it has been observed that some play sessions escalate into serious aggression and, interestingly, during these sessions, no play signals (play faces) were performed (Giada Cordoni, personal communication). Dogs and other animals keep track of what is happening when they play; so, too, should researchers.

Relatedly, play bows also are honest signals, a sign of trust. There is little evidence that social play is a manipulative activity (Bekoff and Pierce 2009). Play signals are rarely used to deceive others in canids or other species. Deceptive signaling is so rare that I cannot recall seeing more than a few occurrences in thousands of play sequences. Cheaters are unlikely to be chosen as play partners because others can simply refuse to play with them and choose others. My long-term field research on coyotes living in the Grand Teton National Park, near Jackson, Wyoming (summarized in Bekoff and Wells 1986), has shown that coyotes who do not play fairly—who invite others to play and then try to dominate them—often wind up leaving their pack because they don't form strong social bonds with other individuals. We also discovered that they suffer higher mortality than those who remain with others. This is a good example in which the data we collected on social play and the dispersal patterns of identified individuals did not make sense to us at the time, however as ideas about fair play emerged they took on significance. The message from research on captive and wild canids is clear: don't bow if you don't want to play. The field data show nicely how the study of the development of play in captive animals that are virtually impossible to gather in the wild can inform what is happening in the wild. (This is not an endorsement of keeping animals in captivity.)

In domestic dogs there is little tolerance for non-cooperative cheaters. Cheaters may be avoided or chased from playgroups. There seems to a sense of what is right, wrong, and fair. For instance, while studying dog play on a beach in San Diego, California, Alexandra Horowitz (2002) observed a dog enter into a play group and interrupt the play of two other dogs. The interloper was chased out of the group and when she returned the playing dogs stopped playing and looked off toward a distant sound. One of the players began moving in the direction of the sound and the interloper ran off following their line of sight. The playmates resumed their game. In rats as well, fairness and trust are important in the dynamics of playful interactions. Sergio Pellis (2002) discovered that sequences of rat play consist of individuals assessing and monitoring one another and then fine-tuning and changing their own behavior to maintain the play mood.

There also are trade-offs in play that help to maintain fair play. Animals engage in two activities that help create an equal and fair playing field: self-handicapping and role-reversing. Self-handicapping (or 'play inhibition') occurs when individuals perform behavior patterns that might compromise them outside of play (Bekoff and Byers 1981; Horowitz 2009). For example, individuals of many species will inhibit the intensity of their bites, thus abiding by the rules and helping to maintain the play mood.

Role-reversing occurs when a dominant animal performs an action during play that would not normally occur during real aggression. For example, during play, a dominant coyote or wolf would not roll over on his back during fighting, but would do so while playing. In one study, Bauer and Smuts (2007) discovered that role-reversals are not always necessary, but they do facilitate play. Giada Cordoni (2009), while studying captive wolves, discovered that "rank distance between conspecifics negatively correlated with play distribution: by playing wolves with closest ranking positions tested each other for acquiring information on skills of possible competitor and gaining hierarchical advantage over it." Young crab-eating foxes, maned wolves, and bush dogs do not show stable hierarchies (Biben 1983). In their study of third-party interventions in play between littermates of dogs Ward et al. (2009) discovered that littermates "use interventions opportunistically to practice offence behaviours directed at littermates already behaving subordinately." These interventions may help structure dominance relationships among littermates.

In many species individuals also show play partner preferences and it is possible that these preferences are based on the trust that specific individuals place in one another or because it is more fruitful for an animal to test its cognitive and physical skills with a particular partner or because it is useful for strengthening the social relationship with a particular partner (Cordoni and Palagi 2011; Palagi and Cordoni 2012). Because social play cannot occur in the absence of cooperation or fairness, it might be a "foundation of fairness" (Bekoff 2004a).

The highly cooperative nature of play has evolved in many species, so there are a number of questions that need to be asked and answered. For most of these questions the database remains scanty. These include: Why might animals continually keep track of what they and others are doing and modify and fine-tune play on the run, while they are playing? Why might they try hard to share one another individual's intentions? Why do animals carefully use play signals to tell others that they really want to play and not try to dominate them? Why do they engage in self-handicapping and role-reversing? Why do animals behave fairly? By "behave fairly" I mean that animals "often have social expectations when they engage in various sorts of social encounters the violation of which constitutes being treated unfairly because of a lapse in social etiquette" (Bekoff and Pierce 2009).

I have also stressed that social morality, in this case behaving fairly, is an adaptation that is shared by many mammals (Bekoff 2004a). By behaving fairly young animals acquire social and other skills they will be needed as they mature into adults (Allen and Bekoff 1997). Without social play individuals and social groups would lose out (Antonacci et al. 2010; Cordoni and Palagi 2011). Morality evolved because it is adaptive, because it helps animals, including humans, survive and thrive in particular environments. And, there is no reason to assume that social morality is unique to humans (Bekoff and Pierce 2009; Brosnan 2012; de Waal 2013; Sussman and Cloninger 2011). "Uncooperative play" is impossible, "an oxymoron, and so it is likely that natural selection weeds out cheaters, those who do not play by the accepted and negotiated rules" (Bekoff 2005).

3.3.3 Future Research: Play and the Ethology of Peace

Detailed observations of social play, a behavior tossed aside because it was supposedly a waste of time to study, can inform the development of "big theories" including those about the evolution of morality. Studies of play also inform theories about the evolution of peaceful behavior (Gray 2014; Verbeek 2008). Furthermore, Niko Tinbergen and renowned field workers including Hans Kruuk and George Schaller (e.g. Schaller and Lowther 1969) have suggested looking to the social carnivores for gaining insights into the evolution of social behavior in humans.

Mammalian social play is a good choice for a behavior to study in order to learn more about the evolution of fairness and social morality, even in humans. This is not to say that animal morality is the same as human morality. If one is a "good"

Darwinian, it is premature to claim that only humans can be empathic and moral beings.

So, where are we? Play is a voluntary activity and individuals have the right to quit when they have had enough or want to do something else (Gray 2014). I still stand by what I have written before (Bekoff 2004a), "Social play may be a unique category of behavior because inequalities between players are tolerated more so than in other social situations. Play cannot occur if individuals choose not to engage in the activity and the equality (or symmetry) needed for play to continue makes it different from other forms of seemingly cooperative behavior such as hunting and care giving. This sort of symmetry, or egalitarianism, is thought to be a precondition for the evolution of social morality in humans" (see also Bekoff and Pierce 2009; Ciani et al. 2012).

Social play is a category of behavior about which we now know quite a bit, but there still is much we have to learn about the details of playful interactions in most of the species in which play has been observed. Much of what we already know about the development, evolution, and social dynamics of play has come from detailed comparative research on domestic dogs and their wild relatives and this information and the methods of study can also be used to learn more about play in other nonhuman animals and the need for "wild play" in human animals (Bekoff 2012). Studying play is challenging and exciting and I hope that additional detailed studies from a wide variety of species will be forthcoming. These data are essential for coming to a further understanding of the evolution of play across diverse species, how ecological variables influence the development of play in individuals of the same species, and how an individual's playful experiences or the lack thereof influence his or her future behavior.

3.4 Ethological Studies of Urination Patterns

The importance of ethological studies is also highlighted in inquiries about urination patterns. I was very surprised four decades ago to discover, when someone asked me some very basic questions about urination patterns of free-running, unrestrained dogs including gender differences in marking rates, what stimuli trigger urinating or scent-marking, and whether seeing another dog urinate stimulated others to do so, sniffing patterns, ground-scratching, and responses to "yellow snow", that there were not any detailed data. There still are not. Based on enduring myths especially in the popular literature about why dogs pee, it was simply assumed that urinating meant scent-marking—although Devra Kleiman (1966) and others pointed out that this is not necessarily so because dogs and other animals do "simply pee." Earlier studies had described in detail the postures dogs use and some general patterns of urination (see Bekoff 1979a and references therein), however little else was known especially about free-running dogs.

To fill the knowledge gap my students and I studied urination patterns in two populations of free-running dogs, one on the campus of Washington University in

St. Louis, Missouri and the other in and around Nederland, Colorado (Bekoff 1979a). Twenty-seven males and Twenty Four anestrous females, all individually identified, were observed. Marking was distinguished from merely urinating in that the urine was aimed at a specific object or area (it had "directional quality", Kleiman 1966) and generally less urine was expelled during marking (see also Palagi et al. 2005; Palagi and Norscia 2009). We also scored the frequency of occurrence of what we called the Raised Leg Display (RLD) that occurred when a dog raised his leg but did not deposit any obvious urine.

The results of this study can be summarized as follows (Bekoff 1979a). Males marked more than females and at a higher rate (males, 71.1 % of urinations; females, 18 %); males ground-scratched significantly more than females after marking and males did it significantly more when other dogs could see him do it (Bekoff 1979b); both males and females marked at the lowest rate in areas in which they spent the greatest amount of time; seeing another male mark was a strong visual releaser for urine marking by males; sniffing did not invariably precede marking by either males or females; the RLD appeared to function as a visual display; and males performed the RLD significantly more frequently when other males were in sight. We concluded that the RLD might be a ploy by which one male gets another male to use his urine because the RLD was a strong visual releaser or trigger for urination by other males. We also concluded then, and the same conclusion obtains now, that we need to pay more attention to the visual aspects of the postures (see also Palagi et al. 2005; Palagi and Norscia 2009) and behavior patterns involved in the deposition of scent, in this case urine. What has been accomplished by observing dogs can serve as a model for studying other species.

3.4.1 What can we Learn from Yellow Snow?

The study of urination and sniffing patterns can also inform ideas about "bigger" questions about cognitive capacities. Various studies have shown that some non-human Great apes, an African elephant, bottlenose dolphins, orcas, and European magpies show "self-recognition", sometimes called "self-awareness" and "self-consciousness", using what is called the "mirror test". Paul Sherman and I (2004) labeled various senses of self as "self-cognizance". In general, in the classic mirror test developed by Gordon Gallup (1970) (for further discussion see Gallup et al. 2002) that has been used, with revisions, on individuals other then land animals (for details about research done on dolphins see Reiss 2011), individuals are habituated to a mirror, anesthetized, and a mark is then placed on their forehead using an odorless dye. When the animal wakes up they are tested to see if the make self-directed movements to the mark. If they do this it is taken as support for self-recognition, self-awareness, or self-consciousness. Individual (but not all) chimpanzees, an elephant, and magpies have passed the mirror test with a good deal of exposure, however, researchers disagree about just what the self-directed movements mean and if the mirror test is really a valid measure of "self-awareness", and they also are concerned with a lack of replication in studies in different laboratories.

A detailed discussion of the mirror test and what it means is beyond the scope of this essay. But what is important is that dogs and wolves do not pass the mirror-test. Michael Fox and I tried to use this method in the early 1970s and tried to publish the negative results, but the paper was repeatedly rejected because of the results were negative. Our negative result, of course, did *not* mean that dogs did not have the capacity for self-cognizance, only that perhaps the mirror test was not the good test to use.

The mirror tests depends on a visual stimulus and for a long while I wondered if perhaps a test using olfactory cues could be designed to see if dogs could discriminate self from others. Once again, following up on Tinbergen's (1951) stress on the importance of conducting simple field experiments I decided to use urine-soaked snow—"yellow snow"—to see if dogs discriminated their own urine from that of others.

To investigate the role of urine in eliciting urinating and marking, in a pilot study that took place over five winters when there was snow on the ground (Bekoff 2001) I moved urine-saturated snow from place-to-place to compare the responses of an adult male domestic dog, Jethro, to his own and others' urine. What I found was that Jethro spent less time sniffing his own urine than that of other males or females, and that while his interest in his own urine waned with time it remained relatively constant for other individuals' urine. Jethro infrequently urinated over or sniffed and then immediately urinated over (scent-marked) his own urine. He marked over the urine of other males more frequently than he marked over females' urine. He clearly had some sense of "self": a sense of "mine-ness" but not necessarily of "I-ness" (Bekoff 2001).

Clearly, as with the study of play, a "simple" ethological approach to urination patterns produced very interesting and useful preliminary results. These data are foundational for the development of hypotheses about, for instance, why animals pee the way they do and where they choose to do it, and the generalizability of the patterns that have been observed in dogs needs to be assessed in other animals.

3.5 Back to Basics: The Ethological Approach and Watching Animals

It is easy to see that the ethological approach is invaluable to the study of animal behavior. Many of the papers in the symposium I organized (Bekoff 1974a) provide excellent examples of just how important it is to watch animals carefully and to develop detailed ethograms. Watching animals is not merely 'stamp collecting' as some pejoratively called it years ago (Jamieson and Bekoff 1993), nor is it just for those interested in natural history or bird watching. We need to know what animals do in order to be able to generate relevant and valid models and theories and explain why animals do what they do. This essential role of ethological inquiries is not as highly prized as it was in the early days of ethology and observational studies are often dismissed as unwelcomed hangovers from the past when natural history accounts were popular even among researchers. Let us hope that funding becomes available for these sorts of foundational studies because in so many areas of animal behavior we really need to go back to the basics, in this case detailed accounts of what animals do in social encounters or when they are on their own. Without these sorts of detailed data attempts to develop wide-ranging, some might say grandiose hypotheses and theories, are similar to trying to build a house without a suitable foundation. The ideas and the home might last for a while but sooner or later someone is going to have to go back to building a firm foundation. Ethological studies do just this.

Furthermore, they are the foundation for important insights about not just behavior but animal experience. Despite some lingering and rapidly declining skepticism about whether or not other animals are conscious or experience deep and enduring emotions (summarized in Bekoff 2013b), it is now time to stop ignoring who these animals really are and to stop pretending that we do not know that they are indeed conscious and feeling beings who experience a wide range of emotions (Bekoff 2013a, b). The minds of other animals are not "all that private" so as to make impossible to know what they want, need, and feel. And, there is no doubt that dogs and other animals love to play and deeply enjoy it. They voluntarily seek it out relentlessly, take certain risks, and will play to exhaustion and seek out playmates with very little rest.

We must also make every effort to use this information on their behalf. Indeed, in July 2012 a group of renowned scientists met at Cambridge University and finally declared that animals are truly conscious and produced a long overdue document they called the Cambridge Declaration on Consciousness (2012). They wrote, "Convergent evidence indicates that non-human animals have the neuroanatomical, neurochemical, and neurophysiological substrates of conscious states along with the capacity to exhibit intentional behaviors. Consequently, the weight of evidence indicates that humans are not unique in possessing the neurological substrates that generate consciousness. Non-human animals, including all mammals and birds, and many other creatures, including octopuses, also possess these neurological substrates." They could also have included fish, for whom the evidence supporting sentience and consciousness is also compelling (Braithwaite 2010).

Although we really have known for a much longer period of time that other animals are conscious, perhaps this highly publicized declaration will be helpful for radically improving animal well being. We can only hope this declaration is not merely gratuitous hand waving. I have proposed a Universal Declaration on Animal Sentience (Bekoff 2013a) that expands the Cambridge Declaration and also notes that we must factor sentience into the decisions we make about how we treat other animals.

Acknowledgments I thank Alexandra Horowitz for inviting me to write this essay and for her insightful comments and also Giada Cordoni and Elisabetta Palagi for taking the time to provide input on an earlier version of this essay. It has been an absolute pleasure to see interest in, and detailed comparative studies of, social play develop over the years.

References

Allen, C., & Bekoff, M. (1997). *Species of mind: The philosophy and biology of cognitive ethology.* Cambridge, MA: MIT Press.

Antonacci, D., Norscia, I., & Palagi, E. (2010). Stranger to familiar: Wild strepsirhines manage xenophobia by playing. *PLoS ONE, 5*(10), e13218.

Barlow, G. (1977). Modal action patterns. In T. Sebeok (Ed.), *How animals communicate* (pp. 98–134). Bloomington: Indiana University Press.

Barrett, L., Blumstein, D. T., Clutton-Brock, T. H., & Kappeler, P. M. (2013). Taking note of Tinbergen, or: The promise of a biology of behavior. *Philosophical Transactions of the Royal Society B, 368,* 20120352.

Bauer, E. B., & Smuts, B. B. (2007). Cooperation and competition during dyadic play in domestic dogs, *Canis familiaris. Animal Behaviour, 73,* 489–499.

Bekoff, M. (1972). The development of social interaction, play, and metacommunication in mammals: An ethological perspective. *Quarterly Review of Biology, 47,* 412–434.

Bekoff, M. (Ed.) (1974a). Social play in mammals. *American Zoologist, 14,* 265–436.

Bekoff, M. (1974b). Social play and play-soliciting by infant canids. *American Zoologist, 14,* 323–340.

Bekoff, M. (1977a). Social communication in canids: Evidence for the evolution of a stereotyped mammalian display. *Science, 197,* 1097–1099.

Bekoff, M. (1977b). Mammalian dispersal and the ontogeny of individual behavioral phenotypes. *American Naturalist, 111,* 715–732.

Bekoff, M. (1979a). Scent marking by free-ranging domestic dogs: Olfactory and visual components. *Biology of Behaviour, 4,* 123–139.

Bekoff, M. (1979b). Ground scratching by male domestic dogs: A composite signal? *Journal of Mammology, 60,* 847–848.

Bekoff, M. (2001). Observations of scent-marking and discriminating self from others by a domestic dog (*Canis familiaris*): Tales of displaced yellow snow. *Behavioural Processes, 55,* 75–79.

Bekoff, M. (1995). Play signals as punctuation: The structure of social play in canids. *Behaviour, 132,* 419–429.

Bekoff, M. (2004a). Wild justice and fair play: Cooperation, forgiveness, and morality in animals. *Biology & Philosophy, 19,* 489–520.

Bekoff, M. (2004b). Social play behavior and social morality. In M. Bekoff (Ed.). *Encyclopedia of Animal Behavior* (pp. 833–845), Westport, CT: Greenwood Press.

Bekoff, M. (2005). *Animal passions and beastly virtues.* Philadelphia, PA: Temple University Press.

Bekoff, M. (2007). *The emotional lives of animals.* Novato, CA: New World Library.

Bekoff, M. (2012). The need for "wild play": Let children be the animals they need to be. *Psychology Today,* http://www.psychologytoday.com/blog/animal-emotions/201202/the-need-wild-play-let-children-be-the-animals-they-need-be

Bekoff, M. (2013a). A universal declaration of animal sentience: No pretending. *Psychology today,* http://www.psychologytoday.com/blog/animal-emotions/201306/ universal-declaration-animal-sentience-no-pretending.

Bekoff, M. (2013b). *Why dogs hump and bees get depressed. The fascinating science of animal intelligence, emotions, friendship, and conservation.* Novato, CA: New World Library.

Bekoff, M., & Byers, J. A. (1981). A critical reanalysis of the ontogeny and phylogeny of mammalian social and locomotor play: An ethological hornet's nest. In K. Immelmann, G. Barlow, M. Main, & L. Petrinovich (Eds.), *Behavioral development: The Bielefeld interdisciplinary project* (pp. 296–337). Cambridge and New York: Cambridge University Press.

Bekoff, M., & Byers, J. A. (Eds.) (1998). *Animal play: Evolutionary, comparative, and ecological perspectives.* New York: Cambridge University Press.

Bekoff, M., & Pierce, J. (2009). *Wild justice: The moral lives of animals.* Chicago: University of Chicago Press.

Bekoff, M., & Sherman, P. W. (2004). Reflections on animal selves. *Trends in Ecology and Evolution, 19*, 176–180.

Bekoff, M., & Wells, M. C. (1986). Social ecology and behavior of coyotes. *Advances in the Study of Behavior, 16*, 251–338.

Berns, G. (2013). *How dogs love us: A neuroscientist and his adopted dog decode the canine brain.* New York: New Harvest/Amazon.

Biben, M. (1983). Comparative ontogeny of social behavior in three South American canids, the maned wolf, crab-eating fox and bush dog: Implications for sociality. *Animal Behaviour, 31*, 814–826.

Braithwaite, V. (2010). *Do fish feel pain?* New York: Oxford University Press.

Brosnan, S. (2012). Justice in animals. *Social Justice Research, 25*.

Burghardt, G. (1997). Amending Tinbergen: A fifth aim for ethology. In R. W. Mitchell, N. Thompson, & L. Miles (Eds.), *Anthropomorphism, Anecdote, and Animals: The Emperor's New Clothes?* (pp. 254–276). Albany, NY: SUNY Press.

Burghardt, G. M. (2005). *The genesis of animal play: Testing the limits.* Cambridge, MA: MIT Press.

Burghardt, G. M. (2010). Play. In M. D. Breed & J. Moore (Eds.) *Encyclopedia of animal behavior,* vol. 2 (pp. 740–744). Oxford: Academic Press.

Cambridge Declaration on Consciousness. (2012). http://fcmconference.org and www.psychologytoday.com/blog/animal-emotions/201208/scientists-finally-conclude-nonhuman-animals-areconscious-beings.

Ciani, F., Dall'Olio, S., Stanyon, R., Palagi, E. (2012). Social tolerance and adult play in macaque societies: A comparison with different human cultures. *Animal Behaviour, 84*, 1313–1322.

Cordoni, G. (2009). Social play in captive wolves (*Canis lupus*): Not only an immature affair. *Behaviour, 146*, 1363-1385.

Cordoni, G., & Palagi, E. (2011). Ontogenetic trajectories of chimpanzee social play: Similarities with humans. *PlosOne, 6*, e27344.

de Waal, F. (2013). *The bonobo and the atheist: In search of humanism among the primates.* New York: Norton.

Dugatkin, L. A., & Bekoff, M. (2003). Play and the evolution of fairness: A game theory model. *Behavioural Processes, 60*, 209–214.

Fagen, R. M. (1981). *Animal Play Behavior.* New York: Oxford University Press.

Gallup, G. G., Jr. (1970). Chimpanzees: Self-recognition. *Science, 167*, 86–87.

Gallup, G. G., Jr., Anderson, J. R., & Shillito, D. J. (2002). In M. Bekoff, C. Allen, G. M. Burghardt (Eds), *The cognitive animal: Empirical and theoretical perspectives on animal cognition* (pp. 325–333). Cambridge, MA: The MIT Press.

Graham, K. L. (2010). A coevolutionary relationship between striatum size and social play in nonhuman primates. *American Journal of Primatology, 73*, 1–9.

Graham, K. L., & Burghardt, G. M. (2010). Current perspectives on the biological study of play: Signs of progress. *Quarterly Review of Biology, 85*, 393–418.

Gray, P. (2014). The play theory of hunter-gatherer egalitarianism. In D. Narvaez, K. Valentino, A. Fuentes, J. McKenna, & P. Gray (Eds.), *Ancestral landscapes in human evolution.* New York: Oxford University Press.

Henry, J. D., & Herrero, S. M. (1974). Social play in the American black bear: Its similarities to canid social play and an examination of its identifying characteristics. *American Zoologist, 14*, 371–389.

Hill, H. L., & Bekoff, M. (1977). The variability of some motor components of social play and agonistic behaviour in Eastern coyotes, *Canis latrans* var. *Animal Behaviour, 25*, 907–909.

Horowitz, A. C. (2002). The behaviors of theories of mind, and A case study of dogs at play. Ph.D. Dissertation, University of California, San Diego.

Horowitz, A. (2009). Attention to attention in domestic dog (*Canis familiaris*) dyadic play. *Animal Cognition, 12*, 107–118.

Jamieson, D., & Bekoff, M. (1993). On aims and methods of cognitive ethology. *Philosophy of Science Association, 2*, 110–124.

Kappeler, P. M., Barrett, L., Blumstein, D. T., Clutton-Brock, T. H. (2013). Constraints and flexibility in social behaviour: Introduction and synthesis. *Philosophical Transactions of the Royal Society B, 368*.

Kleiman, D. (1966). Scent marking in the canidae. *Symposium of the Zoological Society of London, 18*, 167–177.

Lazar, J. W., & Beckhorn, G. D. (1974). Social play or the development of social behavior in ferrets (*Mustela putorius*)? *American Zoologist, 14*, 405–414.

Lewis, K. P., & Barton, R. A. (2006). Amygdala size and hypothalamus size predict social play frequency in non-human primates: A comparative analysis using independent contrasts. *Journal of Comparative Psychology, 120*, 31–37.

Palagi, E. (2006). Social play in bonobos (*Pan paniscus*) and chimpanzees (*Pan troglodytes*): Implications for natural social systems and interindividual relationships. *American Journal of Physical Anthropology, 129*, 418–426.

Palagi, E. (2011). Playing at every age: Modalities and potential functions in non-human primates. In A.D. Pellegrini (Ed.), *The Oxford handbook of the development of play* (pp. 70–82). New York: Oxford University Press.

Palagi, E., Antonacci, D., & and Cordoni, G. (2007). Fine-tuning of social play in juvenile lowland gorillas (*Gorilla gorilla gorilla*). *Developmental Psychobiology, 49*, 433–445.

Palagi, E., & Cordoni, G. (2009). Postconflict third-party affiliation in *Canis lupus*: Do wolves share similarities with the great apes? *Animal Behaviour, 78*, 979–986.

Palagi, E., & Cordoni, G. (2012) The right time to happen: Play developmental divergence in the two *Pan* species. *PLoS ONE, 7(12)*, e52767.

Palagi, E., Cordoni, G., & Borgognini Tarli, S. (2004). Immediate and delayed benefits of play behaviour: New evidence from chimpanzees. *Ethology, 110*, 949–962.

Palagi, E, Dapportto, L., & Borgognini Tarli, S. (2005). The neglected scent: On the marking function of urine in *Lemur catta*. *Behavioral Ecology and Sociobiology, 58*, 437–445.

Palagi, E., & Norscia, I. (2009). Multimodal signaling in wild *lemur catta*: Economic design and territorial function of urine marking. *American Journal Of Physical Anthropology, 139*, 182–192.

Pellis, S. (2002). Keeping in touch: Play fighting and social knowledge. In M. Bekoff, C. Allen, & G.M. Burghardt (Eds.), *The cognitive animal*. Cambridge, MA: MIT Press.

Pellis, S., & Pellis, V. (2010). *The playful brain: Venturing to the limits of neuroscience*. London, UK: Oneworld Publications.

Pierce, J., & Bekoff, M. (2009). Moral in tooth and claw. The Chronicle of Higher Education; http://chronicle.com/article/Moral-in-ToothClaw/48800/.

Pierce, J., & Bekoff, M. (2012). Wild justice redux: What we know about social justice in animals and why it matters. *Social Justice Research, 25*, 122–139.

Reiss, D. (2011). *The dolphin in the mirror: Explaining dolphin minds and saving dolphin lives.* Boston, MA: Houghton Mifflin Harcourt.

Schaller, G. B., & Lowther, G. R. (1969). The relevance of carnivore behavior to the study of early hominids. *southwestern Journal of Anthropology, 25*, 307–341.

Spinka, M., Newberry, R. C., & Bekoff, M. (2001). Mammalian play: Training for the unexpected. *Quarterly Review of Biology, 76*, 141–168.

Sussman, R. W., & Cloninger, C. R. (Eds.) (2011). *Origins of altruism and cooperation*. New York: Springer.
Symons, D. (1978). *Play and aggression: A study of rhesus monkeys*. New York: Columbia University Press.
Tinbergen, N. (1951). *The study of instinct*. New York: Oxford University Press.
Tinbergen, N. (1963). On aims and methods of ethology. *Zeitschrift für Tierpsychologie, 20*, 410–433.
Verbeek, P. (2008). Peace ethology. *Behaviour, 145*, 1497–1524.
Ward, C., Trisko, R., Smuts, B. B. (2009). Third party interventions in dyadic play between littermates of domestic dogs, *Canis lupus familiaris*. *Animal Behaviour, 78*, 1153–1160.
Ward, C., & Smuts, B. B. (2007). Why does carnivore play matter? *Journal of Developmental Processes, 2*, 31–38.

Part II
Behavior and Cognition: Observational and Experimental Results

Chapter 4
Dog Imitation and Its Possible Origins

Ludwig Huber, Friederike Range and Zsófia Virányi

Abstract The question of social learning in dogs is characterized by dispute. Ever since Thorndike more than a century ago, researchers believed that domestic dogs have poor social learning skills. However, recently it has been proposed that dogs have enhanced social cognitive skills due to their selection to live in the human environment and cooperative with humans. Thus, dogs might not just be able to learn through observation from conspecifics but also from humans. The most convincing argument for the latter assumption would be experimental evidence of true imitation, since imitation is considered to be the most complex and also most rare social learning mechanism in the animal kingdom. In this chapter, we will report recent evidence first of social facilitation and social influences on individual learning and then of true social learning in dogs. The latter includes three hallmarks of imitation: faithful copying (of both a human and a conspecific model), deferred imitation, and selective imitation. In the final part we address the potential origins of these remarkable skills of dogs. We propose imitation has been inherited from their ancestors, wolves, which are well known for their advanced social system, including cooperative breeding and hunting. This hypothesis has recently been supported by experimental evidence with wolves outperforming dogs in a manipulative problem-solving task after observation of a skilled conspecific.

4.1 Introduction

Social learning in dogs is a contentious issue. While some believe that dogs are adapted perfectly to the human world and learn massively from their human caretakers, others argue that they may be highly attentive to human cues but do not

L. Huber (✉) · F. Range · Z. Virányi
Messerli Research Institute, University of Veterinary Medicine,
Medical University of Vienna, University of Vienna,
Veterinärplatz 1 1210 Vienna, Austria
e-mail: ludwig.huber@vetmeduni.ac.at

learn anything. At best, they are obedient and follow humans blindly, but they do not modify their behavior, knowledge or skills in a persistent and adaptive manner. Of course, the discussion of whether an animal species is able to learn by observation depends on the definition of 'learning'. Although not aiming at being prisoned with terminological issues, it is important to make things clear before risking too much confusion later on.

According to Thorpe (1956), learning can be considered as a process which manifests itself through adaptive modifications of the behavior of the individual as a result of experiences. While psychologists use the term 'learning' in regard to a relatively long-lasting change in behavior, physiologists use it in regard to a derived change of the neuronal mechanism. Biologists—like Lorenz (1981)—however, do restrict the use of the term 'learning' to only *adaptive* modifications and do not include muscle twitches and motoric stereotypes.

Within contemporary learning theory, learning is regarded as change in an animal that is caused by a specific experience at a certain time, t_1, and that is detectable later, t_2, in the animal's behaviour (Rescorla 1988). More specifically, cognitive psychologists and learning theorists consider learning as a kind of development of internal representations of relationships, which occur between events in the animal's environment (Dickinson 1980). An associative learning mechanism is one that produces, under specified conditions, increments and decrements in the strength of a connection between psychological representations (Dickinson 1980). In other words, learning is the acquisition of knowledge. Knowledge may be represented in two kinds, in a procedural (knowing how) and a declarative (knowing what) form. However, as noted by Heyes (1994), investigators of social learning seldom refer to animal learning theory, even when they are discussing mechanisms. Mostly they refer to the outcomes of social learning and discuss the adaptive value of those outcomes in terms of behavior synchrony, conformity, information transmission, tradition formation, and culture.

Matching one's behaviour to that of a demonstrator is the only widely recognized outcome of social learning. However, simply doing what others do is not necessarily learning. It still may lead to favorable outcomes, because in most cases the behavior of others has already been shaped by consequences. There are many ways in which 'matching behavior' is produced, but only some are based on true learning processes as defined above.

Most researchers would agree that the term 'social learning' refers to learning that is influenced by observation of, or interaction with, another animal (typically a conspecific) or its products (Galef 1988; Heyes 1994). But it is not clear whether it is based on the same mechanism as asocial (or individual) learning, with the only difference being the source of information (a social one in the first case, an asocial one in the second). And the learning phenomenon is blurred by the possibility that species-typical, motivational, or perceptual factors generate the convergence between demonstrator and observer by producing matching behaviors (Zentall 2006). For example, animals may be predisposed to engage in certain behaviors (e.g., eating) when others are seen engaging in those behaviors ('contagious' behaviors). Alternatively, being in the presence of conspecifics may result in an increase in general arousal, which may make certain behaviors more probable (motivational effects). Also, the behavior of others may draw attention to a place or object independently of the behavior itself, and that attention may facilitate subsequent individual learning (perceptual 'enhancement' effects). All these phenomena can be categorized as socially biased learning (Fragaszy and Visalberghi 2004) or socially influenced individual learning in contrast to true social learning (Zentall 2006).

True learning in the above-mentioned sense may occur if observers learn the relation between the effect(s) of the observed behavior on the environment, which subsequently facilitates the performance of the observers. Varieties of social learning can be distinguished according to the role of the demonstrator in generating matching behaviour on the part of the observer (Heyes 1994). The demonstrator's behaviour may act as (i) an unconditioned stimulus eliciting a matching response (observational conditioning), (ii) a discriminative stimulus (matched-dependent behaviour), or (iii) a model within a goal-directed (imitation) or non-goal-directed (copying) process (Galef 1988; Whiten and Ham 1992).

There is wide consensus about the ability and disposition of dogs to benefit from interactions with either conspecifics or humans (e.g. Kubinyi et al. 2009). Species-typical (e.g. contagion), motivational (social facilitation) or perceptual (local or stimulus enhancement) factors have been shown repeatedly to produce matching behavior or to help an observer dog to solve a problem. Much less consensus exists about whether forms of true social learning, like observational conditioning, affordance learning, or emulation and imitation are within the range of the dog's cognitive abilities. Especially imitation, whether animals can "from an act witnessed, learn to perform that act" (Thorndike 1911, p. 79), has produced diverging opinions about the dog's respective capacity. Are these creatures, like humans and possibly great apes, able to use representations to guide their actions? Are they able to selectively switch between imitation and emulation, between copying the observed behavior or using the own, preferred behavior depending on how they interpret the demonstrator's performance in relation to its goals, situational constraints, and possibilities? And if they can, how and when have they gained these abilities?

Fortunately, the last years have produced some empirical answers to these important questions. In this review we proceed in four steps: (1) First, with a critical view differentiating between socially biased individual learning and true social learning, we summarize the evidence of several main forms of social influence on learning in dogs, like social facilitation and stimulus enhancement. (2) Then, in the main part of the chapter, we discuss in more depth the evidence for and against imitation in dogs, which has become a controversial issue in recent years. (3) Subsequently, we address the potential origins of these skills of dogs, elaborating on the social system of canids as well as on the potential effects of domestication and individual learning in a social environment shared with humans. (4) Finally, we highlight a few results of the first study that compared dogs and wolves in an imitative learning task and conclude with a cautious interpretation that begs further research.

4.2 Social Influences and Socially Biased Individual Learning in Dogs

If one looks back into the history of social learning in dogs, one can find both positive and negative reports about the ability of dogs to learn from observation. The two most cited negative examples are those from Thorndike (1898) and Brogden (1942). According to their historic findings, dogs failed to learn to manipulate a latch after observing a human, or to solve a puzzle box faster after watching another dog escaping from it (Thorndike 1898); neither were dogs classically conditioned faster after watching a similarly conditioned dog (Brodgen 1942). On the positive side of the coin, people count the much later reports of Adler and Adler (1977) and Slabbert and Rasa (1997). When given the opportunity to observe their trained littermates, Miniature Dachshund puppies showed enhanced learning in a string-pulling task (Adler and Adler 1977). And the participation in the training exercises of their mother improved the drug search performance of German shepherd puppies (Slabbert and Rasa 1997). In sum, those early reports of social learning studies remained ambiguous regarding whether and what observer dogs learned and, more importantly, were at best suggestive, due to the lack of important experimental controls. Therefore, rigorously controlled experimental studies were needed to further our understanding of the true nature of the dog's social learning abilities.

4.2.1 Species-Typical and Motivational Factors

The (mainly psychological) research of imitation in animals has clearly revealed that methodological improvements have facilitated our understanding of the underlying processes of social learning, especially by uncovering factors that influence the behavioral modification of the observer and the information gained from the model's demonstration (Byrne 2002; Heyes et al. 2009; Heyes and Ray 2000; Zentall 2006). It became obvious that not one but several mechanisms may be involved, most of them simpler on a cognitive account than imitation. For instance, simply the presence of a conspecific may result in changes of behavior of the observer (social facilitation), e.g. by increasing its general arousal or vigilance. Matching behavior is more obvious if specific behaviors of one individual are repeated in more or less the same form by another upon observation. The most popular cases in the dog literature are gaze following (Range and Viranyi 2011; Teglas et al. 2012) and contagious yawning (Harr et al. 2009; Joly-Mascheroni et al. 2008; Madsen and Persson 2013; O'Hara and Reeve 2011; Silva et al. 2012). However, as the controversial issue of contagious yawning has shown, even here, the underlying mechanisms (contagious behaviour, mimicry, or even empathy) have not been clarified (Silva and de Sousa 2011; Yoon and Tennie 2010).

If dogs are not only watching each other but also interact, facilitative effects may multiply. Such a combination of motivation and information transfer was indicated in a study by Heberlein and Turner (2009). Food availability to a demonstrator dog during the demonstration phase increased the willingness of observer dogs to have snout contact with their demonstrators, and this snout contact increased their motivation to search for hidden food. This also increased their search efficiency, as they were then more likely to look at the location where the demonstrator had previously found food. As in rats, snout contact is used by dogs as an important source of information when establishing food preferences. This has been experimentally examined by Lupfer-Johnson and Ross (2007) who tested twelve demonstrator-observer pairs of domestic dogs. Observers exhibited a significant preference for the flavored diet consumed by the demonstrators. In feral dogs it might be important for unsuccessful foraging dogs to get information about food locations by interacting with successful conspecifics.

4.2.2 Perceptual Factors

Watching others provides the opportunity to become aware of interesting objects or interesting places in the environment, something that one would unlikely recognize if alone. Importantly, the observer may have a head start by simply focusing on the object with which the other has interacted before and then may subsequently learn to solve the problem alone (stimulus enhancement). There are numerous examples from dog research showing clear perceptual enhancement effects. Especially when confronted with search problems in complex spatial arrangements, where the baseline probability of finding the desired object (e.g., food) is low, the opportunity to observe a skilful demonstrator bears a high facilitative potential. A series of experiments testing dogs' ability to solve a detour problem showed how effective simple perceptual cues are in leading the naive observer to the target location (Pongrácz et al. 2001, 2003). Dogs with no experience had difficulty obtaining the food when the only solution was to go around (to detour) a V-shaped transparent wire mesh fence. The problem is solved by going farther away from the food first before returning on the other side of the fence. But if demonstrated by a human experimenter, most dogs solved this spatial problem. The most parsimonious explanation for this effect of observation is in terms of stimulus enhancement: the demonstrator's behavior served to direct the attention of the observer to the end points of the fence. The importance of the attention on the side of the observer is also indicated by the fact that ostensive-communicative cues of the human demonstrator (verbal encouragement during the demonstration) had facilitating effects (Pongrácz et al. 2004) (but see Range et al. 2009b).

The question still remained if the human demonstrator only served as means to make important aspects of the environment more salient (like the end points of the fence) rather than as a social model and source of important information to solve the task (see Heyes 1994). A recent study with a box being pulled behind the fence along the detour path suggests that the former effect is indeed sufficient to increase the dogs' performance in this kind of spatial problem (Mersmann et al. 2011). Obviously, the inanimate cues are as effective as real social cues (human or conspecific demonstrator) to draw the observer's attention to otherwise unrecognized parts of the environment and to increase the search space of the dog.

Stimulus enhancement can also be an important mechanism involved in learning to solve manipulative problems. When dogs were confronted with a box from which a ball was released if a protruding lever was manipulated (pushed to the right or left) they could not solve the problem without demonstration (Kubinyi et al. 2003). Only when a human demonstrator (the dog's owner) pushed the lever sideways to release the ball (in contrast to not interacting with the box), did observer dogs solve the problem. Interestingly, it seemed that it was the *pushing* action (with or without resulting in the ball rolling out) that attracted the observers' attention to the lever, since the demonstrator's touching of the lever (tapping on it) had no facilitating effect on the observers' own attempts. However, the observers did not manipulate the lever in the same way as the demonstrator, namely pushing it in the same direction. Given that finding, the most parsimonious interpretation for the subject dogs' behavior is again in terms of stimulus enhancement.

4.3 Social Learning

4.3.1 Behaviour Matching (Perhaps Without Imitation)

A major driving force in the search for imitative learning in non-human animals was the attempt to determine if the observer improved its performance by using the demonstrated actions rather than simply interacting with the same objects or the same parts of the objects as the demonstrator. However, the power of perceptual enhancement effects in social learning tasks was sometimes underestimated when researchers designed experiments to test imitative learning in their study species. Higher cognitive processes like imitation cannot be assessed (and maybe also not found) if these 'lower-level effects' have not been meticulously controlled for. As we will see later, the modification of a motor plan is inhibited if (new) objects are in place, as they are likely more salient than the actions of the demonstrator.

In one study that aimed at testing imitation, when dogs were confronted with a tube mounted on a pole and required to manipulate it so that a ball rolls out, they could do so in different ways, according to what they have seen before from a human demonstrator (Pongrácz et al. 2012). One group of dogs saw the experimenter putting the right hand on the top of the tube (at its end) and pushing it down until the ball rolled out from it. A second group saw the demonstrator tilting the tube by grabbing one of the 27-cm-long, thick ropes that were attached on each end of the tube with their right hand and pulling it down until the ball rolled out. Thus, the matching of the demonstrated action (pushing or pulling) could be achieved by simply manipulating the same (part of the) object (the tube's end or the rope). These two parts of the apparatus were both sufficiently well spatially separated (about 30 cm apart) and visually different to make this distinction. Furthermore, and most importantly, the different parts necessitate different actions when a dog attempts to manipulate them. While the end of the rope cannot be pushed down (but can easily be grasped with the mouth and pulled), the top of the tube cannot be grasped with the mouth and pulled (but can easily be pushed down). Therefore, the most parsimonious explanation of the matching effect the authors found is again stimulus enhancement (but not local enhancement, as the dogs showed no preference for manipulating the same side of the tube as the demonstrator) rather than in terms of action copying (pulling vs. pushing).

A similar argument can be used for most experiments that use the so-called bidirectional control method to test for imitation. By using an object (a rod or a sliding door) that can be moved in two different directions, researchers aimed at controlling for social facilitation and perceptual enhancement effects. Although the movements are essentially the same in terms of the motor pattern (e.g. pushing with the muzzle), they can be considered different in terms of the overall action plan. Two experiments in which the bidirectional control was applied produced contrasting results. While the dogs in the study by Kubinyi and colleagues (2003) failed to show a significant matching effect of lever pushing (see above), a more recent study was successful in this respect. On the first test trial, 11 of 12 dogs that could observe, from a position standing obliquely behind another dog, this dog pushing a sliding door to either the left or the right, pushed the screen in the same direction as demonstrated (Miller et al. 2009).

Why did the dogs perform much better in the second study? Two methodological differences between the studies may have accounted for the different outcomes. First, the overall action was more salient in the second study. The sliding door was much larger and heavier than the protruding lever and thus required larger movements (approximately 16 cm in case of the sliding screen) to the left or right. Second, the demonstrator in the first study was the owner, while in the second study it was a trained dog. The use of a conspecific demonstrator may have been an advantage, as it showed the exactly same action (pushing with the muzzle), while the human demonstrator used her/his hand. Perhaps the dogs in the Kubinyi et al. (2003) study had difficulties in transforming the hand action into the appropriate action of a dog. A human demonstrator was also used in a further group of dogs by Miller et al. (2009), but those dogs showed no significant matching effect (though being only marginally different from chance).

Still there is a third possibility to explain the positive results of Miller and colleagues (2009). Rather than the action itself (pushing left or right), the demonstration may make the edge of the aluminium screen at which it is pushed particularly salient. The two edges were separated by 30.5 cm, which seems to be sufficiently distant to discriminate them. Interestingly, for the training of the demonstrator dog a visual cue was sufficient to indicate which direction to push the screen: a single finger point to one side of the screen from the experimenter located behind the apparatus. Importantly, simply going to the same side of the screen would produce a sufficient degree of demonstrator-observer matching. Manipulation at the left edge can only result in a right-push, manipulation at the right edge can only result in a left-push. A complete control for stimulus or local enhancement effects would have required offering a hole in the screen into which the dog could insert its snout (as we have used in a study with pigs; (Ricke 2013)). From there, each direction of pushing is equally likely.

It is necessary to note that an account in terms of stimulus or local enhancement is not the only alternative to imitation in such an object movement task. Observers may directly learn something about the object—such as its movement or the relationship between the movement and the outcome—even without any intervention by a demonstrator. Instead of matching the behavior of the demonstrator, the observer might produce the action via *object movement reenactment* (copying the direction the object moves (Custance et al. 1999)). When one group of dogs observed the screen moved unobtrusively by the experimenter while another dog was present but did not interact with the apparatus, they also matched the direction in which the screen moved. Therefore, as Miller and colleagues (2009) pointed out, dogs seemed to have learned by observation about the properties of the door (emulation) and their relationship with reward (observational conditioning), rather than about body movements (imitation).

To determine whether animals learn by observation about responses or about changes of state in the environment, a two-action/one-outcome procedure is necessary. Such a procedure would guarantee that only the demonstrator's response topographies (the movements of the demonstrator) differed rather than their effects on the environment (the movements of the objects). Zentall and colleagues pioneered this approach by comparing naïve pigeons (Zentall et al. 1996) and quail (Akins and Zentall 1996) that had observed a demonstrator either pecking at or stepping on a treadle for food (see also McGregor et al. 2006; Saggerson et al. 2005). In a similar manner Huber and colleagues found true imitation effects with marmosets; these monkeys learned by observation to operate a swinging door (Bugnyar and Huber 1997) or to open a lid of a food canister with either their hand or mouth (Voelkl and Huber 2000, 2007). In the following sections we describe experiments with dogs using the two-action/one-outcome procedure.

4.3.2 Automatic Imitation

Copying body movements is at the core of theories about imitation. Current theories assume that imitation is achieved by activation of motor representations through observation of action. More specifically, observing somebody else executing an action leads to an activation of an internal motor representation in the observer because the observed action is similar to the content of the equivalent motor representation (Prinz 2002; Wohlschläger et al. 2003).

This pre-activated motor representation is then used to imitate the observed behaviour (e.g. Brass and Heyes 2005).

Where does this motor representation come from? One influential theory in regard to imitative learning, the associative sequence learning model (ASL) (Heyes 2001; Heyes and Ray 2000), suggests that the development of imitation depends on sensorimotor experience and phylogenetically general mechanisms of associative learning and motor control. The experience consists of correlated (contiguous and contingent) observation and execution of the same body movement, the mechanisms of associative learning are the same that produce Pavlovian and instrumental conditioning in the laboratory. During (early) ontogeny, individuals form 'matching vertical associations', excitatory links between sensory and motor representations of the same action, forged through correlated experience of observing and executing the same action.

We tested the effects of correlated experience of observing and executing the same (matching) or different (nonmatching) actions in dogs (Range et al. 2011). First we trained dogs to open a sliding door of a wooden box using their head and their paw. One group of dogs was rewarded with food for opening the door using the same method (head or paw) as demonstrated by their owner, another for opening the door using the alternative method. The second group, which had to counter-imitate to receive a food reward, was significantly slower to learn the task. This suggests that dogs cannot inhibit online the tendency to imitate using the same body part. In a subsequent transfer test, all dogs were required to imitate their owners' demonstrated sequences of head and paw use for food rewards. Consistent with the training, the second group made a greater proportion of incorrect, counter-imitative responses than dogs of the first group that had previously been rewarded for the matching response.

The findings provided evidence that dogs, like humans (Heyes et al. 2005) and budgerigars (Mui et al. 2008), are subject to 'automatic imitation'. Such automatic imitation is pervasive in everyday human life, where it promotes affiliation and cooperation among social partners (Van Baaren et al. 2009), and compatible with theories that propose a mediating role of the 'mirror neuron system' in response facilitation (Catmur et al. 2009). But most importantly is is the key to solve the 'correspondence problem', i.e., how the imitator knows what pattern of motor activation will make its action look like that of the model (Heyes 2001).

This imitative capacity, which seems to be an emergent property of the motor system, does not, however, mean that imitation is only blind copying. Why don't we imitate all the time? The most recent years of research have been dedicated to the examination of the flexibility of the imitation mechanism. It has addressed the question how the capacity to imitate is brought under intentional control (Heyes et al. 2009); in other words, how voluntary imitation is. Some researchers believe that this ability to facilitate, reorganize, coordinate, and inhibit externally triggered motor representations may be what distinguishes human imitative capacity from that of other species (Rizzolatti 2005).

4.3.3 Voluntary Imitation

Inhibitory control is needed to ensure that imitative behaviour is goal directed rather than compulsive (Brass et al. 2009). Would dogs be able to decide whether to imitate or not in a manipulative problem-solving task? We used a variant of the two-action/one-outcome logic when testing the imitative abilities of pet dogs (Range et al. 2007). A female border collie ('Guinness') was trained to push down with her paw a wooden rod that was dangling from a tree and connected via chains to a food box. If the rod was sufficiently strongly pushed down (not only touched!), the bottom of the food box opened and released a piece of food. To see how naive dogs would manipulate the rod spontaneously, we tested 14 dogs in a control condition without demonstration. Since only two of them used their paws, mouth operation was clearly the preferred method in our sample of dogs. In contrast with this, dogs of two observer groups could observe the demonstrator dog using her paw action ten times, but in two slightly different ways. For one group, Guinness carried a ball in her mouth (mouth-occupied group, MO); for the other, her mouth was free of any object (mouth-free group, MF). To keep the attention of the observers high, the experimenter addressed them using communicative cues and baited the food box ostentatiously. Furthermore, they were allowed to retrieve the food produced by the demonstrator. The owners of the dogs were also present, but one-third of them were blind-folded to control for "Clever Hans" effects.

After the ten demonstrations the observers were encouraged to manipulate the rod themselves. In the very first trial, observer dogs from the MF group showed demonstrator-matching behavior; 13 of 18 dogs used their paws to push down the rod and to retrieve the food. Three more dogs also used their paws, but tentatively, and also their mouths, but never achieved to open the box. Only three dogs used their mouths. In contrast, 15 of 19 members of the MO group used the method of the control group, namely biting into the rod and pulling it down (sometimes pushing it down with the snout). However, this first-trial pattern changed in the course of seven additional test trials. Already from the second trial on they exhibited a tendency to the demonstrated paw use and remained with this method for the rest of the testing.

Especially surprising was this marked difference in the first-trial performance of the observers given the tiny change in conditions. Why has a small blue ball in the mouth of the demonstrator produced such a non-imitative effect? Obviously, it was a salient enough cue for the observer to change the meaning of the demonstration. It prevented them from 'blindly copying' the demonstrated action and instead encouraged them to use their own, preferred one. Probably it rendered Guinness' demonstration less exemplary. Dogs seemed to be choosy in this situation.

In one of the detour tasks of Pongrácz and collaborators (Pongrácz et al. 2003), dogs chose the suddenly open shorter route instead of the long detour to reach the food bowl. So they seem to behave goal-directed and, like other animal species, to optimize their behavior on the basis of efficiency. Given the clear preference of the control group for the mouth method, using the paw in this context is likely to be 'inefficient' for dogs. But it can be turned into efficiency when the other, preferred method is no longer available. The blue ball may have produced such a transformation. Obviously, the performance of the dogs in the MO group was affected by this. The observable combination of the goal (the manipulation of the rod), the inefficient action (paw use) and the situational constraints of the demonstrator (the ball in the mouth) encouraged the observers (without a ball in the mouth) to use their preferred and efficient method on test. We concluded that the clear divergence in the performance of the two experimental groups suggests an ability of dogs—like chimpanzees (Buttelmann et al. 2007; Horner and Whiten 2005)—to imitate 'selectively'. However, their suggested ability to predict the most efficient action to achieve a goal within the constraints of a given situation does not necessarily require the attribution of mental states to others. It may not even reflect the possession of something like the non-mentalistic teleological interpretational system of human infants (Gergely and Csibra 2003), though this seemed to us the most likely interpretation. And although the outcome of the study was strikingly similar to an earlier one with non-verbal infants, after which it was modelled (Gergely et al. 2002), it does not per se lend support to a convergence of dogs and humans in the ability to inferentially evaluate the 'rationality' of others' actions. The much more mundane aim of the study was to examine whether dogs automatically copy a demonstrated action (in whatever situation) or selectively re-enact the demonstrated action depending on the constraints of the situation.

Two studies from Leipzig failed to replicate these findings. One was attempted as a faithful replication, but with the addition of one important control. In addition to the MO group with a demonstrator having a ball inside the mouth, Kaminski et al. (Kaminski et al. 2011) used a control group for which the demonstrator had a ball near the head instead. This tested whether a ball has a distracting effect on the observer's attention (away from the situational constraints of the demonstrator) and instead 'primes' the observer's tendency to grasp things with their mouth. Unfortunately, procedural differences (reversed baseline tendency of using the paw in the control group due to extensive pretraining) and differences in the analysis (not first trial data but first successful trial data) invalidated this study as an attempt at replicating the Range et al. (2007) study (Huber et al. 2012). Furthermore, as the baseline tendency for the demonstrated action (paw action) was already high (62.5 %), the potential for an imitation effect (increasing the baseline tendency) was low. This study found no difference in the number of demonstrated actions between observers and non-observers.

The second study also failed to find evidence for imitation in dogs (Tennie et al. 2009). Observer dogs were confronted with one of two actions performed by a demonstrator dog (to lie down on either the belly or sideways). These actions had been previously trained in the observer dogs by their owners (except in one condition) and were therefore not novel. The critical measure was the difference in frequency of showing these actions in the observer dogs and dogs that had not witnessed the demonstration. But despite high levels of attention during demonstrations, observers did not outperform the control dogs.

The most obvious difference between the studies outlined before and the one by Tennie et al. (2009) was the nature of the demonstrated action(s). While pushing, pulling, pawing etc. are *transitive* actions, i.e. targeted towards an object (a lever, a rod, a sliding door), here a target object is missing and thus the action is *intransitive*. In terms of action understanding this difference is important. In monkeys, mirror neurons have been found in the premotor cortex (F5) responding to transitive, object-related actions, like grasping a piece of food, chewing, sucking it, but none that respond to intransitive actions (except in the mouth region for communicative sounds, like lip smacking (Ferrari et al. 2003)). Thus it seems as if the neuronal substrate for the brain to go into resonance mode when the subject watches intransitive actions is not existent in primates, and maybe also not in dogs.

4.3.4 Do-As-I-Do Imitation

When animals (or humans) are tested in social learning tasks, it is often difficult to determine what exactly has been copied and what information the observer has acquired from the demonstration. The observer may have used the new information to improve its knowledge or to modify its behaviour (productive imitation) or to apply already known actions in the right circumstances (context imitation) (Byrne 2002). In many situations, instead of precisely copying others' actions (and their results on the environment), it is more useful to understand the goal of the demonstrator's actions, and only copy those actions of the demonstrator that are relevant to the task or preferred by the observer. Furthermore, animals may learn through observation how the environment works by learning about the affordances of objects or causal relationships between them (Huber et al. 2001).

There are only a few studies that have measured the fidelity or precision of the copied action. For instance, marmosets proved movement imitation not only at the level of action matching (pushing or pulling) (Bugnyar and Huber 1997) or body part matching (using the mouth or the hand) (Voelkl and Huber 2000), but also at the highest level of movement matching. Marmoset observers showed a very high level of copying fidelity—assessed through movement analysis—when creating a novel opening technique from observation (Voelkl and Huber 2007).

Dogs had initially not been found to achieve such high levels of movement imitation. The situation changed when Topál and colleagues (Topál et al. 2006) started to use a special method to assess copying fidelity and thereby used tasks (among others) without involving objects in the demonstrations. Only if intransitive movements like arbitrary gestures or facial expressions are used does the imitation task require mere movement copying. The researcher can thus determine if the observer uses the demonstrated action elements as a sample against which to match his choice of corresponding action. In contrast to all studies reviewed so far, the animals have first been trained (tutored) to copy ("do it as I do it!") a set of actions demonstrated by the experimenter or the caretaker and then tested with novel ones. This explicit nature of the method has therefore been called 'do-as-I-do' paradigm.

Initially used to test the home-reared chimpanzee Viki's ability to reproduce a variety of actions on command, the do-as-I-do paradigm has later been applied for many other animals (see, for review, Huber et al. 2009). The 4-year-old Belgian Tervueren 'Philip' was the first dog to prove capable of copying human actions with this paradigm (Topál et al. 2006). Philip had been trained as a service dog, that is to assist his disabled owner in tasks such as to open doors, pick up items, switch on/off lights. As in the other do-as-I-do studies, Philip was first tutored to repeat human-demonstrated actions on command ('Do it!') and then to generalize his understanding of copying to untrained action sequences and to actions shown by other people.

Interestingly, the precise topography of the copied movement patterns revealed severe limitations. Compared with children, who showed recognizable matching on all of the actions in the battery used, fidelity in all those studies with great apes was typically low overall (Whiten et al. 2004). The dog Philip was no exception to this. Like the apes, he had difficulty replicating body-oriented actions compared with object-oriented ones and often confused some demonstrated actions with similar ones that were already stored in his action repertoire. A study in our lab with the female Weimaraner named 'Joy' corroborated these findings (Huber et al. 2009).

Like Philip, Joy was first trained (with reward) to perform a sample of (eight) human-demonstrated actions on verbal command ('Do it!') in order to achieve some functional correspondence. As soon as Joy's performance reached a high, asymptotic level, she was tested on different types of novel actions without reinforcement. If confronted with previously untrained actions composed of familiar elements, Joy showed high degrees of matching, irrespective of whether she had to copy object-oriented or body-oriented actions. This indicates that she recognized and encoded a perceived action and then selected the motor response that achieved a match between the observed and performed action (called 'response facilitation' by Byrne (2002)). Deviations only occurred by choosing other trained actions, possibly revealing effects of pro- and retro-active inference. In the few mismatching cases, like in great apes, Joy responded initially with a training action or an action from her repertoire and later approximated what was shown.

Joy's copying fidelity was much poorer when confronted with action sequences (composed of two distinct actions) and with exotic actions. We have considered the latter ones as extremely unusual actions, of which her body should have been capable, but which she has never performed before (according to her caregiver). Among them we tested non-functional, gesture-like movements, because neither action results nor the demonstrator's goal could be used to infer the action.

Joy did not replicate any of the exotic actions on the first trial. However, while she showed a tendency to approximate object-oriented actions in three trials, she completely failed with intransitive (body) actions. This is not only congruent with the findings from great apes, but also with those from autistic children (Heimann et al. 1992). Furthermore, like apes (see Call 2001), dogs seem not be attuned to the details of the actions, but more to a functional replication. They show similar tendencies of perseveration as great apes and quickly fall back into the attractors of training actions.

Two more tests with Joy provided suggestive, if not conclusive evidence, that dogs seem capable of imitation that goes beyond blind copying or simple mimicry. In order to test whether Joy's copying was the result of an enduring representation of the demonstrator's behaviour, a key requirement of true imitation (Zentall 2006), we conducted a so-called *deferred imitation* test. Although Joy's matching degree decreased with increasing delays of the command, she could perform correctly with delays shorter than 5 s and once matched a familiar action even after 35 s. Importantly, Joy was required to fetch a stick between the demonstration and the 'Do it!' command, and not only to wait (Dorrance and Zentall 2001). This positive evidence of deferred imitation was recently confirmed on a bigger sample of dogs. Using a human demonstrator, dogs could reproduce novel actions after 1-min intervals and familiar actions after 10-min intervals, the latter even if distracted by different activities during the retention interval (Fugazza and Miklósi 2013).

A final test with Joy asked whether she would try to make sense of the action and then re-create the most effective or most plausible solution. We therefore devised actions for which the 'target object' was not (or was no longer) present (like in pantomime). For instance, the human demonstrator pretended to jump over a hurdle, but there was no hurdle around. In this first 'invisible hurdle' test, Joy ran in the same direction as the demonstrator before, but did not jump, only stopped and looked back to the demonstrator. Half a year later this test was replicated, again unrewarded, but this time a hurdle was positioned in the garden about 5 m away from the demonstrator's 'jumping position'. This time, Joy ran straight to this real hurdle and jumped over. Possibly she made sense of the observed action by 'completing' it in her own attempt. Altogether, the three 'do-as-I-do' studies with dogs, plus a recent study using dogs as demonstrators (Bentlage 2013), suggest that dogs form a kind declarative (non-procedural) memory for the observed actions and then choose a matching response (action imitation) or one that produces a similar result (goal emulation).

4.4 The Potential Origins of Dog Social Learning

The range of possible origins of dogs' social skills includes short-, mid- and long-term developments. As with the enhanced cognitive performance of primates that grow up in a human environment and regularly interact with humans (Call and Tomasello 1996), enculturation during individual ontogeny, it has been proposed, boosts the cognitive abilities of pet dogs as well (Miklósi et al. 2004; Topál et al. 2009). Specifically, the sensitivity of an individual dog to human actions may depend on its acceptance of (some) humans as social companions (Udell et al. 2010). Alternatively or additionally, a similar alteration of dog behaviour towards a more human-like direction might have taken place in a (mid-term) evolutionary sense during the course of domestication (Hare and Tomasello 2005).

Concerning the attentiveness of dogs towards humans, dogs seem to have evolved a special relationship with and interest towards humans. Already at their age of four months dogs show attachment behaviours to their owners, similar to the relationship between human infants and their mothers (Topál et al. 1998, 2005). Domestication plays a role here since human-raised wolves do not show a similar preference for their caretaker over an unfamiliar person (Topál et al. 2005). This close relationship with their human caretaker has also been shown to result in a higher attentiveness of dogs towards humans in comparison to conspecifics, but also in regard to different people with whom they have different relationships (Horn et al. 2013a, b; Mongillo et al. 2010; Range et al. 2009b). Comparing similarly human-raised young wolves and dogs also confirms that domestication enhanced the attentiveness of dogs towards humans: in a communicative situation, after calling the animal's attention, a human experimenter could establish eye-contact with dogs more easily than with wolves (Gacsi et al. 2009; Miklosi et al. 2003; Viranyi et al. 2008).

There is also evidence that dogs benefit from their increased attentiveness towards humans and are extraordinarily responsive to human-given social cues. Numerous studies using the experimental paradigm of object-choice tasks employing human-given cues have provided convincing evidence that dogs are experts in following human gestures to find hidden food (Miklósi and Soproni 2006; Reid 2009). Neither chimpanzees nor wolves seem to use human communication as flexibly as the domestic dog (Hare et al. 2002; Miklósi et al. 2003; Gácsi et al. 2009) (but see Range and Viranyi 2011). However, as fascinating as this pet dog feature is, it is per se not indicative that the dog learns anything from observing its human 'demonstrator'. In this chapter, we have shown that pet dogs can learn from observing humans—but because comparisons with wolves are missing in this respect, we cannot know what role domestication and growing up in the human environment play here. Only rigorous and fair comparisons between (equally raised) dogs and wolves, pet and stray dogs, more and less enculturated dogs, young and old dogs, etc. can disentangle the different strengths of these long- and short-term influences on social learning in dogs.

A thorough analysis of this question could also answer the question in what way and to what extent humans have altered the social learning skills of dogs and to what extent these reflect the social learning abilities of wolves. Wolves live in packs that are basically one-family units that hunt, rear young, and protect a communal territory as a stable group with the pack members usually being related individuals (Mech and Boitani 2007). Formation and persistence of the pack as a functional unit is based on the strong social bonds among its members. This provides the natural basis of high attentiveness towards and possibly of substantive learning from pack members. Wolves should therefore be well adapted to learn by observation, including the most sophisticated forms like emulation and imitation. This has been supported by anecdotal evidence three decades ago (Frank 1980). First, several wolves locked in a kennel learned by observing a conspecific how to open a door that required two distinct operations to unlatch: pushing the handle toward the door and then rotating it. In contrast, dogs—in this case malamutes— never learned to open the door despite observing humans opening it several times a day for 6 years. However, although several wolves learned the actions by observation, they all used their own distinctive methods to do so.

Feral dog groups differ from wolves in regard to their breeding system and possibly other intraspecific interactions (Boitani and Ciucci 1995; Butler et al. 2004; but see Bonanni et al. 2010b). Although they live in pack-like social groups and display differentiated social relationships with each other including forming an age-graded linear dominance hierarchy (Bonanni et al. 2010a; Cafazzo et al. 2009), female feral dogs raise their pups alone (Boitani and Ciucci 1995; Daniels and Bekoff 1989) or with the help of the fathers that may contribute to the defence of the pups but rarely feed them (Pal 2005). Thus, dogs might lack some of the coordinated activities of wolf packs, which might have consequences for learning from conspecifics.

Experimental studies of social learning in feral or stray dogs are not available. It would be even more speculative to reason about the social attentiveness and learning propensities of pet dogs, but fortunately some experimental data is available in this respect. Though on the above basis one may expect that dogs have lost their abilities to learn from conspecifics during domestication, the studies reviewed above show that dogs readily learn from conspecifics in various social learning tasks and by using different mechanisms. Consequently, we need to seriously consider the hypothesis that the true social learning abilities of dogs originate from their ancestors, wolves.

We addressed this crucial question in a recent experiment at the Wolf Science Center (Ernstbrunn, Austria) by comparing wolves' and dogs' imitative abilities (Range and Virányi 2013). We presented 6-month-old wolves and dogs that were raised and kept under comparable conditions with a box that had food hidden inside. To open the box, the animals had to push down a wooden lever, which released the lid of the box. Each animal observed six demonstrations by one of two familiar dogs dominant over the subjects. One group saw a dog opening the box using its paw, while the other group saw a dog using its mouth to achieve the same action: pressing down the lever. We found that wolves were more successful in opening the box than the dogs and moreover, that they were more likely to match the demonstrated action. These results could not be attributed to delayed development of dogs compared to the wolves, since the dogs could also not solve the problem better one year later, nor could they solely be attributed to a better causal understanding of wolves compared to dogs, since a control group performed significantly worse. These are the first data suggesting that wolves might outperform dogs in regard to their imitative abilities at least with conspecific demonstrators.

4.5 Conclusion

Altogether, the findings of social learning experiments in dogs from the last decade have provided ample evidence that dogs are not only especially gifted for attentiveness to and adjusting their behaviour to humans, but can also learn socially.

Dogs have proved to profit from the demonstration of a skilled model—being either another dog or a human—in solving manipulative problems, by either being more motivated or by using various kinds of information and—in specific circumstances—even forming new representations of the observed actions.

However, this does not mean that dogs are always rational or strategic imitators with outstanding abilities in this respect. It is crucial to distinguish between automatic and voluntary imitation. The development of imitation depends on sensorimotor experience and phylogenetically old mechanisms of associative learning. Domestic dogs are perhaps special insofar as they are known to be highly sensitive to social cues, and their interactions with humans may provide the kind of sensorimotor experience that is required for the development of (automatic) imitation. More complex forms of imitation require not only that the observation of action elements automatically activates a corresponding motor programme based on previous experience, but also the encoding of the demonstrated order of elements in a novel sequence of body movements (sequence learning) and the inhibition of automatic copying tendencies. The findings from the selective imitation experiment and the Do-as-I-do studies suggest that dogs have some capacities in the latter respect as well, and that they can form enduring representations of the demonstrated actions. How flexibly and voluntarily dogs can use these representations in their subsequent attempts to solve the task, and how much dogs are guided by their understanding of the goals and the situational constraints of the demonstrator, are still challenging questions and invite further experiments in the future.

Further studies need to investigate also to what extent the respective propensities and skills of dogs originate from their ontogeny and domestication, and to what extent they lie in their much more distant past. The common ancestor of dogs and wolves seem to have at least the same imitative abilities as dogs, and in manipulative problem solving situations with a conspecific demonstrator wolves may even outperform dogs. Given wolves' fine-grained and complex social life and their dependency on coordinating their actions with pack members during rearing of young and hunting, this is not surprising (Range and Virányi 2013) but also (Range and Viranyi 2012). This does not change the fact, however, that dogs are special animals, both in terms of their evolutionary history of domestication, and the range and intensity of their developmental training by humans. It will be an exciting scientific enterprise to figure out in what way these factors enhance the attentiveness of dogs towards humans and conspecifics and their learning abilities.

References

Adler, L. L., & Adler, H. E. (1977). Ontogeny of observational learning in the dog (*Canis familiaris*). *Developmental Psychobiology, 10,* 167–271.
Akins, C. K., & Zentall, T. R. (1996). Imitative learning in male Japanese quail (Coturnix japonica) using the two-action method. *Journal of Comparative Psychology, 110*(3), 316–320.
Bentlage, J. (2013). "Do as I do" imitation in dogs: Learning of transitive and intransitive actions from a con-specific model. Unpublished master thesis, University of Vienna.

Boitani, L., & Ciucci, P. (1995). Comparative social ecology of feral dogs and wolves. *Ethology Ecology Evolutionary, 7,* 49–72.

Bonanni, R., Cafazzo, S., Valsecchi, P., & Natoli, E. (2010). Effect of affiliative and agonistic relationships on leadership behaviour in free-ranging dogs. *Animal Behaviour, 79*(5), 981–991.

Bonanni, R., Valsecchi, P., & Natoli, E. (2010). Pattern of individual participation and cheating in conflicts between groups of free-ranging dogs. *Animal Behaviour 79*(4), 957–968.

Brass, M., & Heyes, C. (2005). Imitation: Is cognitive neuroscience solving the correspondence problem? Trends. *Cognitive Science, 9*(10), 489–495.

Brass, M., Ruby, P., & Spengler, S. (2009). Inhibition of imitative behaviour and social cognition. *Philosophical Transactions of the Royal Society of London. Series B, 364*(1528), 2359–2367.

Brogden, W. J. (1942). Imitation and social facilitation in the social conditioning of forelimb-flexion in dogs. *American Journal of Psychology, 55,* 77–83.

Bugnyar, T., & Huber, L. (1997). Push or pull: An experimental study on imitation in marmosets. *Animal Behaviour, 54*(4), 817–831.

Butler, J. R. A., du Toit, J. T., & Bingham, J. (2004). Free-ranging domestic dogs (*Canis familiaris*) as predators and prey in rural Zimbabwe: Threats of competition and disease to large wild carnivores. *Biological Conservation, 115*(3), 369–378.

Buttelmann, D., Carpenter, M., Call, J., & Tomasello, M. (2007). Enculturated chimpanzees imitate rationally. *Developmental Science, 10*(4), F31–F38.

Byrne, R. W. (2002). Imitation of novel complex actions: What does the evidence from animals mean? *Advances in the Study of Behaviour, 31,* 77–105.

Cafazzo, S., Valsecchi, P., Fantini, C., & Natoli, E. (2009). Social dynamics of a group of free-ranging domestic dogs living in a suburban environment. *Journal of Veterinary Behavior, 4*(2), 61.

Call, J. (2001). Body imitation in an enculturated orangutan (Pongo pygmaeus). *Cybernet Systems, 32*(1–2), 97–119.

Call, J., & Tomasello, M. (1996). The effect of humans on the cognitive development of apes. In A. E. Russon, K. A. Bard & S. T. Parker (Eds.), *Reaching into thought.* Cambridge University Press, New York, pp 371–403.

Catmur, C., Walsh, V., & Heyes, C. (2009). Associative sequence learning: The role of experience in the development of imitation and the mirror system. *Philosophical Transactions of the Royal Society of London. Series B, 364*(1528), 2369–2380.

Custance, D., Whiten, A., & Fredman, T. (1999). Social learning of an artificial fruit task in capuchin monkeys (Cebus apella). *Journal of Comparative Psychology, 113*(1), 13–23.

Daniels, T. J., & Bekoff, M. (1989). Feralization: The making of wild domestic animals. *Behavioural Processes, 19*(1–3), 79–94.

Dickinson, A. (1980). *Contemporary animal learning theory.* Cambridge University Press, Cambridge, UK.

Dorrance, B. R., & Zentall, T. R. (2001). Imitative learning in Japanese quail depends on the motivational state of the observer at the time of observation. *Journal of Comparative Psychology, 115,* 62–67.

Ferrari, P. F., Gallese, V., Rizzolatti, G., & Fogassi, L. (2003). Mirror neurons responding to the observation of ingestive and communicative mouth actions in the monkey ventral premotor cortex. *European Journal of Neuroscience, 17*(8), 1703–1714.

Fragaszy, D. M., & Visalberghi, E. (2004). Socially biased learning in monkeys. *Learning Behaviour, 32,* 24–35.

Frank, H. (1980). Evolution of Canine information-processing under conditions of natural and artificial selection. *Zeitschrift fur Tierzuchtung, 53,* 389–399.

Fugazza, C., & Miklósi, Á. (2013). Deferred imitation and declarative memory in domestic dogs. *Anim Cogn* published online.

Gacsi, M., Gyori, B., Viranyi, Z., Kubinyi, E., Range, F., Belenyi, B., & Miklosi, A. (2009). Explaining dog wolf differences in utilizing human pointing gestures: Selection for synergistic shifts in the development of some social skills. *PLoS One, 4*(8), e6584.

Galef, B. G. (1988). Imitation in animals: History, definition and interpretation of data from the psychological laboratory. In T. R. Zentall & B. G. Galef (Eds.), Social learning: Psychological and biological perspectives. Erlbaum, Hillsdale, NJ, pp 207–223.

Gergely, G., Bekkering, H., & Kiraly, I. (2002). Rational imitation in preverbal infants. *Nature, 415*(6873), 755.

Gergely, G., & Csibra, G. (2003). Teleological reasoning in infancy: The naive theory of rational action. *Trends in Cognitive Sciences, 7*(7), 287–292.

Hare, B., & Tomasello, M. (2005). Human-like social skills in dogs? *Trends in Cognitive Sciences, 9*, 439–444.

Hare, B., Brown, M., Williamson, C., & Tomasello M (2002). The domestication of social cognition in dogs. *Science, 298*, 1634–1636.

Harr, A. L., Gilbert, V. R., & Phillips, K. A. (2009). Do dogs (*Canis familiaris*) show contagious yawning? *Animal Cognition, 12*, 1435–1448.

Heberlein, M., & Turner, D. C. (2009). Dogs, *Canis familiaris*, find hidden food by observing and interacting with a conspecific. *Animal Behaviour, 78*(2), 385–391.

Heimann, M., Ullstadius, E., Dahlgren, S.-O., & Gillberg, C. (1992). Imitation in autism: A preliminary research note. *Behavioral Neurology, 5*, 219–227.

Heyes, C. M. (1994). Social learning in animals: Categories and mechanisms. *Biological Reviews, 69*, 207–231.

Heyes, C. (2001). Causes and consequences of imitation. *Trends in Cognitive Sciences, 5*(6), 253–261.

Heyes, C., Bird, G., Johnson, H., & Haggard, P. (2005). Experience modulates automatic imitation. *Brain Research. Cognitive Brain Research, 22*(2), 233–240.

Heyes, C., Huber, L., Gergely, G., & Brass, M. (2009). Evolution, development and intentional control of imitation. *Philosophical Transactions of the Royal Society of London. Series B, 364*(1528), 2291–2443 (special issue).

Heyes, C. M., & Ray, E. D. (2000). What is the significance of imitation in animals? In P. J. B. Slater, J.S. Rosenblatt, C. T. Snowdon & T. J. Roper (Eds.), *Advances in the Study of Behaviour*, vol. 29, (pp 215–245). Academic Press, New York.

Horn, L., Huber, L., & Range, F. (2013a). The importance of the secure base effect for domestic dogs - evidence from a manipulative problem-solving task. *PLoS One, 8*(5), e65296.

Horn, L., Range, F., & Huber, L. (2013b). Dogs' attention towards humans depends on their relationship, not only on social familiarity. *Animal Cognition, 16*(3), 435–443.

Horner, V., & Whiten, A. (2005). Causal knowledge and imitation/emulation switching in chimpanzees (Pan troglodytes) and children (Homo sapiens). *Animal Cognition, 8*(3), 164–181.

Huber, L., Range, F., & Viranyi, Z. (2012). Dogs imitate selectively, not necessarily rationally: Reply to Kaminski et al. (2011). *Animal Cognition, 83*(6), e1–e3.

Huber, L., Range, F., Voelkl, B., Szucsich, A., Viranyi, Z., & Miklosi, A. (2009). The evolution of imitation: What do the capacities of nonhuman animals tell us about the mechanisms of imitation? *Philosophical Transactions of the Royal Society of London. Series B, 364*, 2299–2309.

Huber, L., Rechberger, S., & Taborsky, M. (2001). Social learning affects object exploration and manipulation in keas, Nestor notabilis. *Animal Behaviour, 62*(5), 945–954.

Joly-Mascheroni, R. M., Senju, A., & Shepherd, A. J. (2008). Dogs catch human yawns. *Biology Letters, 4*, 446–448.

Kaminski, J., Nitzschner, M., Wobber, V., Tennie, C., Bräuer, J., Call, J. et al. (2011). Do dogs distinguish rational from irrational acts? *Animal Behaviour, 81*(1), 195–203.

Kubinyi, E., Pongrácz, P., & Miklosi, A. (2009). Dog as a model for studying conspecific and heterospecific social learning. *Journal of Veterinary Behavior, 4*(1), 31–41.

Kubinyi, E., Topál, J., Miklosi, A., & Csanyi, V. (2003). Dogs (*Canis familiaris*) learn from their owners via observation in a manipulation task. *Journal of Comparative Psychology, 117*(2), 156–165.

Lorenz, K. Z. (1981). The foundations of ethology. New York: Springer.

Lupfer-Johnson, & G., Ross, J. (2007). Dogs acquire food preferences from interacting with recently fed conspecifics. *Behavioural Processes*, *74*(1), 104–106.

Madsen, E. A., & Persson, T. (2013). Contagious yawning in domestic dog puppies (Canis lupus familiaris): The effect of ontogeny and emotional closeness on low-level imitation in dogs. *Animal Cognition*, *6*(2), 233–240.

McGregor, A., Saggerson, A., Pearce, J., & Heyes, C. (2006). Blind imitation in pigeons, Columba livia. *Animal Behaviour*, *72*, 287–296.

Mech, L. D., & Boitani, L. (2007). Wolves: Behavior, Ecology, And Conservation. University of Chicago Press.

Mersmann, D., Tomasello, M., Call, J., Kaminski, J., & Taborsky, M. (2011). Simple mechanisms can explain social learning in domestic dogs (*Canis familiaris*). *Ethology*, *117*, 675–690.

Miklósi, Á., & Soproni, K. (2006). A comparative analysis of animals' understanding of the human pointing gesture. *Animal Cognition*, *9*, 81–93.

Miklósi, Á., Topál, J., & Csányi, V. (2004). Comparative social cognition: What can dogs teach us? *Animal Behaviour*, *67*, 995–1004.

Miklósi, Á., Kubinyi, E., Topál, J., Gacsi, M., Virányi, Z., & Csányi, V. (2003). A simple reason for a big difference: Wolves do not look back at humans, but dogs do. *Current Biology*, *13*(9), 763–766.

Miller, H. C., Rayburn-Reeves, R., & Zentall, T. R. (2009). Imitation and emulation by dogs using a bidirectional control procedure. *Behavioural Processes*, *80*, 109–114.

Mongillo, P., Bono, G., Regolin, L., & Marinelli, L. (2010). Selective attention to humans in companion dogs, *Canis familiaris*. *Behavioural Processes*, *80*(6), 1057–1063.

Mui, R., Haselgrove, M., Pearce, J., & Heyes, C. (2008). Automatic imitation in budgerigars. In *Proceedings of the Royal Society B: Biological Sciences*, *275*(1651), 2547–2553.

O'Hara, S. J., & Reeve, A.V. (2011). A test of the yawning contagion and emotional connectedness hypothesis in dogs, *Canis familiaris*. *Animal Behaviour*, *81*, 335–340.

Pal, S. K. (2005). Parental care in free-ranging dogs, *Canis familiaris*. *Applied Animal Behaviour Science*, *90*(1), 31–47.

Pongrácz, P., Bánhegyi, & P., Miklósi, Á. (2012). When rank counts—dominant dogs learn better from a human demonstrator in a two-action test. *Behaviour*, *149*, 111–132.

Pongrácz, P., Miklósi, A., Kubinyi, E., Gurobi, K., Topál, J., & Csányi, V. (2001). Social learning in dogs: The effect of a human demonstrator on the performance of dogs in a detour task. *Animal Behaviour*, *62*, 1109–1117.

Pongrácz, P., Miklosi, A., Kubinyi, E., Topál, J., & Csanyi, V. (2003). Interaction between individual experience and social learning in dogs. *Animal Behaviour*, *65*(3), 595–603.

Pongrácz, P., Miklosi, A., Timar-Geng, K., & Csanyi, V. (2004). Verbal attention getting as a key factor in social learning between dog (*Canis familiaris*) and human. *Journal of Comparative Psychology*, *118*(4), 375–383.

Prinz, W. (2002). Experimental approaches to imitation. In A. N. Meltzoff & W. Prinz (Eds.), *The imitative mind: Development, evolution, and brain bases* (pp. 143–162). Cambridge University Press, Cambridge.

Range, F., Heucke, S. L., Gruber, C., Konz, A., Huber, L., & Virányi, Z. (2009a). The effect of ostensive cues on dogs' performance in a manipulative social learning task. *Applied Animal Behaviour Science*, *120*(3–4), 170–178.

Range, F., Horn, L., Bugnyar, T., Gajdon, G. K., & Huber, L. (2009b). Social attention in keas, dogs, and human children. *Animal Cognition*, *12*, 181–192.

Range, F., Huber, L., & Heyes, C. M. (2011). Automatic imitation in dogs. In *Proceedings of the Royal Society B*, *278*, 211–217.

Range, F., & Viranyi, Z. (2011). Development of gaze following abilities in wolves (Canis lupus). *PLoS ONE*, *6*(2), e16888.

Range, F., & Viranyi, Z. (2012). The effect of domestication on cognitive abilities of dogs. *Journal of Veterinary Behavior: Clinical Applications and Research*, *7*(6), e14.

Range, F., Viranyi, Z., & Huber, L. (2007). Selective imitation in domestic dogs. *Current Biology, 17*, 1–5.

Range, F., & Viranyi, Z. (2013). Wolves are better imitators of conspecifics than dogs. *PLoS ONE* in press.

Reid, P. J. (2009). Adapting to the human world: Dogs' responsiveness to our social cues. *Behavioural Processes, 80*(3), 325–333.

Rescorla, R. A. (1988). Behavioral studies of Pavlovian conditioning. *Annual Review of Neuroscience, 11*, 329–352.

Ricke, D. (2013). Zeigen Schweine (Sus scrofa domestica) imitative Anzeichen beim Lernen bei Verwendung eines Two-Action-Tests? Unpublished diploma thesis, Messerli Research Institute, University of Veterinary Medicine Vienna.

Rizzolatti, G. (2005). The mirror neuron system and imitation. In S. Hurley & N. Chater (Eds.), *Perspectives on imitation. From neuroscience to social science* (pp 55–76). MIT Press, MA: Cambridge.

Saggerson, A. L., George, D.N., & Honey, R. C. (2005). Imitative learning of stimulus–response and response–outcome associations in pigeons. *Journal of Experimental Psychology-Animal Behavior Processes, 31*, 289–300.

Silva, K., Bessa, J., & de Sousa, L. (2012). Auditory contagious yawning in domestic dogs (*Canis familiaris*): First evidence for social modulation. *Animal Cognition, 15*(4), 721–724.

Silva, K., & de Sousa, L. (2011). 'Canis empathicus'? A proposal on dogs' capacity to empathize with humans. *Biology Letters, 7*(4), 489–492.

Slabbert, J. M., & Rasa, O. A. E. (1997). Observational learning of an acquired maternal behaviour pattern by working dog pups: An alternative training method? *Applied Animal Behaviour Science, 53*, 309–316.

Teglas, E., Gergely, A., Kupan, K., Miklosi, A., & Topál, J. (2012). Dogs' gaze following is tuned to human communicative signals. *Current Biology, 22*(3), 209–212.

Tennie, C., Glabsch, E., Tempelmann, S., Bräuer, J., Kaminski, J., & Call, J. (2009). Dogs, *Canis familiaris*, fail to copy intransitive actions in third-party contextual imitation tasks. *Animal Behaviour, 77*(6), 1491–1499.

Thorndike, E. L. (1898). Animal intelligence: An experimental study of the associative processes in animals. *Psychological Review Monograph. Supplement, 2*.

Thorndike, E. L. (1911). Animal intelligence. Macmillan, New York.

Thorpe, W. H. (1956). Learning and instinct in animals. Methuen, London.

Topál, J., Byrne, R. W., Miklosi, A., & Csanyi, V. (2006). Reproducing human actions and action sequences: "Do as I Do!" in a dog. *Animal Cognition, 9*(4), 355–367.

Topál, J., Gacsi, M., Miklosi, A., Viranyi, Z., Kubinyi, E., & Csanyi, V. (2005). Attachment to humans: A comparative study on hand-reared wolves and differently socialized dog puppies. *Animal Behaviour, 70*, 1367–1375.

Topál, J., Gergely, G., Erdohegyi, A., Csibra, G., & Miklosi, A. (2009). Differential sensitivity to human communication in dogs, wolves, and human infants. *Science, 325*(5945), 1269–1272.

Topál, J., Miklosi, A., Csanyi, V., & Doka, A. (1998). Attachment behavior in dogs (*Canis familiaris*): A new application of Ainsworth's (1969) strange situation test. *Journal of Comparative Psychology, 112*(3), 219–229.

Udell, M. A., Dorey, N. R., & Wynne, C. D. (2010). What did domestication do to dogs? A new account of dogs' sensitivity to human actions. *Biological Reviews of the Cambridge, 85*(2), 327–345.

van Baaren, R., Janssen, L., Chartrand, T. L., & Dijksterhuis, A. (2009). Where is the love? The social aspects of mimicry. *Philosophical Transactions of the Royal Society B, 364*(1528), 2381–2389.

Viranyi, Z., Range, F., & Huber, L. (2008). Attentiveness toward others and social learning in domestic dogs. In L. S. Röska-Hardy & E. M. Neumann-Held (Eds.), *Learning from animals? Examining the Nature of Human Uniqueness* (pp. 141–153). Hove: Psychology Press.

Voelkl, B., & Huber, L. (2000). True imitation in marmosets. *Animal Behaviour, 60*(2), 195–202.

Voelkl, B., & Huber, L. (2007). Imitation as faithful copying of a novel technique in marmoset monkeys. *PLoS ONE, 2*(7), e611.

Whiten, A., & Ham, R. (1992). On the nature and evolution of imitation in the animal kingdom: Reappraisal of a century of research. In P. J. B Slater, J. S Rosenblatt, C. Beer, & M. Milkinski (Eds.), *Advances in the study of behavior* (Vol. 21, pp. 239–283). New York: Academic Press.

Whiten, A., Horner, V., Litchfield, C. A., & Marshall-Pescini, S. (2004). How do apes ape? *Learning Behaviour, 32*(1), 36–52.

Wohlschläger, A., Gattis, M., & Bekkering, H. (2003). Action generation and action perception in imitation: An instance of the ideomotor principle. *Philosophical Transactions of the Royal Society B, 358*(1431), 501–515.

Yoon, J. M.D., & Tennie, C. (2010). Contagious yawning: Areflection of empathy, mimicry, or contagion? *Animal Behaviour, 79*(5), e1–e3.

Zentall, T. (2006). Imitation: Definitions, evidence, and mechanisms. *Animal Cognition, 9*, 335–353.

Zentall, T. R., Sutton, J. E., & Sherburne, L. M. (1996). True imitative learning in pigeons. *Psychological Science, 7*(6), 343–346.

Chapter 5
Social Looking in the Domestic Dog

Emanuela Prato-Previde and Sarah Marshall-Pescini

Abstract The study of dog social cognition is relatively recent and is rapidly developing, providing an interesting and multi-faceted picture of our "best friend's" sociocognitive abilities. In particular, since Miklósi et al.'s (2003) seminal work "A simple reason for a big difference: wolves do not look back at humans, but dogs do", there has been a surge of interest in the area of dog–human communication. In the current chapter we focus on dogs' comprehension of the human gaze and their ability to use human-directed-gazing as a communicative tool. We first review studies on the social significance of human eye contact for dogs, their understanding of eyes as indicators of attention, and their ability to take another's visual perspective into account. We also consider dogs' understanding of human eye-gaze as a communicative act, in terms of its potentially referential nature and as an ostensive cue signalling the communicative intent of the actor. We then move on to review studies on dogs' human-directed gazing behaviour, discussing whether it may be considered part of an intentional and referential communicative act, what the underlying motivations and contexts in which this behaviour is exhibited may be, and what variables affect its occurrence. Where open questions remains, we outline current debates and highlight potential directions for future research.

E. Prato-Previde (✉) · S. Marshall-Pescini
Dipartimento di Fisiopatologia Medico-Chirurgica e dei Trapianti, Sezione di Neuroscienze, Università degli Studi di Milano, Via F.lli Cervi 93 20090 Segrate, MI, Italy
e-mail: emanuela.pratoprevide@unimi.it

S. Marshall-Pescini
e-mail: sarah.marshall@unimi.it

S. Marshall-Pescini
Comparative Cognition, Messerli Research Institute, University of Veterinary Medicine, University of Vienna, Vienna, Austria

S. Marshall-Pescini
Wolf Science Centre, Ernstbrunn, Austria

5.1 Introduction: Dogs' Scientific Renaissance

Dogs were likely the first animals to be domesticated and they have shared a common environment with humans for longer than any other species (Vilà et al. 1997; Bokyo et al. 2009; Pang et al. 2009; vonHoldt et al. 2010; Ding et al. 2012; Larson et al. 2012; Druzhkova et al. 2013; Wang et al. 2013). At present, they are almost omnipresent in human lives and undoubtedly have a unique relationship with humans, which has been described since ancient times by poets, writers, and artists. Interestingly, the long history of proximity and closeness between dogs and humans has been one of the main reasons why dogs have been widely snubbed and regarded as an ethologically 'uninteresting' species and have only recently become subjects of scientific inquiry in the field of comparative cognition (Miklósi et al. 2004; Miklósi 2007).

In the last 20 years dogs have seen a 'scientific renaissance', with a sudden rise in the number of published studies on canine cognition (Cooper et al. 2003; Miklósi et al. 2004; Bensky et al. 2013). This new wave of canine cognition research has transformed the initially 'uninteresting' dog into a fascinating model for evolutionary cognition research and for the investigation of the building blocks underlying mental abilities in animals, particularly those involving social cognition (Miklósi et al. 2004). From these studies it emerges that dogs' success as domestic animals and their capacity to become "man's best friend" are rooted in a wide range of social skills and competencies that allow them to engage in complex communicative, relational, and cooperative interactions with humans (Miklósi and Topál in press).

Recently, Bensky et al. (2013) reviewed the literature on dog cognition, identifying a number of different areas in which there has been active empirical research. Following the taxonomy proposed by Tomasello and Call (1997), these authors grouped studies on dogs' cognitive abilities into two broad categories, defined mainly in terms of the function of the cognitive processes involved: physical cognition and social cognition. According to this classification, studies on dog physical cognition investigate how dogs perceive and interpret non-social stimuli to make sense of their physical environment, thus focusing on different topics such as discrimination learning, object permanence, object learning, categorization and inferential reasoning, object manipulation, problem-solving, quantitative understanding, spatial cognition, and memory (see Bensky et al. 2013 for a review). On the other hand, studies on social cognition have focused on understanding the dog's social world and social knowledge, investigating what dogs know about others (whether conspecific or human) and how they acquire such knowledge. Therefore, research on dog social cognition covers a wide range of topics dealing with social phenomena ranging from recognition and categorization of conspecifics and humans (Faragó et al. 2010; Huber et al. 2013), perception of their emotions (Nagasawa et al. 2011; Buttleman and Tomasello 2013; Merola et al. 2013b), the development and management of social relationships with humans (e.g. Topál et al. 1998; Prato-Previde et al. 2003; Custance and Mayer 2012; Riemer et al. 2013) and conspecifics (see Bonanni and Cafazzo in press, Smuts in press for recent reviews), the acquisition of new abilities through observation and interaction with others (e.g. Poncrágz et al. 2003, 2004; Range et al. 2007, 2011; Kubinyi et al. 2009; Tennie et al. 2009), and all the different aspects of intraspecific and interspecific communication (e.g. Bradshaw and Nott 1995; Miklósi et al. 2000; Rossi and Ades 2008; Horowitz 2009; Kaminski et al. 2011; Marshall-Pescini et al. 2012; see also Bensky et al. 2013 and Kaminski and Nitzschner 2013 for recent reviews).

Although there is growing evidence that dogs' social skills are in some respects unique and probably contributed to turning dogs into "man's best friend" (Miklósi and Topál in press), how these abilities evolved (Hare et al. 2002, 2010; Miklósi et al. 2003; Udell et al. 2010; Viranyi and Range in press) and develop (Udell and Wynne 2008, 2010; Wynne et al. 2008; Reid 2009; Hare et al. 2010), which are the cognitive mechanisms underlying them (Elgier et al. 2012; Kaminski and Nitzschner 2013), and the relevance of such issues for the understanding of human evolution (Hare and Tomasello 2005; Kubinyi et al. 2007; Topál et al. 2009; Hare et al. 2012) are all matters of lively debate in the field of comparative cognition, with a number of intriguing questions driving current research activities and different hypotheses being put forward.

In the current chapter, we present findings on dog–human communication, with a focus on dogs' ability to *understand* and respond to human gaze and to *use* human-directed-gazing as a communicative tool during interactions; we discuss the potential origins and factors affecting these abilities; and we discuss the evidence regarding the complexity of such behaviour.

5.2 What's in a Gaze? Dogs' Understanding of Gazing

5.2.1 Social Significance of Eye Contact

For most non-human species direct (especially prolonged) eye contact is considered by the receiver as an aggressive threat (Emery 2000), although recent research has suggested that the significance of direct eye contact may change during development: for example, macaque mothers spend time gazing into their infants' face/eyes in the first few weeks of their life (Ferrari et al. 2009; Simpson et al. in press), despite the fact that as adults direct eye contact is mostly used as a threat in this species.

A number of authors have suggested that one of the fundamental changes in human evolution is the different functional role of direct eye-gaze, in that humans may stare in someone's eyes to threaten them, but also to show love and affection. Indeed such a dual function of eye-gazing appears to be somewhat limited to humans and a few other great ape species (Emery 2000), although in fact studies directly investigating the function of eye-gazing in non-primate species are rare.

According to Schenkel, wolves often use gaze to force others into subordination and maintain their position in the group (Schenkel 1967; Fox 1971). Also in dogs, prolonged and direct eye-gazing has been considered a threatening behaviour, associated with agonisitic interactions (Bradshaw and Nott 1995). Hence in both species, extended duration of gazing is often regarded as a form of ritualized aggression, whereas averting the gaze indicates subordination (Scott and Fuller 1965; Fox 1971).

Although no studies have directly investigated if and how dogs (or wolves) use direct eye-gaze in different contexts (i.e. agonisitic and affiliative), there is at least one situation in which prolonged eye-gaze may be interpreted in an affiliative manner i.e. whilst exhibiting a play bow. In this situation the actor normally looks directly in the face of the individual being invited to play and the eye contact may last a few seconds. Hence it would seem that dogs (and potentially wolves) may in fact modulate the significance of a conspecific's prolonged and direct eye-gaze by varying the accompanying postures, and that it can hence be exhibited both in an agonistic and affiliative context.

One potentially important question, therefore, is how dogs perceive a direct and prolonged gaze, not by a conspecific but by a human being. Similarly to the conspecific context, it would seem that a stare can be read as a threat in some cases, since for example, gazing by the owner can trigger aggression in dogs who have dominance-related problems with their owner (Line and Voith 1986). Indeed, in a forced eye-contact test, in which dogs were gently held and a human looked in their eyes attempting to maintain eye contact with them, it was found that they would tolerate their owner's direct eye contact significantly more than a stranger's (Hernádi et al. 2012). The fact that direct eye contact is tolerated more when performed by a person with whom dogs have a strong bond suggests that, with few or no other accopmanying signals, a direct gaze puts dogs in an uncomfortable situation. Indeed, interestingly, it appears that a number of behaviours can affect how a direct gaze is perceived. Vas et al. (2005; see also Vas et al. 2008; Gácsi et al. 2013a, b) were interested in understanding how dogs would modify their own behaviours in response to a human partner changing their behaviour from exhibiting a threatening to a friendly approach towards them (or vice versa). In both cases, the experimenter approached the dog whilst maintaining full eye contact with it as much as possible, but what varied were the accompanying behaviours: in the 'friendly' approach the experimenter walked at a normal speed and talked to the dog in a friendly tone of voice; whereas in the 'threatening' situation she approached haltingly with the upper body bent forward, and in complete silence. The dogs responded differently, with a more fearful (and in some cases aggressive) response to the experimenter and increased heart rate (Gácsi et al. 2013a, b) when she approached in a threatening way, but rapidly switching to a more relaxed stance once the experimenter exhibited the friendlier behaviours.

Hence, it seems that dogs like many other species, may indeed consider direct eye-gaze to be a threatening social stimulus, also when exhibited by humans, however the interpretation of this cue appears to be dependent on which other behaviours occur together with it.

Considering the above, the next important issue is whether dogs' flexibility in understanding human eye-gaze is a specific adaptation to life with humans, selected for during the course of domestication. In multiple studies involving different tasks, young hand-reared wolf pups were significantly slower or more reluctant to establish eye contact with a human than similarly-raised dog pups (Gácsi et al. 2005, 2009a; Virányi et al. 2008; see also Sect. 5.3.4 below). However, a wolf-dog comparison as regards the *understanding* of human eye-gaze is at present difficult to evaluate since it is based on only one study. Adult wolves presented with a stranger approaching either in a friendly or threatening manner whilst maintaining eye contact (the same procedure adopted by Vas et al. 2005) looked away from the approaching threatening stranger significantly faster than dogs. However, unlike dogs they showed no signs of fearfulness or aggression (Gácsi et al. 2013a, b), making any interpretation of the avoidance of eye contact difficult. In the friendly approach, wolves and dogs behaved in largely similar ways and both species behaved differently in the two contexts, which would suggest that at the very least wolves were also capable of perceiving direct eye contact differently in the two context. However, a more systematic comparison of how direct eye-gaze is understood and used in different context both with conspecifics and humans would help to tell us more about the potential changes brought about by the domestication process to this basic, but important behaviour.

Aside from the emotional and social valence, eye-gazing can also deliver important information as regards the gazer's attentional state, the focus of their attention, and, potentially, their communicative intent.

5.2.2 Do Dogs Understand that Eye-gazing Can Reveal Something of the Gazer's State of Attention? Do They Know we can 'See' Them and Are They Sensitive to Others' Visual Perspective?

Only one study, to our knowledge, has investigated dogs' understanding of their conspecific's attentional stance. Horowitz (2009) looked at whether dogs would exhibit and modulate potential attention-getting behaviours such as paw, bump, or bark in accordance with their partners' attentional state in the context of play. Indeed, the author found that play signals were sent nearly exclusively to forward-facing conspecifics, whereas attention-getting behaviors were used most often when a playmate was facing away. Furthermore, the forcefulness of the attention-getter varied in accordance with the degree of disattention in the audience: stronger attention-getters were used when a playmate was looking away or distracted, less forceful ones when the partner was facing forward or laterally. So from this study it appears that dogs perceive something of the attentional state of their conspecifics, although whether eye contact is used as a cue to attention has not so far been investigated.

In contrast to the paucity of studies with conspecifics, a number of studies have been carried out investigating dogs' understanding of human attention. In an experimental paradigm borrowed from the primate literature (Povinelli and Eddy 1996; Kaminski et al. 2004), researchers investigated whether dogs would beg food more from a person who is looking at them or one who is distracted in different ways (e.g. back-turned, reading a book, etc.). In most cases dogs, efficiently chose the 'attending' experimenter (Gácsi et al. 2004; Virányi et al. 2004). A second approach to investigating dogs' understanding of human attention relied on dogs' obedience to human commands. In two studies, dogs were told not to take food from the floor; following the command, the human either kept looking at the dogs, or she was distracted, turned her back, or closed her eyes. Dogs obeyed the commands more when the human was attentive and, impressively, they were also capable of distinguishing between the eyes open versus eyes closed condition (Call et al. 2003; Schwab and Huber 2006). Similarly, Virányi et al. (2004) asked dogs to perform a set of training exercises—however, they imparted the commands whilst directing the attention either to the dog, to another person in the room, to neither, or whilst actually being on the other side of a barrier. Again the likelihood of obedience was directly related to the attentiveness of the owner towards the dog.

In a twist to the 'commands' test, Bräuer et al. (2004) required dogs to refrain from taking food from the floor. However, in this case the food was placed behind three different barriers: a small, a large, or a transparent one. If dogs are sensitive to the human's visual access to the food they should choose to approach the food behind the large barrier (which would be effective in avoiding detection), and, indeed, this is what dogs did. This study suggests that dogs are not only sensitive to whether humans are attending to them or not, but potentially also to the direction of their attention—that is, to their visual perspective.

To investigate the latter issue in more detail Kaminski et al. (2009) set up a familiar toy-retrieval game with dogs who normally engaged in these games at home. However, in the experimental room in which the game took place, there were always two toys placed in different locations behind barriers so that although both toys were visible to the dogs, only one toy was visible to the experimenter. When asking the dog to 'fetch', the experimenter gave no behavioural cues as to which toy she was referring to. Results suggest that dogs appreciated the experimenter's visual perspective despite the fact that it was different from their own, since they mostly retrieved the toy which was visible to her.

In a final recent study dogs' understanding of human eye-gaze was investigated by attempting to address whether dogs actually understand what 'seeing' means. Kaminski et al. (2013) presented a forbidden-food-on-the-floor test but with another twist, since in this case what varied was the direction of the illumination: in some cases the lamp illuminating the person was switched on, in other cases the light was focused on the food, in yet further situations either both or none were in the light. Control conditions were also carried out to check what dogs would do in the different illumination conditions if the experimenter was not in the room. Dogs hesitated longer before taking the food when the food was illuminated than when it was not—irrespective of seeing the human. This result suggests that dogs may have some understanding that when the food and the area around it is illuminated humans can see them, providing some evidence that dogs take into account the human's visual access to the desired food while making their decision to steal it.

As has been suggested by some authors (Bräuer et al. 2004; Bräuer in press), dogs may 'solve' a number of these tasks by simply inhibiting their behaviour when they see their human partner's eyes. Indeed, considering that eye gaze in a social context may be read as a 'threat', in a 'command test' such as the food-on-the-floor situation, such an avoidance strategy may come quite naturally to dogs. However, what is interesting to note here is that in 'begging' paradigms, by contrast, dogs choose the human partner whose eyes are in fact *visible*, suggesting that they are indeed capable of reading the social message of human eye gaze in different ways, depending on the contextual information. Furthermore, this particular criticism is not applicable to the Kaminski et al. (2013) study in which dogs stole food if the food was in the dark irrespective of whether they could see the person or not.

Another criticism raised against a number of these studies is that in most cases results can be explained by dogs' having learned from daily experience with humans how to best obtain what they want in a specific context (Udell et al. 2011). For example, in the begging paradigm the scenarios which were more likely to be familiar to dogs (e.g. the experimenter reading a book) were easier for them to solve correctly than those which were likely to have been less familiar (e.g. a person with a bucket over their head) (Udell et al. 2011). Although little is as yet known of the mechanisms or the ontogeny of dogs' comprehension of human attention, a learned component is likely to play at least a part in its acquisition. Notably, however, while this criticism is valid for experimental paradigms such as the begging and obedience tests, it is less likely to explain the studies carried out by Kaminski et al. (2009) involving the familiar 'retrieval' context but with multiple barriers or the Kaminski et al. (2013) study, since it is unlikely that dogs would have extensive experience of the different illumination or barrier combinations presented in those studies.

Indeed what appears striking when putting all the evidence together is the multitude of contexts and hence the flexibility that dogs seem to exhibit in their understanding of human attention (Virányi and Range 2011); perhaps this suggests that rather than simply responding to single cues or triggering stimuli which may have been reinforced in the past, dogs have the capacity to extract knowledge from their experiences and use this knowledge to solve novel social problems (Call 2001).

5.2.3 Can Dogs Follow Eye-gazing and Read this Behaviour as a Referential Gesture?

As well as its use as indicator of whether another is attending or not to a given situation, eye-gazing may also be used to infer the *direction* of another's attention. The functional relevance of this ability may be particularly important for social species, since it would allow an individual to gain valuable information about both its social and physical environment (Emery 2000).

Multiple studies have been carried out on the abilities of various non-human species to follow eye-gaze, and the list of species that appear to be capable of such gaze following is constantly growing. At least three different types of gaze-following contexts can be identified: (1) gaze following into distant space; (2) gaze-following around barriers (which has been shown only in the great apes—Bräuer et al. 2005—and in two corvid species—Bugnyar et al. 2004 and Schloegl et al. 2008); (3) gaze-following to a specific target object, which appears to be the hardest to master by non-human species (Call and Tomasello 2005). Considering that the ability to gaze-follow in one context does not necessarily carry over to others, there is growing evidence that these abilities actual tap into different underlying mechanisms (Povinelli and Eddy 1996; Tomasello et al. 1999; Triesch et al. 2007).

Unfortunately, gaze-following in dogs has been investigated almost exclusively with a human as a demonstrator. The only report of conspecific gaze-following is a conference abstract describing a comparative study carried out in dogs and wolves living in packs where it was found that both species followed their partner's gaze into distant space (Werhahn et al. 2013). With human-reared wolves, gaze-following into distant space has been shown to occur both with conspecifics and familiar humans from 14 weeks of age (Range and Viranyi 2011), although this has not as yet been tested with dogs.

As regards gaze-following into distant space with humans as demonstrators, a few more studies have been carried out with dogs, with mixed results. Agnetta et al. (2000) tested dogs in a gaze-following task where a human experimenter turned her head and looked at one of three predetermined locations (straight up, directly to the left, or directly to the right of the dog) for approximately five seconds. The results found that dogs did not reliably follow human gaze into distant space. However, in the same study authors also used a two choice task—i.e. food was hidden under one of two bowls equidistant from the human signal—and in one of the conditions the demonstrator turned his head and looked at the correct bowl for 5 s. Interestingly, in this case the dogs' performance was above chance at the group level, with 7 of the total 16 dogs also performing above chance at the individual level.

In a similar study, Soproni et al. (2002) also adopted a two-way choice task with the demonstrator either turning her head and gazing toward the correct bowl, turning her head and gazing above the baited bowl to the upper corner of the room, or orienting her head and body to the midline facing the dog and turning only her eye-gaze toward the correct bowl. Results showed that whereas dogs followed the referential head- and eye-gaze directed to the bowl above chance, they did not do so in the other two conditions.

Overall, it appears that whereas dogs have difficulty in following eye-gaze into distant space, in the two-way choice task, they appear to be able to use eye-gaze to infer where the hidden food is located. Indeed, even puppies of between 9 and 24 weeks of age appear to be able to use gazing as a referential cue in a two-way choice task, even if they find this cue much more difficult to follow than, for example, pointing, or a combination of pointing and gazing (Hare et al. 2002). In fact, although it seems that dogs can use gazing as a referential cue, it appears that the ease with which they do so is contingent on how it is presented, since Bräuer et al. (2006) found that a continuous, prolonged gaze directed at the bowl was easier for dogs to follow than more rapid alternating glances between the object and the subject.

The importance of contingent cues is confirmed by a more recent study, using an innovative eye-tracking methodology, in which researchers showed that dogs could reliably follow human eye-gazing in a two-way object choice task (see also Rossi et al., this volume). Importantly, dogs did so significantly more if the referential gesture was preceded by direct eye contact between the demonstrator and the dog, and accompanied by the use of the dog's name in a high-pitched tone of voice (Teglas et al. 2012). According to these authors, the direct eye contact exhibited by the demonstrator prior to the referential gestures set the communicative context between the human and dog, hence alerting the latter that the subsequent actions were intended for them. In other words the eye contact acted as an ostensive cue (see Sect. 5.2.4).

Taken together, dogs' gaze-following abilities in a referential context, coupled with the preliminary report that dogs can follow conspecific eye gaze into distant space, suggests that, this ability is present in dogs. It is more likely that methodological issues in the experimental setup of the two studies finding no human gaze-following affected the results. Future studies on this topic will be necessary to draw final conclusions on this issue. Still, results suggest that dogs' understanding of eye-gaze includes a perception of its directionality and potentially an understanding that eye-gazing can be referential.

5.2.4 Do Dogs Understand that Eye-gazing May be Used as an 'Ostensive Cue' to Signal a Communicative Context?

One of the most interesting aspects emerging from recent research on dog–human communication is the possibility that like human infants, dogs may be sensitive to 'ostensive cues'. Ostensive cues in adult-child dyads are communicative signals such as direct eye contact and body orientation as well as the use of motherese and the child's name, which have the function of alerting infants that a 'learning context' is being set up, and hence assisting them in the social learning process (Nielsen 2006; Brugger et al. 2007; Southgate et al. 2009; Csibra and Gergely 2011). Indeed, the use of ostensive cues by a demonstrator has been shown to override childrens' own perceptions and lead them into making inefficient choices by, for example, 'over-imitating': i.e., copying redundant actions (Nielsen 2006; Brugger et al. 2007; Lyons et al. 2007, 2011) or carrying out 'obvious' errors in their decision making process (Topál et al. 2008).

It has been proposed that dogs, potentially uniquely amongst non-human species, also respond to human ostensive cues in similar contexts—although the functional relevance of these behaviours may be different (see Topál et al. in press for a review). Support for this idea has been growing, as results have found that dogs: show more search errors in A-not-B task when the demonstrator exhibits ostenisve cues rather than other non-social attention getters (Topál et al. 2009a, 2010; Marshall-Pescini et al. 2010; Kis et al. 2012; Sümegi et al. in press); make more counterproductive choices in a food quantity discrimination task when a combination of communicative cues are used (Marshall-Pescini et al. 2012); and follow pointing in a two-way choice task more accurately when they are preceded by communicative cues (Kaminski et al. 2012).

What is striking is that in some cases dogs appear to rely more on the communicative cues delivered by a complete stranger (the experimenter) than their own experience (Szetei et al. 2003; Erdőhegyi et al. 2007; Topál et al. 2009a; Kupán et al. 2011). For example, in a food-quantity discrimination task, although the different size food plates are continuously visible during the demonstration, dogs tend to choose the smaller one when their owner or the experimenter communicate a preference for it (Prato-Previde et al. 2008; Marshall-Pescini et al. 2011; Horowitz et al. 2013).

Exactly what mechanisms are responsible for the powerful social influence effect shown by humans' communicative cues on dogs' choices is still a matter of debate (Topál et al. 2009a, 2010; Marshall-Pescini et al. 2010; Kis et al. 2012), however what is perhaps more interesting for the purpose of this chapter is to note that various studies converge in suggesting that direct eye contact appears to be one of the most potent communicative cues that dogs rely on (Kaminski et al. 2012; Marshall-Pescini et al. 2012; Teglas et al. 2012).

Overall, it would seem that dogs have a rather sophisticated understanding of eye-gazing. Although there are relatively few studies focusing on dogs' understanding of conspecific eye-gazing—this area needs more attention in the future—their understanding of human eye-gazing is rather impressive. Dogs appear to perceive that eye contact may be used to convey emotional and socially relevant information in the dyadic context, and that its valence as a threatening or affiliative signal may change depending on the behaviours that accompany it. Furthermore, dogs also seem to be able to use human eye-gaze referentially (although with more difficulty than pointing), suggesting that they may understand that eye-gaze can be used to direct attention to a target. Finally, a few studies suggest that human eye-gazing is also an important attention-getter in dog–human communication and that it may serve to alert dogs when human communication is intended for and directed at them.

5.3 Looking at Humans: How, When, and Why Dogs Engage in Looking Behaviour

In humans and other animal species intraspecific communication occurs in a variety of different situations, involves the use of a range of observable behaviours, and in general takes place between conspecifics. However, dogs' natural habitat has particular features which requires members of this species to communicate with human beings regularly and effectively. During everyday social interactions with humans, in addition to responding to human communication, dogs also spontaneously and actively communicate with humans through a variety of signals, including gazing, different types of vocalizations, and other behavioural actions (e.g., running back and forth, touching, and assuming specific body postures). These behaviours, besides expressing internal emotional and motivational states, seem to be aimed at achieving specific goals, such as initiating play, going for a walk, or obtaining a person's attention, help, or comfort.

In the past dog–human communication received little scientific attention (e.g. Warden and Warner 1928; McConnell and Baylis 1985; Mitchell and Thompson 1993), but recently this issue has been more systematically investigated. Research has considered different topics including: which signals dogs use and in which contexts (e.g. Hare et al. 1998; Miklósi et al. 2000; Gaunet 2008, 2010; Kaminski et al. 2011; Merola et al. 2012a, b, 2013a); to what extent evolutionary factors, lifetime developmental experiences, and current living conditions affect dogs' communicative skills (Miklósi et al. 2003, 2005; Marshall-Pescini et al. 2009; Barrera et al. 2011; Gaunet and Deputte 2011; Passalacqua et al. 2011, 2013); and how complex the mechanisms underpinning dogs' human-directed communication may be (Bentosela et al. 2008, 2009; Elgier et al. 2009a, b). Concerning the latter, specific attention has been given to the issue of whether dogs' communication towards humans might be considered referential (i.e. about an event, an agent, or a place) and intentional (i.e. in line with the social or spatial context) (Miklósi et al. 2000; Gaunet and Deputte 2011; Marshall-Pescini et al. 2013).

5.3.1 Looking Behaviour: Is it Part of a Referential Communicative Act?

During communication, *looking at others* and *alternating gaze* between an observer (whether conspecific or not) and a specific target are considered ways to initiate communication by attracting and directing the audience's attention towards a specific object or location (Gómez 1990, 1996). There is agreement that in apes, gazing and even more so gaze alternation, in combination with oriented vocalizations and different forms of "pointing" behaviours (gestures with whole hand and arm, body orientation, and positioning), represent functionally referential and intentional communicative behaviours (Call and Tomasello 1994; Leavens et al. 1996, 1998, 2004, 2005a, b; Gómez 2007). Similarly, the capacity to adjust gaze and other communicative signals towards the recipient is interpreted as an indication that the subject understands the partner's role and importance in the communication process (Cheney and Seyfarth 1990; Call and Tomasello 1994; Tomasello et al. 1994).

It has been proposed that for a communicative act to be 'referential' and 'intentional' a number of operational criteria should be satisfied (e.g. Leavens et al. 2004, 2005a, b). In particular, there is agreement that the communicating subject should (1) exhibit gaze alternation between its social partner(s) and distant or inaccessible objects; (2) engage in behaviours clearly aimed at obtaining the partner's attention (e.g. vocalisations, body movements); (3) exhibit communicative signals only if in the presence of an audience; (4) adapt communicative behaviours to the attentional state of the audience; (5) show persistence in, and (6) elaboration of, communicative behaviour when the partner is not attending or responding.

So far, the above-mentioned criteria have been established in human infants (Tomasello 2008) and in apes (Leavens et al. 2004, 2005a, b; Cartmill and Byrne 2007, 2010; Roberts et al. 2013), but, as outlined below, a number of recent studies suggest that at least some of these may also be fulfilled in dogs' communicative interactions with humans.

Most evidence relative to human-directed looking behaviour and gaze alternation has been reported in two different situations: when dogs are unable to obtain a reward located in an out-of-reach position (e.g. Hare et al. 1998; Miklósi et al. 2000; Gaunet 2008, 2010; Gaunet and Deputte 2011), and when they are presented with a difficult or unsolvable problem (e.g. Miklósi et al. 2003; Marshall-Pescini et al. 2009, 2013; Passalacqua et al. 2011, 2013).

For instance, Miklósi et al. (2000) addressed the issue of whether and how dogs use gaze and gaze alternation to communicate. They grouped dogs' indicative behaviours (e.g. looking at an external referent, gaze alternation, jumping, running back and forth, etc.) and introduced the term "showing behaviour" as a communicative act with a directional component related to an external target and an attention-getting component aimed at attracting the social partner's attention. The authors tested dogs in three different conditions in which the presence of the human and/or the hidden object was manipulated to assess under what conditions 'showing' behaviour emerged and to differentiate between motivational and referential components of dogs' signals (Marler et al. 1992). It emerged that when both the food and a naive owner were present, 'looking' behaviour towards the owner and the location of food were more frequent, and gaze alternation between them emerged. Vocalisations were also observed to be an integral part of gazing behaviour, since they were always associated with gazing (at the owner or at the location of the hidden food).

Similar results have been recently reported by Gaunet and Deputte (2011) who found that besides using gaze and gaze alternation dogs can provide information about the location of a desired object by using the position of their body: namely, by standing in close proximity to it while signalling to the human. Interestingly, this shows that apart from using an individual (human or dog)'s body location as a local enhancement cue (Udell et al. 2008a), dogs may be able to use their own body position and orientation as an intentionally communicative cue. How flexibly dogs can use this behaviour, and in what contexts, remains to be tested.

Another situation in which dogs have been observed to use gaze alternation also accompanied by different attention-getters, is the 'unsolvable task' paradigm, in which an initially accessible apparatus containing food becomes impossible to access (Miklósi et al. 2003; Gaunet 2008; Marshall-Pescini et al. 2009). Taken together, these studies provide coherent evidence that dogs use gaze and gaze alternation to communicate with their human partners when confronted with a distant or inaccessible object or food source they desire; it also emerges that dogs can combine looking behaviour with other 'indicative' cues, and that these cues can be preceded by apparent attention-getting signals, fulfilling at least the first two criteria for intentional referential communication (Leavens 2005a).

Recently Gaunet (2010) also tested whether dogs would show elaboration and persistence in their communication (criteria 5 and 6). As with other studies, dogs were given the possibility of 'showing' where their favourite toy had been hidden. However, on some occasions the receiver of their communicative act would retrieve and give the dog an unfamiliar and uninteresting object instead of their favourite toy. Hence in the latter case the dogs' communication was unsuccessful, since it did not result in the desired outcome. In this study, dogs exhibited persistence in (although not elaboration of) their "showing" behaviour when their attempts to 'manipulate' the human partner failed (therefore fulfilling criterion 5). To our knowledge this is the only study investigating the possibility of elaboration and persistence of these behaviours in a communicative settings; hence, results can only be considered preliminary.

Significantly more attention has been given to whether dogs can also modulate their communicative behaviours in accordance with the presence or absence of an audience and perhaps more importantly taking into account the audience's attentional state.

5.3.2 Do Dogs Adjust Gazing (and Showing) in Accordance with Their Human Audience?

A key issue in social cognition research is the relationship between communication and cognitive skills, and in particular the extent to which communication is influenced by the presence of an audience and varies according to its characteristics (Marler et al. 1986; Cheney and Seyfarth 1990; Evans 1997; Tomasello and Call 1997; see operational criteria 3 and 4, Sect. 5.3.1). In particular, the 'audience effect' provides evidence that an individual has some understanding that others play a role in the communication process, by recognizing for example that: (1) there must be an audience to which the behaviours can be directed; (2) the audience must be attending to the message; and (3) information provided should take into account the state of knowledge of the audience, which may vary according to circumstances (Cheney and Seyfarth 1990; Call and Tomasello 1994; Tomasello et al. 1994).

So far a number of studies have evaluated whether and to what extent dogs' gazing and communicative behaviour is influenced by a human partner's presence, their attentional state, and their state of knowledge (Hare et al. 1998; Miklósi et al. 2000; Gaunet 2010; Topál et al. 2006; Virányi et al. 2006; Gaunet and Deputte 2011; Marshall-Pescini et al. 2013), providing mixed results.

The first study looking at this question (Hare et al. 1998), with only two dogs, showed that these dogs exhibited more communicative acts as regards the location of the hidden food in the presence than in the absence of an audience, but found no evidence that they would adapt their communicative behaviour to the audience's attentional state (i.e. facing the dog, back turned, eyes covered). With a slightly larger sample (n = 10), Miklósi et al. (2000) found that, in a food hiding situation, dogs used gaze and gaze alternation significantly more when in the presence of both a human partner and the hidden food than when left alone in the presence of the food. However, the authors did not test whether dogs could modify their behaviour in accordance with the audience's attentional state.

Gaunet (2010) adopted another approach to investigate the same issue, comparing the communicative behaviour of guide dogs for the blind and pet dogs when tested with their blind or sighted owners in a play session based on a 'fetch' task. The authors suggested that if guide dogs appreciated something of the visual status and abilities of their human partners, they would behave differently from pet dogs and adapt their behaviour in accordance with the visual abilities of their audience (e.g. reducing gazing behaviour to attract attention and increasing the emission of sounds and contacts). They found no group differences in communicative behaviour between guide dogs and pet dogs towards their respective owners, thus concluding that in their experimental setting guide dogs did not show sensitivity to the blind owner's visual attentional state (criterion 4, Sect. 5.3.1). Though this finding is in line with previous ones in showing a very limited sensitivity of guide dogs to their human partner visual status (Gaunet 2008; Ittyerah and Gaunet 2009), a number of explanations are possible (Ittyerah and Gaunet 2009; Gásci et al. 2004) and, among these, the fact that guide dogs are raised and in general continue to be surrounded by sighted people may have affected their performance.

In a later study the same author and colleagues (Gaunet and Deputte 2011) tested pet dogs in a hidden object paradigm, varying the location of the audience by placing barriers in the room in which the object was hidden. Results showed that besides signalling the presence of the out-of-reach object, dogs also adjusted their position so as to adopt the optimal location which allowed them to alternate their gaze between the hidden object and the owner, and insured that the owner would be in a position to direct their gaze to the hidden object. In sum, dogs adjusted their behaviour to the human recipient's point of view.

The possibility that dogs adapt their behaviour according to the state of attention of a human partner was also investigated recently by Marshall-Pescini et al. (2013). In particular, these authors evaluated whether dogs and toddlers adjusted their human-directed gazing behaviour and gaze-alternation depending on the attentional state of the audience using an unsolvable task paradigm and varying, in the crucial unsolvable trial, the attentional stance of the audience (facing versus back turned). Both dogs and toddlers increased their gaze alternation between the apparatus and the caregiver when the task became unsolvable. Both also preferentially directed their gazing behaviour towards the attentive audience, suggesting a basic understanding in both species that, for their requesting gesture to be effective, the audience needed to be looking towards them and the object of interest.

Only two studies, to our knowledge, have evaluated whether dogs take into account the state of knowledge of the audience (which may vary according to circumstances) when providing information, by testing whether dogs are capable of adapting their behaviour to what their audience knows when looking at humans and exhibiting "showing behaviour" (Topál et al. 2006; Virányi et al. 2006). Both studies used an adapted nonverbal knowledge-attribution paradigm, originally used with nonhuman primates—the "Ignorant–Helper" paradigm (Kaminski et al. 2008)—in which dogs must communicate with a human "helper" to obtain an out-of-reach reward, and the helper's state of knowledge is manipulated. Virányi et al. (2006) presented dogs with conditions in which they had to discriminate between what a person had, or had not, seen being hidden in a specific situation, and compared their performance with that of two-and-one-half-year-old children. Both a toy and a stick (necessary to retrieve the toy) were hidden in various out-of-reach locations in a room; however, depending on the experimental condition, the dog's helper (i.e. the person who retrieved the object for the dog) could witness the hiding of both, none, or only one of the two objects. The question was whether dogs would be sensitive to what the helper had witnessed, and thus adjust their communicative behaviour accordingly (i.e. looking at the location of the toy only when the helper had not witnessed it being baited).

Results showed that infants were capable of discriminating between situations, indicating the appropriate object in accordance with what the helper had or had not witnessed. On the other hand, dogs hardly ever indicated the stick location, preferentially indicating the toy location (Virányi et al. 2006). The dogs' behaviour could be explained by the fact that, despite a pre-training phase, dogs failed to appreciate the functional connection between the stick and the toy, thus considering the former barely relevant in the situation (whereas children would by this age have had multiple experience with tools). Species difference in performance could be also attributed to working memory problems in dogs or to a different 'motivational value' of the two goal objects (stick and toy), with dogs being over-motivated to get the toy.

Interestingly, dogs, like children, signalled the place of the toy more frequently if the helper had been absent during toy-hiding compared to those conditions when she had participated in the hiding: thus, dogs apparently discriminated between when the helper had or had not witnessed the toy being hidden and adapted somewhat to the helper's knowledge state, but there was less sophistication in communication compared to infants. Again, though, as outlined by Virányi et al. (2006), this performance could be explained at different levels, and a number of more parsimonious explanations need to be ruled out before concluding that dogs are able to attribute knowledge and ignorance to their human partners.

Overall, these findings indicate that dogs are sensitive to the presence or absence of an audience when communicating, but provide mixed results on their ability to take into account human attentional states. In particular, more studies are needed before it can be concluded that dogs are capable (or incapable) of attributing knowledge and ignorance to their human partners.

Taken together, the experimental evidence suggests the existence of some behavioural flexibility in communication between dogs and humans, supporting the idea that gaze alternation and showing behaviour may be considered intentional and referential communicative acts. However, the relative paucity of studies on dogs' ability to adapt to their audience's state of attention and knowledge, and the mixed results emerging from this literature, mandates a note of caution as to the depth of dogs' underlying comprehension of their communicative actions.

5.3.3 What Do Dogs Want When They Look at us?

Given all the studies mentioned above, perhaps the most intriguing outstanding question is what dogs want, or in other words, what dogs might be 'saying' through looking at us? There are number of possible answers to this question, although there has been little systematic research trying to tease them apart.

The simplest explanation is that dogs use 'looking' as a way to obtain attention from their human partners. Indeed, in a questionnaire asking what kind of behaviours dogs used to get their attention, owners reported that 'looking' was one of the more prominent ones (Mills et al. in press). However, the available evidence, based on different experimental procedures, indicates that dogs also engage in active communication with humans and look at them to request their intervention when unable to obtain a desired goal. Interestingly, in these cases, dogs seem to use gaze alternation (and not just direct gazing) to direct humans to the object of their desire, suggesting there is both intentionality and referentiality in the exhibition of this behaviour (see Sect. 5.3.1).

Another important function of looking is that of monitoring one's own environment to gain information prior to making a decision. Studies on free-ranging dogs show that when a pack encounters another pack, a number of individuals will confront the strangers by barking and moving forward, but they will look back often to check the status of their companions (i.e. whether they are following or not), and their decision to engage in a confrontation will be based on their partners' movements as well as the size of the two packs (Bonanni and Cafazzo in press; Bonanni et al. 2011). Thus, monitoring each other's action is likely an important behaviour for pack animals, since it allows behavioural synchronization, which is necessary for cohesive action to occur. Something similar might also occur during dogs' interaction with humans.

Moreover, dogs are in general strongly dependent on their human partners, establishing close bonds with them (Topál et al. 1998; Prato-Previde et al. 2003) and, like infants and hand reared chimpanzees, dogs appear to use their caregivers as a 'secure base' from which to explore the environment (Bard 1991; Palmer and Custance 2008; Gácsi et al. 2013a; see Prato-Previde and Valsecchi in press for a review). It is possible that dogs may especially look at humans to seek information about external objects or events in a context of *uncertainty*. There is evidence that human infants at around twelve months of age (Mumme et al. 1966; Vaish and Striano 2004; de Rosnay et al. 2006), and in some cases chimpanzees (Itakura 1995; Russell et al. 1997 but see Tomonaga et al. 2004), look at other individuals more when facing unfamiliar situations that are difficult to interpret, and act in accordance with the informer's positive or negative emotional reactions. This process has been dubbed 'social referencing'. Social referencing, as other social learning processes, is considered to be a useful and safe way to learn about the outside world (Feinman 1982; Roberts et al. 2008). In fact, emotional cues not only provide important information about the emotional states of others and the likelihood of their future behaviour, but also about the nature of the environmental events leading to such states.

To investigate the possibility that dogs may look at humans when facing an ambiguous situation, and take into account the informer's emotional cues and behaviour, we borrowed a social referencing paradigm from the infant literature in developing a series of studies (Merola et al. 2012a, b, 2013a). In the first study, adult dogs were tested with their owner delivering either a positive or negative message from a distance and then either approaching or moving away from a potentially scary object (i.e. a noisy fan with ribbons flying around when activated). We found that the majority of dogs looked back to their human partner when confronted with the strange object and chose to move forward or away from it depending on the person's movements, thus mirroring their owner's behaviour. Hence, dogs looked referentially towards their owner and there was evidence that they synchronized their behaviour with him or her (Merola et al. 2012a).

In a subsequent study, we tested dogs with either their owner or a stranger, acting as the informant and delivering either a positive or negative emotional message using only facial and vocal expressions (rather than also approaching or withdrawing from the object). As is the case with human infants, most dogs looked referentially at the human informant regardless of her identity; however, they based their decision on whether to approach the potentially scary object on their owner's, and not the stranger's, emotional message (Merola et al. 2012b). According to a number of authors (e.g. Stenberg 2003; Walden and Geunyoung 2005; Stenberg and Hagekull 2007), looking at a stranger as much as at a familiar caregiver in a social referencing paradigm indicates that looking behaviour under ambiguous conditions cannot be considered just a form of comfort seeking (due to the activation of the attachment system); rather, it indicates a search for information about the specific context. The recognition of the valence of the owners' message is probably based on dogs' daily experience with the owner. This interpretation is supported by another recent study in which dogs showed recognition of positive but not negative emotions, and did so only when the owner, not a stranger, was expressing it (Merola et al. 2013b).

Overall, these social referencing studies suggest that in a context of uncertainty, most dogs look at their owners (but also a stranger) and monitor what humans do before deciding on their own course of action. However, dogs' reduced synchronization to the stranger's emotional cues remains open to interpretation, since it is unclear whether it is due to their inability to comprehend the stranger's emotional communication or whether the dog's willingness to modify their own behaviour is tied to their relationship with the person. This will be an interesting avenue for future studies.

Finally, at least in principle dogs might engage in communication with humans to provide information. In fact communication can be initiated also with the aim of informing others—providing them with information when they need it (Tomasello 2008). Human infants, unlike chimpanzees, engage in cooperative communication from a very early age, even when doing so has no direct benefit for themselves (Liszkowski et al. 2006, 2008; Bullinger et al. 2011).

The possibility that dogs might use gaze to provide information to humans even without any direct benefit for themselves has been investigated in a study by Kaminski et al. (2011), who tested the occurrence and the flexibility of "showing" behaviour in situations in which the hidden or out-of-reach object was of interest either only to the dog, only to the human, or both. Based on the dog's showing behaviour, the human partner found the target object more frequently in situations where dogs requested their own preferred object than in situations where the information was relevant only to the human. This confirms that dogs can show communicative behaviour, including gazing and gaze alternation, to request access to a toy they themselves desire, but provides no evidence that dogs are capable of informing a human of the location of the object that only the human desires.

The lack of flexibility exhibited by dogs in this study may support the notion that gaze alternation is a behaviour elicited by specific trigger situations as a way to use humans as social tools to obtain a desired goal, and that dogs have learned to do so during their daily interactions with people. However, further research will be necessary to probe the flexibility of dogs' showing behaviour in different contexts, and hence draw conclusions on whether it is in fact exhibited only in a 'requesting' context.

5.3.4 Nature and Nurture in Dogs' 'Looking' Behaviour

It has been suggested that human-directed gazing represents a foundation on which dog–human communication evolved, and that dogs' propensity to look at humans, or to quickly learn to do so, represents a behavioural feature that distinguishes dogs from wolves, emerging during the course of domestication (Miklósi et al. 2003; Kubinyi et al. 2007; Virányi et al. 2008; Gácsi et al. 2009a, b).

On a number of tasks—for example in two-way choice pointing tasks (Virányi et al. 2008; Gácsi et al. 2009a, b), and a task in which animals were reinforced for looking into a person's face—young hand-raised wolf pups were significantly slower to initiate or maintain eye contact with humans compared to similarly-raised dog pups (Gácsi et al. 2005). This difference was maintained also as juveniles (Gácsi et al. 2009a, b). Furthermore, when young hand-reared wolves and dogs were presented with a task which after being accessible became unsolvable, wolves tended to work independently and tried out different strategies, whereas dogs looked towards their human partner sooner and for longer than wolves (Miklósi et al. 2003).

Recently, Smith and Litchfield (2013) used the same unsolvable task to test dingoes (*Canis dingo*). Dingoes are interesting subjects within the framework of the 'domestication debate', since they arrived in Australia between 3500 and 500 years ago, are thought to have evolved from very early domestic dogs of East Asiatic origin (Savolainen et al. 2002, 2004), and, since their arrival in Australia, have had no further direct selection pressures from indigenous Australians (Smith and Litchfield 2009). Hence, dingoes may be able to tell us something of the effects of domestication without the subsequent 'interference' of direct selection for specific breed characteristics.

In Smith and Litchfield's (2013) study, based on Miklósi et al. operational definition of "looking back", dingoes looked back at the experimenter earlier compared to both dogs and wolves, though they used fleeting glances more similar to those used by wolves than by dogs. However, when applying a restrictive and more context specific definition of "looking back", no evidence of looking back in dingoes emerged. This finding raises important methodological issues on the measurement of "looking back" behaviours but leaves open the question of whether it was the process of domestication or the subsequent systematic process of breed selection that had a stronger effect on the emergence of looking towards humans for information. Indeed a number of studies have shown significant breed differences in gazing behaviour, suggesting that more recent artificial selection has had a fundamental impact on its emergence (Jakovcevic et al. 2010; Passalacqua et al. 2011).

Jakovcevic et al. (2010) compared three breeds of dogs, retrievers, German shepherds and poodles, in the acquisition and extinction of gazing behaviour in a situation in which food was in sight but out of the dogs' reach. They found that, with or without any previous training, retrievers took longer to extinguish the gazing response compared to the other groups.

Passalacqua et al. (2011) used the 'unsolvable task' paradigm to investigate the effect of breed (and also age and experience) on human-directed gazing behaviour in three different breed groups: 'primitive', 'hunting/herding', and 'molossoid' (i.e. mastiff-types). They found no evidence of breed group differences at 2 months, but breed group differences started to emerge at 4.5 months and were clearly evident in adult dogs, with dogs in the hunting/herding group showing significantly more human-directed gazing behaviour than dogs in the other two breed groups.

Taken together these findings suggest that genetic changes occurred during both domestication and artificial selection which have shaped dog's human-directed gazing behaviour, predisposing this species, and certain breeds selected for close work with people in particular, for effective communication with humans. The role of genes in human-directed gazing behaviour is further supported by a study by Hory et al. (2013), providing evidence for an association between owner-directed gazing behaviour in an unsolvable task and polymorphisms in the dog DRD4 gene.

Looking behaviour in dogs and its occurrence are also influenced by ontogenetic factors and environmental and life experiences (Udell and Wynne 2010; Passalacqua et al. 2011) including learning opportunities and training and living conditions (Marshall-Pescini et al. 2009; Bentosela et al. 2008, 2009; Barrera et al. 2011). For example, Passalacqua et al. (2011) found that, independently from breed group, at two months of age only about 50 % of the pups tested in an unsolvable task showed looking behaviour towards humans. This tendency increased with age, as 4.5-month-old pups used gaze more than two-month-old ones, but less than adults, the latter being the fastest at looking back towards humans for longer periods. These findings, like those that have emerged in the "response-to-human-cues" literature (Gácsi et al. 2009b), indicate that exposure to a household environment and to humans may be a crucial factor for this behaviour to emerge.

Perhaps not surprisingly, learning also plays an important role in the looking behaviour of dogs, with a number of studies clearly showing that gazing is shaped by reinforcement contingencies (Bentosela et al. 2008, 2009; Barrera et al. 2011). For example, Bentosela et al. (2008), using a test situation involving food in sight but out of the dogs' reach, showed that gaze duration toward the experimenter's face significantly increased with just three reinforcement trials, and also quickly diminished when it was no longer reinforced ('extinction'). However, there were differences between dogs living in shelters and pet dogs on this task, with shelter dogs showing a shorter gaze duration and a faster extinction process when reinforcement was no longer provided (Barrera et al. 2011). This suggests that apart from reinforcement schedules, different living conditions and life experiences may affect this behaviour.

The latter point emerges also from a study comparing highly trained and untrained dogs on the unsolvable task. Marshall-Pescini et al. (2009) found that agility-trained dogs gazed longer at humans than dogs trained for search-and-rescue, which in turn gazed longer than untrained dogs. Interestingly, dogs trained for search-and-rescue were more prone to vocalize towards the owner when the task became unsolvable. This pattern of results reveals that, although in neither of the two training contexts dogs were specifically trained to look at their owner, the type of experiences that dogs have whilst in the training context affect this behaviour. Indeed, agility and search-and-rescue dogs differ in the amount of physical closeness and independence and in the decision-making processes required to carry out their respective tasks, as nicely reflected in the results.

Finally, a couple of recent studies have shown a relationship between gazing behaviour towards humans and both sociability levels and anxiety in dogs. More sociable dogs are more prone to use gaze to obtain out of reach food (Jakovcevic et al. 2012), whereas dogs with anxiety-related problems exhibit different patterns of human-directed gazing behaviour and gaze alternation compared to non-anxious dogs when faced with an unsolvable task (Passalacqua et al. 2013). These studies provide preliminary evidence that in dogs, just as is reported in humans (e.g. Iizuka 1994; Schneier et al. 2011; Wieser et al. 2009) gazing behaviour may be related to personality traits. This intriguing aspect could have practical applications but is in need of further testing.

Taken together, these results clearly show that human-directed gazing is the result of a complex combination of genetic and environmental factors. The predisposition for an increased likelihood of gazing was probably inadvertently selected for during the process of domestication, since looking at humans forms the basis of any communicative act. Breed selection, especially in breeds required to perform a close working or cooperative activity with humans, further shaped this behaviour, resulting in noticeable differences between breeds. However, the genetic predisposition is greatly affected by the social environment dogs are exposed to. Exposure to a human household for just a couple of months significantly increases the occurrence of this behaviour (Passalacqua et al. 2011) and experiences such as specific training regimes can also modulate its frequency (Marshall-Pescini et al. 2009). Future studies combining genetic and environmental variables in systematic ways may reveal much of the flexibility of this behaviour.

5.4 Conclusions

In the last 10 years, since Miklósi et al.'s (2003) seminal work "A simple reason for a big difference: wolves do not look back at humans, but dogs do", there has been an explosion of studies focusing on dogs' communicative abilities. In this chapter we reviewed studies focused on dogs' understanding of the human gaze, and how they themselves use gazing to communicate, especially with humans. As regards dogs' understanding of gazing, gathering evidence suggests that this signal, whether exhibited by conspecifics or humans, has no fixed meaning, and can carry different meanings depending on the context and what other cues accompany it. Dogs appear to be able to use human gazing as a referential cue, although there is mixed evidence as regards their ability to follow human gaze into space and around barriers. There is also evidence that dogs can use human eyes as a cue to their attentional state, and in some (but not all) cases use them to perceive another's perspective. There is also some support for the idea that dogs use human gazing as an ostensive cue, i.e. as a signal that a communicative interaction is about to take place.

Considering the multitude of studies using different paradigms and the mostly consistent results emerging from these, what is striking is the flexibility exhibited by dogs in their understanding of the human gaze. Of course, there are still open questions needing further investigation, especially as regards the origin and development of such understanding: for example, to what extent dogs understand and use their conspecifics' gaze? Is the dog's understanding of the human gaze a recent adaptation or can it be found equally in human-raised wolves?

As regards dogs' use of gazing behaviours towards humans, there is growing evidence that dogs use this behaviour as a form of communication, first and foremost to gain their human partners' attention. However, dogs' use of gaze alternation in a variety of contexts, showing elements of persistence if the desired outcome is not achieved, exhibiting it whilst taking into account the line of sight and in some cases the attentional state of the audience (and possibly their level of knowledge), suggests that this behaviour can be considered both an intentional and referential communicative act. Exactly in what contexts this is more likely to occur and what the underlying motivations for social looking are, awaits further investigation. Most studies converge on the fact that dogs show gaze alternation in a requesting context when they seek to achieve a desired goal. However, dogs also look to their human partners before approaching a potentially scary object and they take into consideration their partner's emotional and behavioural action in deciding how to act. This suggests that looking towards humans may also function as a way to monitor and synchronize their own response to the environment with that of their partner. Yet dogs do not seem to be able to exhibit gaze alternation to help their partner obtain their goal. Whether this is for a lack of understanding of the other's goal or a lack of motivation remains an open question.

Finally, studies comparing identically-raised wolves and dogs, different dog breeds, using both behavioural and genetic methods and analysing the different experiential factors, which may affect human-directed gazing behaviour, suggest that this behaviour is the product of a combination of genetic and environmental factors. Clearly the investigation of these aspects is still relatively limited and future studies simultaneously analysing both will be needed to better understand the interplay between them.

Acknowledgements We are grateful to all the owners and their dogs that participated as volunteers in our studies. A special thanks to Paola Valsecchi, Isabella Merola, Chiara Passalacqua, Lara Tomaleo, Martina Lazzaroni, and Elisa Colombo for their contribution in many of the studies cited above. Thanks to Alexandra Horowitz for inviting us to contribute to this book. Sarah Marshall-Pescini was supported from grants from the University of Milan and funding from the European Research Council under the European Union's Seventh Framework Programme (FP/2007-2013)/ERC Grant Agreement n. [311870]. Last but not least we thank our dogs Laika, Mago and Tika for participating as pilot subjects in planning our studies at the *"Canis sapiens Lab"*.

References

Agnetta, B., Hare B., & Tomasello, M. (2000). Cues to food location that domestic dogs (*Canis familiaris*) of different ages do and do not use. *Animal Cognition, 3,* 107–112.

Barrera, G., Mustaca, A., & Bentosela, M. (2011). Communication between domestic dogs and humans: Effects of shelter housing upon the gaze to the human. *Animal Cognition, 14,* 727–734.

Bard, K. (1991). Distribution of attachment classifications in nursery chimpanzees. *American Journal of Primatology*, 24–88.

Bensky M. K., Gosling, S. D., & Sinn, D. L. (2013). The world from a dog's point of view: A review and synthesis of dog cognition research. *Advances in the Study of Behavior, 45,* 209–406.

Bentosela, M., Barrera, G., Jakovcevic, A., Elgier, A. M., & Mustaca, A. E. (2008). Effect of reinforcement, reinforcer omission and extinction on a communicative response in domestic dogs (*Canis familiaris*). *Behavioural Processes, 78*, 464–469.

Bentosela, M., Jakovcevic, A., Elgier, A. M., Mustaca, A. E., & Papini, M. R. (2009). Incentive contrast in domestic dogs (*Canis familiaris*). *Journal of Comparative Psychology, 123*, 125–130.

Bonanni, R., Natoli, E., Cafazzo, S., & Valsecchi, P. (2011). Free-ranging dogs assess the quantity of opponents in intergroup conflicts. *Animal Cognition, 14*, 103–115.

Bonanni, R., & Cafazzo, S. (in press). The social organization of a population of free-ranging dogs in a suburban area of Rome: A reassessment of the effects of domestication on dogs' behaviour. In J. Kaminski & S. Marshall-Pescini (Eds.), *The social dog: Behaviour and cognition*. San Diego, CA: Elsevier.

Boyko, A. R., Boyko, R. H., Boyko, C. M., Parker, H. G., Castelhano, M., Corey, L., et al. (2009). Complex population structure in African village dogs and its implications for inferring dog domestication history. *Proceedings of the National Academy of Sciences USA, 106*(33), 13903–13908.

Bradshaw, J. W. S., & Nott, H. M. R. (1995). Social and communication behaviour of companion dogs. In J. Serpell (Ed.), *The domestic dog: Its evolution, behaviour and interactions with people* (pp. 115–130). Cambridge: Cambridge University Press.

Bräuer, J., Call, J., & Tomasello, M. (2004). Visual perspective taking in dogs (*Canis familiaris*) in the presence of barriers. *Applied Animal Behaviour Science, 88*, 299–317.

Bräuer, J., Call, J., & Tomasello, M. (2005). All Great Ape species follow gaze to distant locations and around barriers. *Journal of Comparative Psychology, 119*, 145–154.

Bräuer, J., Kaminski, J, Riedel, J, Call, J., & Tomasello, M. (2006). Making inferences about the location of hidden food: Social dog-causal ape. *Journal of Comparative Psychology, 120*, 38–47.

Bräuer, J. (in press). What dogs understand about humans. In J. Kaminski & S. Marshall-Pescini (Eds.), *The social dog: Behaviour and cognition*. San Diego, CA: Elsevier.

Brugger, A., Lariviere LA, Mumme DL, & Bushnell E.W. (2007). Doing the right thing: Infants' selection of actions to imitate from observed event sequences. *Child development, 7*, 806–824.

Bugnyar, T., Stowe, M., & Heinrich, B. (2004). Ravens, Corvus corax, follow gaze direction of humans around obstacles. *Proceedings of the Royal Society B, 271*, 1331–1336.

Bullinger, A., Zimmermann, F., Kaminski, J., & Tomasello, M. (2011). Different social motives in the gestural communication of chimpanzees and human children. *Developmental Science, 14*, 58–68.

Buttelmann, D., & Tomasello M (2013) Can domestic dogs (*Canis familiaris*) use referential emotional expressions to locate hidden food? *Animal Cognition, 16*, 137–145.

Call, J. (2001). Chimpanzee social cognition. *TRENDS in Cognitive Sciences, 5*, 388–393.

Call, J., Bräuer, J., Kaminski, J., & Tomasello, M. (2003). Domestic dogs (*Canis familiaris*) are sensitive to the attentional state of humans. *Journal of Comparative Psychology, 117*, 257–263.

Call, J., & Tomasello, M. (1994). Production and comprehension of referential pointing by orangutans (Pongo pygmaeus). *Journal of Comparative Psychology, 108*, 307–317.

Call, J., & Tomasello, M. (2005). What chimpanzees know about seeing, revisited: An explanation of the third kind. In N. Eilan, C. Hoerl, T. McCormack & J. Roessler (Eds.), *Joint attention: Communication and other minds* (pp. 45–64). Oxford: Oxford University Press.

Cartmill, E. A., & Byrne, R. W. (2007). Orangutans modify their gestural signaling according to their audience's comprehension. *Current Biology, 17*, 1345–1348.

Cartmill, E. A., & Byrne R. W. (2010). Semantics of primate gestures: Intentional meanings of orangutan gestures. *Animal Cognition, 13*, 793–804.

Cheney, D. L., & Seyfarth, R. M. (1990). *How monkeys see the world*. Chicago: University of Chicago Press.

Cooper, J. J., Ashton, C., Bishop, S., West, R., Mills D. S., & Young R. J. (2003). Celver hounds: Social cognition in the domestic dog (*Canis familiaris*). *Applied Animal Behaviour Science, 81,* 229–244.

Csibra, G., & Gergely, G. (2011). Natural pedagogy as evolutionary adaptation. *Philosophical Transactions of The Royal Society B, 366,* 1149–1157.

Custance, D., & Mayer, J. (2012). Empathetic-like responding by domestic dogs (*Canis familiaris*) to distress in humans: An exploratory study. *Animal Cognition, 15,* 851–859.

de Rosnay, M., Cooper, P. J., Tsigaras, N., & Murray, L. (2006). Transmission of social anxiety from mother to infant: An experimental study using a social referencing paradigm. *Behaviour Research and Therapy, 44,* 1164–1165.

Ding, Z. L., Oskarsson, M., Ardalan, A., Angleby, H., Dahlgren, L. G., Tepeli, C., et al. (2012). Origins of domestic dog in Southern East Asia is supported by analysis of Y-chromosome DNA. *Heredity, 108,* 507–514.

Druzhkova, A. S., Thalmann, O., Trifonov, V. A., Leonard, J. A., Vorobieva, N. V., Ovodov, N. D., et al. (2013). Ancient DNA analysis affirms the canid from Altai as a primitive dog. *PLoS ONE, 8,* 57754.

Elgier, A. M., Jakovcevic, A., Barrera, G., Mustaca, A. E., & Bentosela, M. (2009a). Communication between domestic dogs (*Canis familiaris*) and humans: Dogs are good learners. *Behavioural Processes, 81,* 402–408.

Elgier, A. M., Jakovcevic, A., Mustaca, A. E., & Bentosela, M. (2009b). Learning and owner–stranger effects on interspecific communication in domestic dogs (*Canis familiaris*). *Behavioural Processes, 81,* 44–49.

Elgier, A. M., Jakovcevic, A., Mustaca, A. E., & Bentosela, M. (2012). Pointing following in dogs: Are simple or complex cognitive mechanisms involved? *Animal Cognition, 15,* 1111–1119.

Emery, N. J. (2000). The eyes have it: The neuroethology, function and evolution of social gaze. *Neuroscience and Biobehavioral Reviews, 24,* 581–604.

Erdőhegyi, Á., Topál, J., Virányi, Zs., & Miklósi Á. (2007). Dog-logic: Inferential reasoning in a two-way choice task and its restricted use. *Animal Behaviour, 74,* 725–737.

Faragó, T., Pongrácz, P., Miklósi, Á., Huber, L., Virányi, Z., & Range, F. (2010). Dogs' expectation about signalers' body size by virtue of their growls. *PLoS ONE, 5,* e15175.

Feinman, S. (1982). Social referencing in infancy. *Merrill-Palmer Quarterly, 28,* 445–470.

Ferrari, P. F., Paukner, A., Ionica, C., & Suomi, S. J. (2009). Reciprocal face-to-face communication between rhesus macaque mothers and their newborn infants. *Current Biology, 19,* 1768–1772.

Fox, M. W. (1971). *Behaviour of wolves, dogs and related canids.* London: Jonathan Cape.

Evans, C. S. (1997). Referential signals. In D. Owings, M. Beecher & N. Thompson (Eds.), *Perspectives in ethology* (pp. 99–143). New York: Plenum.

Gácsi, M., Miklósi, A., Varga, O., Topál, J., & Csányi, V. (2004). Are readers of our face readers of our minds? Dogs (*Canis familiaris*) show situation-dependent recognition of human's attention. *Animal Cognition, 7,* 144–153.

Gácsi, M., Gyori, B., Miklósi, Á., Virányi, Z., & Kubinyi, E. (2005). Species-Specific Differences and Similarities in the Behavior of Hand-Raised Dog and Wolf Pups in Social Situations with Humans. *Developmental Psychobiology, 47,* 111–122.

Gácsi, M., Györi, B., Virányi, Z., Kubinyi, E., Range, F., Belényi, B., & Miklósi, Á. (2009a). Explaining dog wolf differences in utilizing human pointing gestures: Selection for synergistic shifts in the development of some social skills. *PLoS ONE, 4*(8), e6584.

Gácsi, M., Kara, E., Belenyi, B., Topál, J., & Miklósi, Á. (2009b). The effect of development and individual differences in pointing comprehension of dogs. *Animal Cognition, 12,* 471–479.

Gácsi, M., Vas, J., Topál, J., & Miklósi, Á. (2013a) Wolves do not join the dance: Sophisticated aggression control by adjusting to human social signals in dogs. *Applied Animal Behaviour Science, 145,* 109–122.

Gácsi, M., Maros, K., Sernkvist, S., Farago, T., & Miklósi, Á. (2013b). Human analogue safe haven effect of the owner: Behavioural and heart rate response to stressful social stimuli in dogs. *PLoS ONE, 8*, e58475.

Gaunet, F. (2008). How do guide dogs of blind owners and pet dogs of sighted owners (*Canis familiaris*) ask their owners for food? *Animal Cognition, 11*, 475–83.

Gaunet, F. (2010). How do guide dogs and pet dogs (*Canis familiaris*) ask their owners for their toy and for playing? *Animal Cognition, 13*, 311–323.

Gaunet, F., & Deputte, B. L. (2011). Functionally referential and intentional communication in the domestic dog: Effects of spatial and social contexts. *Animal Cognition, 14*, 849–860.

Gómez, J. C. (1990). The emergence of intentional communication as a problem-solving strategy in the gorilla. In S. Parker & K. Gibson (Eds.), *"Language" and intelligence in monkeys and apes: Comparative developmental perspectives* (pp. 333–355), Cambridge: Cambridge University Press.

Gómez, J. G. (1996). Non-human primate theories of (non-human primate) minds: Some issues concerning the origins of mindreading. In P. Carruthers & P. K. Smith (Eds.), *Theories of theories of mind* (pp. 330–343). Cambridge: Cambridge University Press.

Gómez, J. G. (2007). Pointing Behaviors in Apes and Human Infants: A Balanced Interpretation. *Child Development, 78*, 729–734.

Hare, B., Brown, M., Williamson, C., & Tomasello, M. (2002). The domestication of social cognition in dogs. *Science, 298*, 1634–1636.

Hare, B., Call, J., & Tomasello, M. (1998). Communication of food location between human and dog. *Evolution of Communication, 2*, 137–159.

Hare, B., & Tomasello, M. (2005). Human-like social skills in dogs? *Trends in Cognitive Sciences, 9*, 439–444.

Hare, B., Rosati, A., Kaminski, J., Brauer, J., Call, J., & Tomasello, M. (2010). The domestication hypothesis for dogs' skills with human communication: A response to Udell et al., 2008 and Wynne et al., 2008. *Animal Behaviour, 79*, e1–e6.

Hare, B., Wobber, V., & Wrangham, R. (2012). The self-domestication hypothesis: Evolution of bonobo psychology is due to selection against aggression. *Animal Behaviour, 83*, 573–585.

Hernádi, A., Kis, A., Turcsán, B., & Topál, J. (2012). Man's Underground Best Friend: Domestic Ferrets, Unlike the Wild Forms, Show Evidence of Dog-Like Social-Cognitive Skills. *PLoS ONE, 7*, e43267.

Horowitz, A. (2009). Attention to attention in domestic dog (*Canis Familiaris*) dyadic play. *Animal Cognition, 12*, 107–118.

Horowitz, A., Hecht, J., & Dedrick, A. (2013). Smelling more or less: Investigating the olfactory experience of the domestic dog. *Learning and Motivation, 44*, 207–217.

Hory, Y., Kishi H., Murayama, M. I., Inoue, M., & Fujita, K. (2013). Dopamine receptor D4 gene (DRD4) is associated with gazing toward humans in domestic dogs (*Canis familiaris*). *Open Journal of Animal Sciences, 3*, 54–58.

Huber, L., Racca, A., Scaf, B., Virányi, Z., & Range, F.(2013). Discrimination of familiar human faces in dogs. *Learning and Motivation, 44*, 258–269.

Jakovcevic, A., Elgier, A. M., Mustaca, A. E., & Bentosela, M. (2010). Breed differences in dogs' (*Canis familiaris*) gaze to the human face. *Behavioural Processes, 84*, 602–607.

Jakovcevic, A., Mustaca, A., & Bentosela, M. (2012). Do more sociable dogs gaze longer to the human face than less sociable ones? *Behavioural Processes, 90*, 217–222.

Kaminski, J., Call, J., & Tomasello, M. (2004). Body orientation and face orientation: Two factors controlling apes' begging behavior from humans. *Animal Cognition, 7*, 216–223.

Kaminski, J., Call, J., & Tomasello, M. (2008) Chimpanzees know what others know, but not what they believe. *Cognition, 109*, 224–234.

Kaminski, J., Bräuer, J., Call J., & Tomasello, M. (2009). Domestic dogs are sensitive to a human's perspective. *Behaviour, 146*, 979–988.

Kaminski, J., Neumann M., Bräuer, J., Call J., & Tomasello, M. (2011). Domestic dogs communicate to request and not to inform. *Animal Behaviour, 82*, 651–658.

Kaminski, J., Pitsch, A., & Tomasello, M. (2013). Dogs steal in the dark. *Animal Cognition, 16*, 385–394.
Kaminski, J., & Nitzschner, M. (2013). Do dogs get the point? A review of dog–human communication ability. *Learning and Motivation, 44*, 294–302.
Kaminski, J., Schulz L., & Tomasello, M. (2012). How dogs know when communication is intended for them. *Developmental Science, 15*, 222–232.
Kis, A., Topál, J., Gácsi, M., Range, F., Huber, L., Miklósi, Á., & Virányi, Z. (2012). Does the A-not-B error in adult pet dogs indicate sensitivity to human communication? *Animal Cognition, 15*, 737–743.
Kubinyi, E., Pongrácz, P., & Miklósi, Á. (2009). Dog as a model for studying conspecific and heterospecific social learning. *Journal of Veterinary Behavior: Clinical Applications and Research, 4*, 31–41.
Kubinyi, E., Virányi, Z., & Miklósi, Á., (2007). Comparative social cognition: From wolf to dog to humans. *Comparative Cognition and Behavior Reviews, 2*, 26–46.
Kupán, K., Miklósi, Á., Gergely, G., & Topál, J. (2011). Why do dogs (Canis familiaris) select the empty container in an observational learning task? *Animal Cognition, 14*, 259–268.
Iizuka, Y. (1994). Gaze during speaking as related to shyness. *Perceptual and Motor Skills, 78*, 1259–1264.
Itakura, S. (1995). Snakes as agents of evolutionary change in primate brains. *Journal of Human Evolution, 51*, 44–48.
Ittyerah, M., & Gaunet, F. (2009). The response of guide dogs and pet dogs (*Canis Familiaris*) to cues of human referential communication. *Animal Cognition, 12*, 257–265.
Larson, G., et al. (2012). Rethinking dog domestication by integrating genetics, archeology, and biogeography. *Proceedings of the National Academy of Sciences USA, 109*(23), 8878–8883.
Leavens, D. A., Hopkins, W. D., & Bard, K. A. (1996). Indexical and referential pointing in chimpanzees (*Pan troglodytes*). *Journal of Comparative Psycholology, 110*, 346–353.
Leavens, D. A., & Hopkins, W. D. (1998.) Intentional communication by chimpanzees (*Pan troglodytes*): A cross-sectional study of the use of referential gestures. *Developmental Psychology, 34*, 813–822.
Leavens, D. A., Hopkins, W. D., & Thomas, R. (2004). Referential communication by chimpanzees (*Pan troglodytes*). *Journal of Comparative Psycholology, 118*, 48–57.
Leavens, D. A., Hopkins, W. D., & Bard, K. A. (2005a) Understanding the point of chimpanzee pointing: Epigenesis and ecological validity. *Current Directions in Psychological Science, 14*, 185–189.
Leavens, D. A., Russell, J. L., & Hopkins, W. D. (2005b) Intentionality as measured in the persistence and elaboration of communication by chimpanzees (*Pan troglodytes*). *Child Development, 76*, 291–306.
Line, S., Voith, V. L. (1986). Dominance aggression of dogs towards people: Behavior profile and response to treatment. *Applied Animal Behaviour Science, 16*, 77–83.
Liszkowski, U., Carpenter, M., Striano, T., & Tomasello, M. (2006). 12- and 18-month-olds point to provide information for others. *Journal of Cognition and Development, 7*, 173–187.
Liszkowski, U., Carpenter, M., & Tomasello, M. (2008). Twelve-month-olds communicate helpfully and appropriately for knowledgeable and ignorant partners. *Cognition, 108*, 732–739.
Lyons, D. E., Young, A. G., & Keil, F. C. (2007). The hidden structure of overimitation. *PNAS, 104*(19), 1158–1167.
Lyons, D. E., Damrosch, D. H., Lin, J. K., Macris, D. M., & Keil F. C. (2011). The scope and limits of overimitation in the transmission of artifact culture. *Philosophical Transactions of the Royal Society B, 366*, 1158–1167.
Marler, P., Dufty, A., & Pickert, R. (1986). Vocal communication in the domestic chicken: II. Is a sender sensitive to the presence and nature of a receiver? *Animal Behaviour, 34*, 194–198.
Marler, P., Evans, C.H., & Hauser, M. (1992). Animal signals: Motivational, referenial or both? In H. Papousek, U. Jürgens & M. Papousek (Eds.), *Nonverbal vocal communication:*

Comparative and developmental approaches (pp. 66–86). Cambridge: Cambridge University Press.

Marshall-Pescini, S., Passalacqua, C., Barnard, S., Valsecchi, P., & Prato-Previde, E. (2009) Agility and search and rescue training differently affects pet dogs' behaviour in socio-cognitive tasks. *Behavioural Processes, 81,* 416–422.

Marshall-Pescini, S., Passalacqua, C., Valsecchi, P., & Prato-Previde, E. (2010). Comment on "Differential sensitivity to human communication in dogs, wolves and human infants." *Science, 329,* 142-c.

Marshall-Pescini, S., Prato-Previde, E., & Valsecchi, P. (2011). Are dogs (Canis familiaris) misled more by their owners than by strangers in a food choice task? *Animal Cognition, 14,* 137–142.

Marshall-Pescini, S., Passalacqua, C., Miletto Petrazzini, M. E., Valsecchi, P., & Prato-Previde, E. (2012). Do dogs (*Canis lupus familiaris*) make counterproductive choices because they are sensitive to human ostensive cues? *PLoS ONE, 7,* e35437.

Marshall-Pescini, S., Colombo, E., Passalacqua, C., Merola, I., & Prato Previde, E. (2013) Gaze alternation in dogs and toddlers in an unsolvable task: Evidence of an audience effect. *Animal Cognition, 16,* 933–943.

McConnell, P., & Baylis, J. (1985). Interspecific communication in cooperative herding. Acoustic and visual signals from human shepherds and herding dogs. *Z. Tierpsychol, 67,* 302–328.

Merola, I., Prato-Previde, E., & Marshall-Pescini, S. (2012a). Social referencing in dog-owner dyads? *Animal Cognition, 15,* 175–185.

Merola, I., Prato-Previde, E., & Marshall-Pescini, S. (2012b). Dogs' social referencing towards owners and strangers. *PLoS ONE, 7,* e47653.

Merola, I., Marshall-Pescini, S., D'Aniello, B., & Prato-Previde, E. (2013a). Social referencing: Water rescue trained dogs are less affected than pet dogs by the stranger's message. *Applied Animal Behaviour Science, 147,* 132–138.

Merola, I., Prato-Previde, E., Lazzaroni, M., & Marshall-Pescini, S. (2013b). Dogs' comprehension of referential emotional expressions: Familiar people and familiar emotions are easier. *Animal Cognition, 16*(6), 1–13.

Miklósi, Á. (2007). Dog behaviour, evolution, and cognition. Oxford: Oxford University Press.

Miklósi, Á., Kubinyi, E., Topál, J., Gácsi, M., Virányi, Z., & Csányi, V. (2003). A simple reason for a big difference: Wolves do not look back at humans, but dogs do. *Current Biology, 13,* 763–766.

Miklósi, Á., Polgárdi, Topál, J., & Csányi, V. (2000). Intentional behaviour in dog-human communication: An experimental analysis of "showing" behaviour in the dog. *Animal Cognition, 3,* 159–166.

Miklósi, Á., Pongrácz P., Lakatos G., Topál J., & Csányi, V. (2005). A comparative study of the use of visual communicative signals in interaction between dogs (*Canis familiaris*) and humans and cats (*Felis catus*) and humans. *Journal of Comparative Psychology, 119,* 179–186.

Miklósi, Á., & Topál, J. (in press). What does it take to become 'best friends'? Evolutionary change in canine social competence. *Trends in Cognitive Science.*

Miklósi, Á., Topál, J., & Csányi, V. (2004). Comparative social cognition: What can dogs teach us? *Animal Behaviour, 67,* 995–1004.

Mills, D., van der Zee, E., & Zulch, H. (in press) When the bond goes wrong: Problem behaviours in the social context. In J. Kaminski & S. Marshall-Pescini (Eds.), *The social dog: Behaviour and cognition.* San Diego, CA: Elsevier.

Mitchell, R. W., & Thompson, N. S. (1993). Familiarity and the rarity of deception: Two theories and their relevance to play between dogs (*Canis familiaris*) and humans (*Homo sapiens*). *Journal of Comparative Psychology, 7,* 291–300.

Mumme, D. L., Fernald, A., & Herrera, C. (1966). Infants' responses to facial and vocal emotional signals in a social referencing paradigm. *Child Development, 67,* 3219–3237.

Nagasawa, M., Murai, K., Mogi, K., & Kikusui, T. (2011). Dogs can discriminate smiling faces from blank expression. *Animal Cognition, 14,* 525–533.

Nielsen, M. (2006). Copying actions and copying outcomes: Social learning through the second year. *Developmental Psychology, 42*, 555–565.

Palmer, R., & Custance, D. (2008). A counterbalanced version of Ainsworth's Strange Situation Procedure reveals secure-base effects in dog–human relationships. *Applied Animal Behaviour Science, 109*, 306–319.

Pang, J., et al. (2009). mtDNA Data indicate a single origin for dogs south of Yangtze River, less than 16,300 years ago, from numerous wolves. *Molecular Biological Evolution, 26*, 2849–2864.

Passalacqua, C., Marshall-Pescini, S., Barnard, S., Lakatos, G., Valsecchi, P., & Prato-Previde, E. (2011). Human-directed gazing behaviour in puppies and adult dogs, Canis lupus familiaris. *Animal Behaviour, 82*, 1043–1050.

Passalacqua, C., Marshall-Pescini, S., Merola I., Palestrini C., & Prato-Previde, E. (2013). Different problem-solving strategies in dogs diagnosed with anxiety-related disorders and control dogs in an unsolvabletask paradigm. *Applied Animal Behaviour Science, 147*, 139–148.

Pongrácz, P., Miklósi, Á., Kubinyi, E., Topál, J., & Csányi, V. (2003). Interaction between individual experience and social learning in dogs. *Animal Behaviour, 65*, 595–603.

Pongrácz, P., Miklósi, A., Timár-Geng, K., & Csányi, V. (2004). Verbal attention getting as a key factor in social learning between dog (*Canis familiaris*) and human. *Journal of Comparative Psychology, 118*, 375–383.

Povinelli, D. J., & Eddy, T. J. (1996). What young chimpanzees know about seeing. *Monographs of the Society for Research in Child Development, 61*, 1–152.

Prato-Previde, E., Custance D. M., Spiezio, C., & Sabatini, F. (2003). Is the dog-human relationship an attachment bond? An observational study using Ainsworth's Strange Situation. *Behaviour, 140*, 225–254.

Prato-Previde, E., Marshall-Pescini, S., & Valsecchi, P. (2008). Is your choice my choice? The owners' effect on pet dogs' (Canis familiaris) performance in a food choice task. *Animal Cognition, 11*, 167–174.

Prato-Previde, E., & Valsecchi, P. (in press) The immaterial cord: The dog human attachment bond. In J. Kaminski & S. Marshall-Pescini (Eds.), *The social dog: Behaviour and cognition*. San Diego, CA: Elsevier.

Range, F., Viranyi, Z., & Huber, L. (2007). Selective imitation in domestic dogs. *Current Biology, 17*, 868–872.

Range, F., & Virányi, Z. (2011). Development of gaze following abilities in wolves (*Canis lupus*). *PLoS ONE, 6*, e16888.

Range, F., Huber, L., & Heyes, C. (2011). Automatic imitation in dogs. *Proceedings of the Royal Society B: Biological Sciences, 278*, 211–217.

Reid, P. J. (2009). Adapting to the human world: Dogs responsiveness to our social cues. *Behavioural Processes, 80*, 325–333.

Riemer, S., Müller, C., Virányi, Z., Huber, L., & Range, F. (in press). Choice of conflict resolution strategy is linked to sociability in dog puppies. *Applied Animal Behaviour Science*.

Roberts, S. G., McComb, K., & Ruffman, T. (2008). An experimental investigation of referential looking in free-ranging Barbary macaques (*Macaca sylvanus*). *Journal of Comparative Psychology, 122*, 94–99.

Roberts, A. I., Vick, S. J., & Buchanan-Smith H. M. (2013). Communicative intentions in wild chimpanzees: Persistence and elaboration in gestural signalling. *Animal Cognition, 16*, 187–196.

Rossi, A. P., & Ades, C. (2008). A dog at the keyboard: Using arbitrary signs to communicate requests. *Animal Cognition, 11*, 329–338.

Russell, C. L., Bard, K. A., & Adamson, L. B. (1997). Social referencing by young chimpanzees (*Pan troglodytes*). *Journal of Comparative Psychology, 111*, 185–193.

Savolainen, P., Zhang, Y., Luo, J., Lundeberg, J., & Leitner, T. (2002). Genetic evidence for an east Asian origin of domestic dogs. *Science, 298*, 1610–1613.

Savolainen, P., Leitner, T., Wilton, A., Matisoo-Smith, E., & Lundeberg, J. (2004). A detailed picture of the origin of the Australian dingo, obtained from the study of mitochondrial DNA. *Proceedings of the National Academy of Sciences USA, 101*, 12387–12390.

Schenkel, R. (1967). Submission: Its features and function in the wolf and dog. *American Zoologist, 7*, 319–329.

Schloegl, C., Schmidt, J., Scheid, C., Kotrschal, K., & Bugnyar, T. (2008). Gaze following in non-human animals: The corvid example. In F. Columbus (Ed.), *Animal behaviour: New research* (pp. 73–92). New York: Nova Science Publishers.

Schwab, C., & Huber, L. (2006). Obey or not obey? Dogs (*Canis familiaris*) behave differently in response to attentional states of their owners. *Journal of Comparative Psychology, 120*, 169–175.

Scott, J. P., & Fuller, J. L. (1965). *Genetics and the social behavior of the dog*. Chicago: University of Chicago Press.

Simpson, E. A., Paukner, A., Suomi, S. J., & Ferrari, P. F. (in press). Visual attention during neonatal imitation in newborn macaque monkeys. *Developmental Psychobiology*.

Schneier, F. R., Rodebaugh, T. L., Blanco, C., Lewin, H., & Liebowitz, M. R. (2011). Fear and avoidance of eye contact in social anxiety disorder. *Comprehensive Psychiatry, 52*, 81–87.

Smith, B. P., & Litchfield, C. A. (2013). Looking back at 'looking back': Operazionalizing referential gaze for dingoes in an unsolvable task. *Animal Cognition, 16*, 961–971.

Soproni, K., Miklósi, A., Topál, J., & Csányi, V. (2002). Dogs' (*Canis familiaris*) responsiveness to human pointing gestures. *Journal of Comparative Psychology, 116*, 27–34.

Southgate, V., Chevallier, C., & Csibra, G. (2009). Sensitivity to communicative relevance tells young children what to imitate. *Developmental Science, 12*,1013–1019.

Smith, B., & Litchfield, C. (2009) Review of the relationship between Indig- enous Australians, dingoes (Canis *dingo*), and domestic dogs (Canis *familiaris*). *Anthrozoös, 22*, 111–128.

Smuts, B. (in press) Social behavior among companion dogs with an emphasis on play. In J. Kaminski & S. Marshall-Pescini (Eds.), *The social dog: Behaviour and cognition*. San Diego, CA: Elsevier.

Sümegi, Z., Kis, A., Miklósi, Á., & Topál, J. (in press). Why do adult dogs (*Canis familiaris*) commit the A-not-B search error? *Journal of Comparative Psychology*.

Szetei, V., Miklósi, Á., Topál, J., & Csányi, V. (2003). When dogs seem to lose their nose: An investigation on the use of visual and olfactory cues in communicative context between dog and owner. *Applied Animal Behaviour Science, 83*, 141–152.

Stenberg, G. (2003). Effects of maternal inattentiveness on infant social referencing. *Infant Child Development, 12*, 339–41.

Stenberg, G., & Hagekull, B. (2007). Infant looking behavior in ambiguous situation: Social referencing or attachment behavior? *Infancy, 11*, 111–129.

Teglas, E., Gergely, A., Kupan, K., Miklósi, Á., & Topál, J. (2012). Dogs' gaze following is tuned to human communicative signals. *Current Biology, 22*, 209–212.

Tennie, C., Glabsch, E., Tempelmann, S., Bräuer, J., Kaminski, J., & Call, J. (2009). Dogs, *Canis familiaris*, fail to copy intransitive actions in third-party contextual imitation tasks. *Animal Behaviour, 77*, 1491–1499.

Tomasello, M. (2008). *Origins of human communication*. Cambridge, Massachusetts: MIT Press.

Tomasello, M., & Call, J. (1997). *Primate cognition*. New York: Oxford University Press

Tomasello, M., Call, J., Nagell, K., Olguin, R., & Carpenter, M. (1994). The learning and use of gestural signals by young chimpanzees: A trans-generational study. *Primates, 35*, 137–154.

Tomasello, M., Hare, B., & Agnetta, B. (1999). Chimpanzees, *Pan troglodytes*, follow gaze direction geometrically. *Animal Behaviour, 58*, 769–777.

Tomonaga, M., Tanaka, M., Matsuzawa, T., Myowa-Yamakoshi, M., Kosugi, D., & Mizuno, Y., et al. (2004). Development of social cognition in infant chimpanzees (Pan troglodytes): Face recognition, smiling, gaze and the lack of triadic interactions. *Japanese Psychological Research, 46*, 227–235.

Topál, J., Erdohegyi, Á., Mányik, R., & Miklósi, Á. (2006). Mindreading in a dog: An adaptation of a primate "mental attribution" study. *International Journal of Psychology and Psychological Therapy, 6*, 365–379.

Topál, J, Gergely, G, Miklósi Á, Erdohegyi, Á., & Csibra, G. (2008). Infants' perseverative search errors are induced by pragmatic misinterpretation. *Science, 321*, 1831–1834.

Topál, J., Gergely, G., Erdohegy, Á., Csibra G., & Miklósi, Á. (2009a). Differential sensitivity to human communication in dogs, wolves and human infants. *Science, 325*, 1269–1272.

Topál, J., Miklósi, Á., Csanyi, V., & Doka, A. (1998). Attachment behavior in dogs (Canis familiaris): A new application of Ainsworth's (1969) strange situation test. *Journal of Comparative Psychology, 112*, 219–229.

Topál, J., Miklósi, Á., Gácsi, M., Dóka, A., Pongrácz, P., & Kubinyi, E., et al. (2009). The dog as a model for understanding human social behavior. *Advances in the Study of Behavior, 39*, 71–116.

Topál, J., Miklósi, Á., Sumegi, Z., & Kis, A. (2010). Response to comments on "differential sensitivity to human communication in dogs, wolves and human infants." *Science, 329*, 142-d.

Topál, J., Kis, A., & Katalin' O. (in press). Dogs' sensitivity to human ostensive cues: A unique adaptation? In J. Kaminski & S. Marshall-Pescini (Eds.), *The social dog: Behaviour and cognition*. San Diego, CA: Elsevier.

Triesch, J., Jasso, H., & Deák, G. O. (2007). Emergence of mirror neurons in a model of gaze following. *Adaptive Behavior, 15*(2), 149–165.

Udell, M. A. R., Dorey, N. R., & Wynne, C. D. L. (2010). What did domestication do to dogs? A new account of dogs' sensitivity to human actions. *Biological Reviews, 85*, 327–345.

Udell, M. A. R., Dorey, N. R., & Wynne, C. D. L. (2011). Can your dog read your mind? Understanding the causes of canine perspective taking. *Learning and Behavior, 39*, 289–302.

Udell, M. A. R., Giglio, R. F., & Wynne, C. D. L. (2008a). Domestic dogs (Canis familiaris) use human gestures but not nonhuman tokens to find hidden food. *Journal of Comparative Psychology, 122*, 84–93.

Udell, M. A. R., & Wynne, C. D. L. (2008b). A review of domestic dogs' (Canis familiaris) human-like behaviors: Or why behavior analysts should stop worrying and love their dogs. *Journal of the Experimental Analysis of Behavior, 89*, 247–261.

Udell, M. A. R., & Wynne, C. D. L. (2010). Ontogeny and phylogeny: Both are essential to human-sensitive behaviour in the genus Canis. *Animal Behaviour, 79*, e9–e14.

Vaish, A., & Striano, T. (2004). Is visual reference necessary? Contributions of facial versus vocal cues in 12-month-olds' social referencing behavior. *Developmental Science, 7*, 261–269.

Vas, J., Topál, J., Gácsi, M., Miklósi, A., & Csányi, V. (2005). A friend or an enemy? Dogs' reaction to an unfamiliar person showing cues of threat and friendliness at different times. *Applied Animal Behaviour Science, 94*, 99–115.

Vas, J., Topál, J., Gyori, B., & Miklósi, A. (2008). Consistency of dogs' reactions to threatening cues of an unfamiliar person *Applied Animal Behaviour Science, 112*(3–4), 331–344.

Vilà, C., Savolainen, P., Maldonado, J. E., Amorin, I. R., Rice, J. E., & Honeycutt, R. L., et al. (1997). Multiple and ancient oigins of the domestic dog. *Science, 276*, 1687–1689.

Virányi, Z., Gácsi, M., Kubinyi, E., Topál, J., Belényi, B., Ujfalussy, D., & Miklósi, Á. (2008). Comprehension of human pointing gestures in young human-reared wolves and dogs. *Animal Cognition, 11*, 373–387.

Virányi, Z., Topál, J., Gácsi, M., Miklósi, A., & Csányi, V. (2004). Dogs respond appropriately to cues of humans' attentional focus. *Behavioural Processes, 31*, 161–172.

Virányi, Z., Topál, J., Gácsi, M., Miklósi, Á., & Csányi, V. (2006). A nonverbal test of knowledge attribution: A comparative study on dogs and children. *Animal Cognition, 9*, 13–26.

Virányi, Z., & Range, F. (2011). Evaluating the logic of perspective-taking experiments. *Learning & Behavior, 39*, 306–309.

Virányi, Z., & Range, F., (in press). On the way to a better understanding of dog domestication: Aggression and cooperativeness in dogs and wolves. In J. Kaminski & S. Marshall-Pescini (Eds.), *The social dog: Behaviour and cognition*. San Diego, CA: Elsevier.

vonHoldt, B. M., et al. (2010). Genome-wide SNP and haplotype analyses reveal a rich history underlying dog domestication. *Nature, 464*, 898–902.

Walden, T. A., & Geunyoung, K. (2005). Infants' social looking toward mother and stranger. *The International Journal of Behavioral Development, 29*, 356–360.

Wang, G., et al. (2013). The genomics of selection in dogs and the parallel evolution between dogs and humans. *Nature Communication, 4*, 1860.

Warden, C., & Warner, L. (1928). The sensory capacities and intelligence of dogs. *The Quarterly Review of Biology, 3*, 1–28.

Werhahn, G., Barrera, G., Viranyi, Z., & Range, F. (2013). Wolves and dogs follow their conspecifics gaze into distant space. In *Proceedings of the 3rd ToK Conference of Comparative Cognition*, Vienna, Austria, 3–5 July 2013.

Wieser, M. J., Pauli, P., Alpers, G. W., & Mühlberger, A. (2009). Is eye to eye contact really threatening and avoided in social anxiety?–an eye-tracking and psychophysiology study. *Journal of Anxiety Disorders, 23*, 93–103.

Wynne, C. D. L., Udell, M. A. R., & Lord, K. A. (2008). Ontogeny's impacts on human dog communication. *Animal Behaviour, 76*, 1–4.

Chapter 6
Visual Attention in Dogs and the Evolution of Non-Verbal Communication

Alejandra Rossi, Daniel Smedema, Francisco J. Parada and Colin Allen

Abstract The common history of *Homo sapiens* and *Canis lupus familiaris* dates back to between 11,000 and 32,000 years ago, when some wolves (*Canis lupus*) started living closely with humans. Although we cannot reach back into the past to measure the relative roles of wolves and humans in the ensuing domestication process, it was perhaps the first involving humans and another animal species. Yet its consequences for both species' history are not completely understood. One of the puzzling aspects yet to be understood about the human–dog dyad is how dogs so readily engage in communication in the context of a social interactions with humans. To be sensitive to the meaning of human speech and gestures, dogs need to attend to various visual and vocal cues, in order to reconstruct the messages from patterns of human behavior that remain stable over time, while also generalizing to unfamiliar, novel contexts. This chapter will discuss this topic in light of some of the recent findings about dogs' perceptual capacities for social cues. We describe some of the new technologies that are being used to better describe these perceptual processes, and present the results of a preliminary experiment using a portable eye-tracking system to gather data about dogs' visual attention in a social interaction with humans, ending with a discussion of the possible cognitive mechanisms underlying dogs' use of human social cues.

A. Rossi (✉) · D. Smedema · C. Allen
Cognitive Science Program, Indiana University, Bloomington, IN, USA
e-mail: alejandrarossic@gmail.com

A. Rossi · F. J. Parada
Program in Neuroscience, Indiana University, Bloomington, IN, USA

F. J. Parada
Psychological and Brain Sciences, Indiana University, Bloomington, IN, USA

6.1 Introduction

The eyes, according to proverb, are the windows to the soul. The more mundane truth is that humans and many other social animals cannot help but indicate their attentional and intentional states with their eyes. The eyes, of course, are not the only source of such information; other cues such as head movements, changes in body posture, and pointing gestures provide essential information for the everyday lives of many social animals. Because vision is a primary mode by which the members of many gregarious species detect social cues, as well as mediating their goal-directed interactions with the environment, it is possible for observers to use the overt eye movements of others as a proxy for their intentions and as cues toward the location of significant objects (Shepherd 2010). As scientific observers, we may exploit the importance of eye movements to improve our understanding of the mechanisms of social perception by studying where species look for relevant information when performing a given task. Advances in the study of comparative cognition can be achieved by combining behavioral tasks with techniques coming from the cognitive and neural sciences, and in this chapter we describe the work of our group in applying eye-tracking technology to the study of how dogs use cues provided by humans.

Because of their long history of domestication, dogs present a very important model for making progress in understanding the evolution of social cognition. Technological progress toward capturing the dynamics of dog–human interactions could provide a very productive means of furthering this understanding, as well as furthering the aims for integrative ethological work described by Tinbergen (1963). We have been developing a fully portable eye-tracking system for use with dogs that provides excellent opportunities for gathering data about the abilities of dogs to read human social gestures. In this chapter we describe the technological and methodological challenges presented by our attempt to develop and use the eye-tracking system, and we describe some preliminary results, which should encourage further exploration of these techniques.

Before getting into the details of our methods and results, it is useful to highlight the significance of studying the relationship between humans and dogs for understanding the evolution of social cognition by outlining the likely events that led to this long-standing relationship. The history between dogs (*Canis lupus familiaris*) and humans (*Homo sapiens*) represents what is almost certainly the longest, closest ongoing relationship humans have ever had with another species of mammal. Dogs were the first species to be domesticated by humankind (Clutton-Brock 1995; Davis and Valla 1978; Hemmer 1990; Vila et al. 1997). A proto-dog (an early version of dogs) lived in a close relationship with humans even before they were a 'properly' domesticated species. That is, some members of *Canis lupus* were most probably cohabiting with humans, even before humans started exerting conscious control over their diet and reproduction. This close relationship resulted in genetic, behavioral, morphological and physiological changes that were passed on to subsequent generations (Diamond 2002). The relationship has been so close that dogs have evolved in ways that range from acquiring the ability to metabolize carbohydrates, as a result of being permitted to eat human food (Axelsson et al. 2013), to the ability to use human gestures in way that has no equivalent in the animal kingdom (e.g. Hare and Tomasello 2005; Miklósi et al. 2005; Soproni et al. 2001, 2002) (see also Prato-Previde and Marshall-Pescini, this volume).

Only recently have researchers started paying attention to the full range of implications of the long relationship between dogs and humans. One reason for this may be that dogs were often dismissed as "artificial" or "unnatural", thus not worthy of serious attention from behavioral biologists. However, researchers have started to realize that the domestic dog is a unique yet informative case in nature and have begun to investigate its particularities.

Our focus is on the implications of this relationship for social behavior and cognition. Dogs have become keen observers of human movements and gestures. But importantly, in order for dogs (or any species) to be able to understand or make use of another species' movements and gestures, they have to selectively allocate perceptual and cognitive resources to the detection and processing of particular social aspects of the environment (a process known as *orienting of attention*). Dogs are especially capable of utilizing human pointing gestures and of following our eye-gaze direction. To be able to use these cues efficiently, individuals must be able to exploit their referential nature. Pointing with hands or looking in a particular direction does not have an intrinsic value or meaning in the world; rather these actions allow perceivers to orient their attention to specific environmental cues within the particular context of a social interaction.

To arrive at an understanding of how humans and dogs have managed to achieve such tight social coordination, it is necessary to talk about the process of domestication. The exact period of time, the location or locations of dogs' domestication, and the reasons behind it are matters of substantial debate in the scientific community (Driscoll et al. 2009; Hare and Tomasello 2005; Larson et al. 2012; Serpell 1995). Nevertheless, one thing that most researchers do agree on is that dogs are so closely related to gray wolves (*C. lupus*) that they most probably come from them (Clutton-Brock 1995; Driscoll et al. 2009) (but see Koler-Matznick 2002, for an argument that *Canis familiaris* did not descend from *C. lupus*). There is widespread acceptance, along the way, of the likelihood of the existence of several independent centers of dog domestication beginning in the Late Pleistocene (126,000–11,000 years ago) and early Holocene (11,000–5,000 years ago) (Crockford and Iaccovoni 2000). This is especially relevant if we take into account that the emergence of dogs as an identifiable subspecies is estimated to have occurred between 32,000 and 12,000 years ago. The later part of this range (from 12,000 years ago) coincides with the Neolithic revolution in which humans started to settle down, slowly changing the hunter-gatherer lifestyle to a more sedentary one (Bar-Yosef 1998; Weisdorf 2005). In that respect, as the first domesticated animal species, dogs were probably a strategic factor in this process, since it has been suggested that dogs helped modern humans in hunting and herding, as well as in defending territory, searching and guiding—in addition to being eaten in some cultures (Lupo 2011; Ruusila and Pesonen 2004). Humans, in turn, might have provided a secure source of food to dogs.

The fact that dogs adapted to the human environment in such a distinctive way allows us to explore whether the basis of human–dog social interactions can be informative about the evolution of relevant aspects of sociality such as communication and cooperation. Because of the long history of domestication, dogs' particular responsiveness to human words (Fukuzawa et al. 2005; Kaminski et al. 2004; Pilley and Reid 2011; Ramos and Ades 2012) and their apparent understanding of human emotional expressions (Buttelmann and Tomasello 2013; Merola et al. 2012), it becomes possible to ask to what extent, if any, these capacities are a case of convergent evolution, the process by which species that are not closely related evolve similar traits as they adapt over time to similar ecological and social environments. During the past 12–30,000 years dogs and humans have helped to shape each others' environments in ways that might also have facilitated a co-evolutionary process. For example, playful interactions between puppies and human children could scaffold developmental processes that have strong implications for the evolution of both species' social capacities.

There are several reasons to postulate why it would have been necessary for organisms to develop a new set of social skills enabling them to predict and to manipulate other agents' behaviors (Shultz and Dunbar 2007). It is reasonable to suppose that individuals benefit from creating, discovering, and taking advantage of others' solutions to ecological challenges (Barton 1996; Reader and Laland 2002). Thus, the particularly social way in which wolves (the most probable ancestors of dogs) solve ecologically relevant issues such as foraging could be one of the sources in which dogs' communicative abilities are rooted. The evolution of social and communication skills, and related cognitive abilities, has been related to hunting patterns of a species (Bailey et al. 2013). The gray wolf is well-known as a cooperative hunter. Hence, it has been suggested that the evident communicative skills of dogs with respect to humans originates from wolves being able to cooperate among each other to hunt (Hare and Tomasello 2005).

Wolves belong to the category of pack-living canids with gregarious but independent natures (Koler-Matznick 2002). Though they live in packs of eight to twelve animals (Clutton-Brock and Wilson 2002) it has been reported that they can also live and survive independently of the group (Mech 1970, 2012; Sullivan 1978). Wolves and primates share a highly hierarchical social structure, but unlike primates, only the dominant pair of the pack reproduces (Mech 1970) and most of the time they do so for life (Clutton-Brock and Wilson 2002). Wolves hunt both small and large prey (Clutton-Brock and Wilson 2002). As mentioned before, they hunt large prey cooperatively (Clutton-Brock 1999; Olsen 1985) and they learn to do so in their adolescence, which last up to 2 years (Koler-Matznick 2002).

Although it is difficult to determine the exact role of cooperative hunting, it is believed that such hunting carried significant implications for the evolution of sociality and cognition (Bailey et al. 2013). For instance, wolves have been shown to coordinate their actions in order to fan out and encircle prey (Muro et al. 2011).

The gray wolf also has highly developed reconciliation behaviors, very similar to those seen in non-human primates (Palagi and Cordoni 2009). These reconciliation behaviors would help to decrease aggression between group members and preserve cooperative relationships and social cohesiveness (Cordoni and Palagi 2008; Palagi and Cordoni 2009), which in turn would increase the efficiency of cooperative hunting (Bailey et al. 2013). Some biological models indicate, however, that hunting both large and small prey provides evidence that wolves engage in both cooperation and cheating (Packer and Ruttan 1988).

Cooperation and cheating both require a level of vigilance directed toward the behavior of other group members. Therefore, the production of intentional signals (i.e., intended communication) and unintentional signals (e.g. involuntary behavioral cues) signals, and the ability to recognize such signals, probably plays a crucial role in the decision-making process of group members (Bailey et al. 2013).

Taking this evidence as a whole, it is reasonable to conclude that for social species that hunt cooperatively, a well-developed understanding of others' behaviors is extremely important and probably advantageous (Bernstein 1970; Moore and Dunham 1995).

Capacities such as responding differentially to facial expressions and paying attention to another's focus of attention can help an individual recognize, for example, when another group member is seeing a predator or a prey approaching. Similarly, attention to such cues allows organisms to predict each other's behavior, emotional state or intentions (Shepherd 2010).

Effective prediction of other organisms requires individuals to integrate different sensory information automatically through time. Understanding how the information is integrated over time, therefore, is relevant to understanding the basis of social cognition in different species. It is possible to study this issue in a detailed way by making use of new technologies. Technological advances allow us to study how salient environmental cues (e.g. information provided by movements and gestures) are being used in many different modalities. Some of these advances aim to integrate sensory information, getting us closer to studying sociality as it occurs day to day. One example of such a technology is eye-tracking.

Eye-tracking technology allows researchers to access participants' overt visual attention (Duchowski 2007). Research using eye-tracking techniques has revealed much about the cognitive processes underlying human behavior (e.g. Dalton et al. 2005; Felmingham et al. 2011; Gredebäck et al. 2009; Holzman et al. 1974; Yarbus et al. 1967), and recently researchers in animal behavior have started to make use of this method to answer questions regarding animals' overt attention (Kano and Tomonaga 2013; Teglas et al. 2012; Williams et al. 2011), with greater accuracy than third-person perspectives (e.g. video-cameras and inferences from subjects' head orientation) (Duchowski 2007). It is worth noting that the use of eye-tracking methods to study animal cognition is still quite preliminary. There are a number of unresolved issues in expanding a technique developed for humans to non-human animals. These include constraints related to calibration of the eye-tracker system, and the lack of development of non-invasive and appropriate gaze-estimation models for non-human species (Kjaersgaard et al. 2008).

Other difficulties include getting a suitable and ecologically valid task for the species, and being aware of possible anthropocentric interpretations of results.

Notwithstanding these issues, eye-tracking systems have been recently used in comparative cognition research with nonhuman animals, primarily primates (Hattori et al. 2010; Hirata et al. 2010; Kano and Tomonaga 2010, 2013; Machado and Nelson 2011; Zola and Manzanares 2011) but also recently with dogs (Somppi et al. 2011; Teglas et al. 2012; Williams et al. 2011). This research has produced intriguing results. Using a video-based eye-tracker with a wide-angle lens, Kano and colleagues (Kano et al. 2011; Kano and Tomonaga 2009, 2013) recorded the eye movements of great apes and humans. Their research suggests that these species show similar visual scanning patterns for scenes and faces (particularly eyes and mouth) with respect to both conspecifics and members of other species. Differences were seen in fixation duration: shorter and stereotyped in apes, and longer and variable in humans. Humans also present more frequent and longer fixations to the eye region compared to apes. Similarly, Hirata et al. (2010) showed that when presented with pictures of conspecifics' faces, chimpanzees focus more frequently and for a longer time on the eye region of the faces, but only for upright faces in which chimpanzees had their eyes open (vs. inverted faces or pictures of chimpanzees with closed eyes). By comparison, eye-tracking research on dogs is still very preliminary. Somppi et al. (2011) trained their participants to stay still on a cleverly designed apparatus that allows them to study a dog's visual scanning without the use of more constrictive methods. They showed that dogs are able to focus their attention on informative regions of images displayed on a screen without any task-specific pre-training. Interestingly, they also showed that dogs seem to prefer looking at faces of conspecifics over other images (e.g. human faces, objects) and that they fixate on a familiar images longer than on novel ones. Similarly, adopting a system widely use in human and infant research measuring responses to images on a screen, Teglas et al. (2012) showed that dogs' use of human cues is highly sensitive to the ostensive and referential use of the signals, supporting according to the authors, the existence of a functionally infant-analog social competence in both dogs and humans.

However, existing approaches to studying dogs' visual attention suffer from various drawbacks if one is interested in studying the naturalistic behavior of these animals. For example, maintaining the animal's natural mobility is important when researchers aim to study behaviors in a naturalistic fashion; however, all eye-tracking systems currently in use with dogs either inhibit the dogs' movements in some way, or limit the location and type of stimuli that can be presented to the dogs (e.g. to those that can be presented on a video monitor).

Thus, although there is plenty of evidence supporting dogs' skillfulness at reading human gestures, there are still many open questions about how dogs visually process human social cues. Such questions include: Where do dogs focus their overt visual attention when presented with human gestures? How much does familiarity with the human signaler modulate dogs' visual behavior? How differently do dogs scan pointing with the hand to an object versus head-gaze cues directed at an object? Answering these questions will support several scientific objectives. First, it will allow detailed comparison of the visual behavior of dogs to that of primates and thus to explore further the influence of domestication in the evolution of the species' social skills. Second, knowing whether dogs scan familiar versus unfamiliar persons differently might influence the design of experiments that tap into communicative skills of dogs. Finally, taking a step forward in understanding the salient aspects of our own behavior for dogs might allow scientists and engineers to tap into the properties that make the comprehension of human commands and intentions possible for nonhuman forms of intelligence. Thus, for example, a better understanding of the social-cognitive capacities of dogs could support appropriate developments in robotics, where robots must detect and respond to human social cues.

To start exploring these questions we made use of our head-mounted eye-tracking system (described below) and the so-called object choice paradigm, a well-known experimental paradigm that has been widely used with dogs (e.g. Hare et al. 2002; Miklósi et al. 2005; Soproni et al. 2002). The object choice task consists of giving a dog a choice between two cups—one of them containing a piece of food or a preferred toy—and then directing the dog to the correct container by human pointing gestures. We aimed to investigate, first, what are the most salient regions for dogs when scanning human gestures, second, whether familiarity modulates visual tracking in dogs and third, whether dogs' abilities to follow a pointing gesture is related to the kind of motion (head vs. arms). For this preliminary study a familiar person (owner of the dog) and an unfamiliar person (experimenter) displayed static distal pointing and head-gazing to signal to the subject where a piece of food was hidden, while the dog's overt visual attention was monitored by a head-mounted eye-tracker system.

With fewer spatial and visual constraints than other methods, the use of head-mounted systems such as ours might be especially relevant to the investigation of the communicative and social behavior of canines and other animals in a naturalistic setting.

Since our system allows the full eye-tracking system to be carried in a backpack worn by the dog, the dog is able to move relatively freely. Mobility has the benefit of capturing most spontaneous behaviors of dogs, albeit within the confined circumstances of a room in the lab (but see Fugazza and Miklósi, this volume, for discussion of the naturalism of a lab for dogs). Despite the fact that movements are confined to "indoor behavior"—no rough-and-tumble social play, for example—the freely-moving dogs in our experiments often changed direction abruptly, affecting the stability of the system. It is important, nevertheless, to work with freely-moving dogs because mobility brings the opportunity to study the full range of dogs' visual behavior before they make a choice. Decision-making in dogs involves actions, which translate to movements. The system must be recalibrated after every movement, contributing a source of noise that is sometimes impossible to clean (thus decreasing the number of usable trials). Re-calibrating the system is not difficult per se, though dogs do seem to lose motivation over time. Thus, sessions have to be short in duration, including only few trials per session.

In this preliminary study, the number of dogs providing usable data was limited because of problems with camera and recording quality. More subjects need to be run with the next-generation equipment, but we offer this study as a proof of concept and illustration of the capacity to collect interesting new kinds of data bearing on canine cognitive processes.

6.2 Eye-Tracking Study

6.2.1 Methods

Participants. Six privately owned pet dogs (5 females and 1 male, age range: 2–5 years old; breeds: mixed, American pitbull terrier, Akita mix, Shiba/terrier mix and two schnauzer/poodle mixes) took part in this study. Dogs and their owners were asked to visit the lab twice on different days, one for each experimental session. All experiments were conducted at the Canine Behavior and Cognition (CBC) Laboratory of Indiana University. The Institutional Animal Care and Use Committee and the Institutional Review Board of Indiana University approved the experimental protocol.

For an average period of two weeks prior to coming to the lab, dogs were trained at home by their owners to wear an off-the-shelf set of goggles specially designed for different breeds of dogs. The criteria for inclusion in the study were, first, that the dog could successfully wear the goggles for about 10 min without disruption of the dog's natural behavior (according to their owner's opinion). Second, the dog should not try to take the goggles off by pawing at them or by any other means during those 10 min.

Three people were involved in running the experiment: Experimenter 1 (E1 hereafter), who gave instructions to the owner and carried out the warm-up and the unfamiliar person (UP) trials; Experimenter 2 (E2 hereafter), who assisted E1 throughout the session; and the owner of the dog, who carried out the familiar person (FP) trials and helped in handling her/his dog when required.

Apparatus. A head-mounted dog eye tracker system designed in the CBC Lab was used. The system consists of two small lightweight cameras, mounted on a set of commercially available dog goggles (the same ones worn by the dogs in the pre-training phase), slightly modified to improve stability. One of the cameras is strategically located, pointing at the eye of the participant [eye camera: Sony high-resolution ultra-low-light black-and-white snake camera], while the second camera sits on top of the goggle frame, right above the participant's eye, collecting images of the world [scene camera: RMRC-MINI 12 V (6–24 V) compact camera PAL] (Fig. 6.1). Both streams of data were recorded and stored at 60 frames per second using two lightweight portable video recorders for further processing and analysis.

Warm-up trials. Before each session ('familiar' and 'unfamiliar person' sessions) dogs took part in four warm-up trials that were intended to both familiarize

Fig. 6.1 Digital image of the eye-tracking system worn by one of our participants

dogs and owners with the procedures and also to motivate dogs to participate in the study by increasing their interest in the plastic cups used in the experiment. Owners were carefully instructed about their participation before each session. Specifically, after a verbal explanation of the owners' role in the experiment (without mentioning the purpose of it), E1 demonstrated several times the two types of pointing involved in the experiment. It is worth noting that if at some point during a trial an owner confused or misapplied the gestures, E1 stopped the trial to inform the owner and to again demonstrate how the gesture should be displayed. In that case, since the trial was disrupted it was not included in the analysis. The floor in the experimental room was marked in order to facilitate consistent location of the dogs during the experiment (2.5 m in front of the signaler) and the placement of the cups (0.5 m either side of the signaler). Before each warm-up trial, E2 led the dog and asked him to sit 2.5 m in front of E1. After E1 made sure that the dogs were calm and paying attention to her (making eye-contact) she proceeded with the two static distal pointing (1 left and 1 right) and two head-gaze (1 left and 1 right) trials (described below). The order of presentation of these trials was counterbalanced across participants. The warm-up trials were conducted in the same room as the experimental trials, but the participants were not wearing the eye-tracking system. Participants successfully selected the food container at a rate of 100 % for distal pointing and 60 % for head gazing.

Calibration. After the warm-up trials the calibration phase took place. When the dog was calm, E1 mounted the eye-tracker system on the dog's head. When the system was correctly placed (with the eye camera correctly pointing to and capturing the whole eye), the eye-tracker was calibrated for each participant. In order to calibrate the eye-tracker, E1 stood directly in front of the dog and after calling the dog's name and making eye contact, immediately showed the dog a piece of food, slowly moved it to one of nine Velcro-marked points on the wall, making sure the dog was following the food with his gaze. When the piece of food was in the middle of the calibration dot (a black cardboard circle attached to the Velcro marks on the wall), E1 held it there for approximately 2 s, indicating to the cameras when the calibration trial was successful (i.e., when the dog stared at the food for the entire 2 s). The same sequence was executed for the other eight marked dots on the wall. The threshold for an adequate calibration was accomplished if participants fixated their gaze at eight out of nine calibration points. If for any reason participants did not reach the criterion the experimenter proceeded to take a break before the second and last attempt. One participant out of the six did not meet the threshold and was consequently not considered for the experimental trials.

This calibration procedure was used for the experimental trials, though extra calibration trials (same procedures) were conducted if the goggles moved from the original position. In order to correct for small movements, additional drift-correction was done at arbitrary moments throughout the study.

Experimental trials. Immediately after the calibration procedure the experimental trials took place. Familiar and unfamiliar sessions consisted of exactly the same procedure as in the warm-up, with the only difference being the identity of the signaler (experimenter or owner).

Trials started with the signaler (whether E1 or FP) standing in front of the dog and calling the dog's name to draw his attention; when the dog was attentive the signaler proceeded to hide a piece of food (i.e. dried treats previously approved by the owner) in one of two plastic cups. To control for olfactory cues, both cups were pre-baited with dried food that was taped to the bottom inside the cups (bubble wrap was used to avoid possible interference of auditory cues). Immediately after hiding the food, E1 shuffled the cups behind her back so as to avoid giving the dog visual access to the target cup. Then, the signaler placed the two cups at their corresponding marked locations from a standing position midway between them. Following that, after calling the dog's name again and making eye-contact, the signaler displayed one of two types of pointing gestures: a static distal point or momentary head gaze. Static distal (SD) pointing consisted of signaler extending her arm and hand to the target cup (the one containing the food) and stayed in the same position until participants made their choice by approaching one of the cups, or a maximum of 15 s. The tip of the signaler's finger was approximately 50 cm from the closest edge of the target cup. Head gaze (HG) pointing consisted of the signaler turning her head and gaze to the target cup and remaining looking at it until the participants made their choice, or a maximum of 15 s. Each session consisted of six SD and four HG trials. The low number of trials was intended to avoid fatigue and lack of interest in the dogs. The exact sequence of trials was determined before the session and written on a whiteboard visible to the signaler throughout the experiment. Each sequence was pseudo-randomized for each participant, with the only restriction that no one side was baited more than twice in a row.

Trials were considered passed and food retrieved if the participant touched or approached the target cup within 10 cm, prior to any contact with the other cup. In order to discount the effects of possible extraneous variables, participants also went through two control trials in which no signal was given to the dog by the signaler. These control trials were randomly positioned within the experimental session. For the control trial the signaler, after calling the dog's name to draw her/his attention, maintained a neutral position facing forward and looking directly at the dog for about ten seconds.

Data analysis. A two-way repeated-measure ANOVA was used to analyze the dog's behavior on the choice task: Familiarity (familiar, unfamiliar) versus Type of Gesture (HG, SD) using SPSS for MAC 18.0 (SPSS Inc). Significant effects were identified at p values of less than 0.05. For eye-tracking data, the language archiving tool, ELAN (Brugman and Russel 2004) was used to annotate the movie of each participant's session with the appropriate fixation location and duration. In order to analyze the eye-tracking data we used the ExpertEyes software package that was developed to perform human eye-tracking studies under demanding conditions (Parada et al. submitted). ExpertEyes permits video to be analyzed in an offline mode. Using this mode we were able to export movies containing a fixation cross indicating the subject's overt focus of attention. Movies were imported to ELAN and, making use of the fixation cross, were coded frame-by-frame. Because the experiment was run in naturalistic conditions, it was impossible to control for factors such as consistency of duration for all trials. Thus, we only coded the first 2000 ms immediately after the signaler started making the gesture, independent of dogs' final choice (correct or incorrect). In order to analyze the trials we first divided each image into several features (areas of interest: AOI) to quantitatively analyze participants' viewing patterns. The AOI to be coded in all trials were: face, torso (the arm region was combined with torso for the HG trials, since the signalers stood with their arms behind their backs and elongated the respective arm only to point with the hand), legs, correct cup, and incorrect cup. For the distal pointing trials only, correct hand area and incorrect hand area were also coded. After analyzing the data, legs were discarded since no dog looked at this area during the analyzed temporal window.

We used two dependent variables indicating attention: gaze time (duration of gaze points) and the frequency with which dogs directed their gaze to the AOIs—calculated by using the proportion of frames in which any AOI was the target of the gaze points. A gaze point was scored if the gaze remained stationary for at least 70 ms. Otherwise, the recorded sample was considered as part of a saccade or noise. In order to limit analysis to the visual information actually available to the participants, we excluded the samples recorded after the first 2000 ms post-gesture. Likewise, we did not code samples recorded during saccades.

Raw data from each participant were converted to fixation duration and fixation frequency for each participant taking into account the number of usable trials per participant per session using custom in-house routines running under MATLAB 7 environment (The Mathworks).

For statistical analyses, differences in fixation frequency and gaze time (duration) were separately evaluated using a three-way repeated-measures ANOVA: Familiarity (unfamiliar vs. familiar person) X Location (face, torso, correct cup and incorrect cup) X Type of Gesture (SD vs. HG). Additionally, a two-way repeated measures ANOVA: Familiarity (unfamiliar vs. familiar person) X Location (correct hand vs. incorrect hand) was used to analyze within-subject differences with respect to correct versus incorrect hand visual behaviors for SD trials only. All statistical analyses were run using SPSS for MAC 18.0 (SPSS Inc.). Significant main effects were identified at p values of less than 0.05 (when sphericity did not hold, Greenhouse-Geisser correction was used). Post-hoc comparisons were evaluated using the Bonferroni criterion to correct for multiple comparisons with p values of less than 0.05 identifying significant effects.

6.2.2 Results

Behavioral data. Dogs were above chance for all the gestures except head gazing by FP, where they performed at chance. Specifically, the proportion of correct choices per conditions was 0.77 for UP SD, 0.5 for UP HG, 0.8 for FP SD and finally 0.48 for FP HG (Fig. 6.2a). Statistical analysis showed no significant differences in accuracy to type of gestures ($F(1,4) = 0.433$, $p = 0.127$) or familiarity ($F(1,4) = 0.003$, $p = 0.962$). No interaction effect was seen ($F(1,4) = 0.003$, $p = 0.687$).

In the control trials, for most of the trials participants did not move from the starting point (no choice response were seen 60 % of the times for FP and 70 % for UP); on the few occasions when they did move, they could not accurately predict where the food was hidden (incorrect cup responses: 30 % for FP, 20 % for UP) (Fig. 6.2b).

Frequency. The repeated measures ANOVA showed a significant main effect of location ($F(3,12) = 9.14$, $p = 0.002$), and a significant interaction effect of location by gesture ($F(3,12) = 10.44$, $p = 0.001$). No other significant effects or interaction were seen.

A post hoc analysis comparing the effects of locations did not show any significant differences. This might be due to the effect size for this analysis ($d = 1.5$), which is found to exceed Cohen's convention (1988) for a large effect ($d = 0.8$). However, post hoc tests comparing the effect of location by gesture revealed that

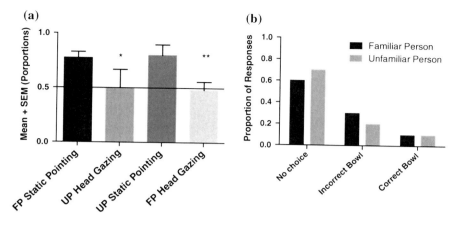

Fig. 6.2 a Proportion of correct choices mean (+sem) proportions of participant's correct choices in each condition. Chance performance (0.5) noted. The *x-axis* depicts trial mean proportions for the four conditions: *UP* (unfamiliar person, E1) *static distal pointing, UP* (unfamiliar person, E1) *head gazing, FP* (familiar person) *static distal pointing* and *FP* (familiar person) *head gazing*. **b** Responses to control proportions of behavioral responses to control trials (no gestures). The *x-axis* depicts the choices our participants took: *No choice* (dog did not approach to any of the cups), *Incorrect Cup* (dog approached to the incorrect cup), *Correct Cup* (dog approached the cup that contained the food)

dogs looked more frequently at faces when presented with head gazing gestures compared to static distal pointing. Similarly, they looked more often at the torso area when presented with distal pointing gestures compared to gazing gestures (HG).

As expected, analysis exclusively for distal pointing gestures (correct vs. incorrect hand) showed that dogs looked only to the correct hand area compared to the incorrect area ($F(1,4) = 80.35$, $p = 0.001$). No significant effect of familiarity was seen.

Gaze time (duration). Analysis showed a significant main effect of location ($F(3,12) = 6.61$, $p = 0.007$) and a significant interaction was found between location and gesture [$F(3,12) = 6.01$, $p = 0.01$). As for the frequency analysis, due to a small sample size, a post hoc test did not show significant differences between locations. Nevertheless, post hoc tests for the interaction effect of location by gesture showed the dogs looked longer at the face region compared to other regions for the head gaze trials (Figs. 6.3 and 6.4). Similarly, dogs looked marginally longer to the incorrect bowls for the head gaze trials.

As for gaze time, analysis exclusively for distal pointing gestures (correct vs. incorrect hand) showed that since dogs did not look at the incorrect hand area, the duration of the looks to the correct hand area was also significantly longer ($F(1,40) = 23.16$, $p = 0.009$) (Figs. 6.3 and 6.4). There was no significant effect of familiarity.

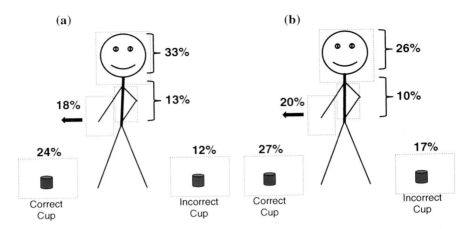

Fig. 6.3 Data depicting the total average duration (gaze time) that dogs spent looking at each of the six AOIs for static distal pointing trials as a function of familiarity. AOIs (roughly represented by *dashed lines*): face area—torso—correct hand area—incorrect hand area (not depicted because dogs did not look on that direction during the 2000 ms after the gesture)—correct cup—incorrect cup. **a** Familiar person (owner), **b** Unfamiliar person (experimenter)

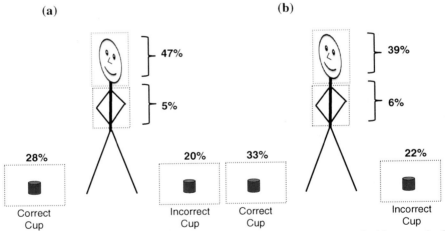

Fig. 6.4 Data depicting the total average duration (gaze time) that dogs spent looking at each of the five AOIs for head gazing trials as a function of familiarity. AOIs (roughly represented by *dashed lines*): face area—torso—correct cup—incorrect cup. **a** Familiar person (owner), **b** Unfamiliar person (experimenter)

6.2.3 *Discussion of Results*

By presenting our eye-tracking system we aimed to extend the limited literature on dog's visual scanning of human gestures and extend an invitation to researchers to make use of this and other technologies widely employed in the human cognitive sciences.

Previous work with eye-tracking on dogs had already suggested that dogs focus their attention on informative regions of images without any task-specific pre-training. This work had also shown that dogs make use of human ostensive and referential signals. In the study presented here we wanted to further explore dogs' visual behavior in a relatively natural situation, which allowed them to move freely. The data acquired in the current study raise some interesting issues regarding the use of eye-tracking on dogs in communicative tasks. We discuss three: first, dogs' focus of visual attention during human gestures; second, the influence of familiarity on dogs' visual behaviors; third, pointing with the hand versus pointing head-gaze cues. We then conclude with a more general discussion.

6.2.3.1 Dogs' Focus of Visual Attention During Human Gestures

Our eye-tracking data show that dogs looked longer and gazed more frequently at the face area than the rest of the locations for the HG trials and at the torso area for SD trials. This can be understood if we consider that dogs' overt visual attention might be functioning, similar to humans, in a reflexive way. Human infants as young as 6 months old shift their attention in the direction of a pointing hand (Bertenthal and Boyer 2012). By this age they also turn their gaze after a model starts turning hers in order to attend to a toy (Gredebäck et al. 2009), and show longer looking times to the attended objects. The apparent similarity of dogs' visual scanning behavior to human infant behavior is something that should be further explored.

6.2.3.2 Influence of Familiarity on Dogs' Visual Behaviors

Our preliminary study did not find any effect of dog gaze related to familiarity of the signaler. Communicative behaviors of dogs toward an unfamiliar person have received some attention. Elgier et al. (2009) studying extinction to a human pointing gesture in an object choice task showed that for dogs, extinction of a pointing gesture took longer when the signaler was the owner; likewise, reverse learning (following the not-pointed-to container) was faster when the owner gave the cue compared with when a stranger did it (Elgier et al. 2009). This suggests that familiarity does matter for dogs. However, Kaminski et al. (2011) showed that dogs differentially communicated the location of a hidden object according to whether the item was an object of their own interest or an object of the person's interest irrespective of whether the person was their owner or a stranger (Kaminski et al. 2011). Merola et al. (2012) had owners and strangers display emotional information (facial and vocal expressions of either happiness or fear) with respect to an electric fan. The authors found that this behavior elicited both referential looking and gaze alternation in dogs irrespective of the informant's identity. In their study, dogs did make different use of the emotional information when coming from their owner compared to the stranger: the owners' emotional reactions influenced the dogs' subsequent behavior, whereas the strangers' emotional reaction did not (Merola et al. 2012).

Rossi et al. (submitted) found that dogs show referential looking between an unfamiliar person and inaccessible food, matching the result of a previous study that had been conducted with familiar people only (Miklósi et al. 2005). But whereas Miklósi and colleagues also demonstrated that, in an apparent effort to communicate with their owners, dogs increased their vocalizations when engaged in referential looking between food and an inattentive (book-reading) owner, the dogs in the study by Rossi et al., when paired with an unfamiliar person who was being similarly inattentive, engaged in similar amounts of referential looking but did not vocalize more. Taken together, these studies suggest that the vocalization rate of dogs is modulated by familiarity when dogs are trying to get a person's attention, even though referential looking is unmodulated.

The preliminary eye-tracking study reported in this chapter found that familiarity did not modulate dogs' behavior in this particular experimental task. This might be related to the nature of the task, which may have been approached by the dogs in our study as a foraging task. Because the dogs in our study, as in the majority of studies using the object-choice task, were pre-trained with warm-up trials before the experimental session, the dogs might have generated some level of expectation about the receipt of food and taken the signaler, regardless of her identity, as a cooperative partner. Further experimentation is needed in order to investigate this possibility.

6.2.3.3 Pointing with the Hand Versus Head-Gaze Cues

As previously mentioned, dogs looked longer and gazed more often at the face area than other locations during the HG trials. Interestingly, there was a trend towards dogs looking marginally longer at the incorrect cup in the HG trials compared to the time spent looking at the incorrect cup in the SD trials (Fig. 6.4). This suggests that dogs might have scanned the scene more thoroughly during the HG trial. One possible reason is that dogs actually perceived that the signaler was signaling to one of the cups—i.e., recognizing communicative intent—but they failed when deciding which cup to approach. Perhaps this was because the means of communicating information (head-gaze) did not rise above their threshold for the decision. Similarly, a general lack of experience with gaze-following should not be discarded as an explanation for the subjects' failure to use the cue to find the hidden food.

Our data showed that for both SD and HG dogs did not fail to attend to the source of information. Rather, they failed more frequently at the moment of making the decision during HG, being practically at chance—taking both FP and UP together—at retrieving the food from the cups. That is, for dogs in our study the provided gaze cues were probably ambiguous signals that did not provide enough information about the environment. Similar effects have been shown in other studies with dogs (Hare et al. 2002; Ittyerah and Gaunet 2009). This might be related to the kind of gesture and the motion involved. Pointing with the arm and hand is different from pointing with the head or eyes. Although pointing with the arm and hand may be harder to follow, because the relationship to the intended target is more indirect and abstractly geometrical than a direct look at the target (Cappuccio and Shepherd 2013), it also occupies more of the visual field than pointing with the eyes, likely drawing more attention and thus facilitating recognition of its target. This, along with the possibility that pet dogs might have more experience following arm and hand pointing gestures than following eye and head gazing makes a case for eye- or head-gaze being more difficult to use than arm-hand gestures.

An interesting possibility is suggested by taking a dynamical systems perspective (Spencer et al. 2001) towards our preliminary data. In the SD trials, adding the time spent looking to the arm/hand and to the correct cup yields 42 % (with the familiar person) and 47 % (unfamiliar) of the total time budget directed towards the side that the dogs should choose to obtain the food. In contrast, for the HG trials, only 28 % (familiar person) and 33 % (unfamiliar) of the total time is spent oriented towards the direction the dog needs to go. Thus, the total time in which the dog is physically oriented towards the "correct" direction in the SD trials versus the HG trials could activate an embodied action trace which lasts long enough to support the decision in the SD case but not the HG case. Although our data are currently insufficient to ground or test a dynamical model we believe that this an important future goal that will be facilitated by the use of cameras and other technology allowing the dog's gaze to be tracked in real time. It will also be useful to use data from eye-tracking to look at the sequencing of the saccades the dogs make. In this way, it will become possible to connect comparative psychology to similar work in human developmental psychology (Smith et al. 2011; Spencer et al. 2001) in a more rigorous fashion than the all-too-common statements concerning the mental equivalence of animal subjects to toddlers of a specific age [dogs are frequently described as having the mental capacity of two-year-old children (e.g. Coren 2005; Lakatos et al. 2009)].

If one considers gaze-following from a developmental perspective, it is likely to be the result of a developmental interplay between inherited dispositions to attend to gaze cues and experience with the significant use of such cues (Gomez 2005). Accordingly, it makes sense to think that there are shared underlying mechanisms between species that are capable of following gaze. However, underlying differences in the phylogenetic roots of this ability may help explain why attentional orientation to movement is different in different species.

6.3 General Discussion

Primates and canids both live in a complex social system. Though the idiosyncrasies of each social structure are different, the convergences between the two, plus the long history of domestication of dogs might have facilitated orientation responses to humans. This can be seen in the data coming from the small number of eye-tracking studies reported above. Among the great apes there seem to be shared visual scanning patterns toward scenes and both conspecific and non-conspecific faces (with eyes and mouth being particularly salient). Consequently, furthering our knowledge of visual scanning patterns of dogs (and, if possible, of human-reared highly socialized wolves) will be extremely informative for assessing several factors that might be playing a role in the typical human preference for looking at the eye region.

We have shown that it is possible to accurately track the way in which dogs allocate attention to human faces, bodies, and locations when humans are providing social cues relevant to the dog as indicators of food location. However, more fine-grained analyses than we have provided are certainly desirable. For instance, how are dogs allocating attention within components of the face, such as eyes and mouth? How important is the finger in a pointing gesture, or is the dog simply extrapolating from the direction of the whole arm? Such questions are answerable in principle given further improvements in the head-mounted eye-tracking system that we have developed. Our present study cannot answer these questions due to our small sample size and due to the fact that, ironically, the mobility that is one of the main advantages of our eye-tracking system was found to be a consistent source of noise as well, which translated into having to drop several trials from the analysis.

Answering questions such as the aforementioned, and tying them to issues such as prior experience of the dogs with particular people could help tease apart ontogenetic differences. Furthermore, physiological work pairing eye-tracking with hormonal assays could help integrate findings about cognitive processes with proximal mechanisms. Although it is difficult to imagine strong comparative methods that could use eye-tracking with (unsocialized) wolves, comparative work with eye-tracking in humans will help us to quantify the degree of convergence between dogs and humans with respect to visual processing of social cues. Of particular interest might be to investigate the differences in visual behavior of dogs performing different tasks; this may help reveal underlying cognitive processes, making it possible to relate the processes to questions about the adaptive value of their interactions with humans in specific contexts.

The field of comparative animal cognition has grown enormously in the past couple of decades, and yet it still lags behind human cognitive science in many ways. Here we hope to have convinced readers that eye-tracking in particular, and the adoption of technologies used to gather data in human cognitive psychology more generally, can help advance an integrative understanding of the evolution, development, mechanisms, and adaptiveness of canine cognition in its natural environment: the dog–human social group.

Acknowledgments This research and the first author were supported by the Office of Vice Provost for Research of Indiana University. We wish to thank our participants and their owners for participating in this study. The authors would also like to thank to Mr. Jeff Sturgeon for technical advice, Dr. Bennett Bertenthal for helpful comments and discussion, Dr. Nicholas Port and the IU School of Optometry for their generous help and last but not least to the Indiana Statistical Consulting Center, particularly to Stephanie Dickinson for her guidance.

References

Axelsson, E., Ratnakumar, A., Arendt, M. L., Maqbool, K., Webster, M. T., Perloski, M., et al. (2013). The genomic signature of dog domestication reveals adaptation to a starch-rich diet. *Nature, 495*(7441), 360–364.

Bailey, I., Myatt, J. P., Wilson, A. M. (2013). Group hunting within the Carnivora: Physiological, cognitive and environmental influences on strategy and cooperation. *Behavioral Ecology and Sociobiology, 67*(1), 1–17.

Bar-Yosef, O. (1998). On the nature of transitions: The middle to upper Palaeolithic and the Neolithic revolution. *Cambridge Archaeological Journal, 8*(2), 141–163.

Barton, R. A. (1996). Neocortex size and behavioural ecology in primates. *Proceedings of the Royal Society B, 263*(1367), 173–177.

Bernstein, I. S. (1970). Primate status hierarchies. *Primate Behavior, 1*, 71–109.

Bertenthal, B., & Boyer, T. (2012). Developmental changes in Infants' visual attention to pointing. *Journal of Vision, 12*(9), 480.

Brugman, H., & Russel, A. (2004). Annotating multimedia/multi-modal resources with ELAN. *Proceedings of the Fourth International Conference on Language Resources and Evaluation* (pp. 2065–2068). Citeseer.

Buttelmann, D., & Tomasello, M. (2013). Can domestic dogs (Canis familiaris) use referential emotional expressions to locate hidden food? *Animal Cognition, 16*(1), 137–145.

Cappuccio, M. L., & Shepherd, S. V. (2013). 13 Pointing Hand: Joint Attention and Embodied Symbols. The Hand, an Organ of the Mind: What the Manual Tells the Mental, 303.

Clutton-Brock, J. (1995). Origins of the dog: Domestication and early history. In: J. Serpell (Ed.), *The domestic dog: Its evolution, behaviour, and interactions with people* (pp. 7–20). Cambridge: Cambridge University Press.

Clutton-Brock, J. (1999). *A natural history of domesticated mammals* (2nd ed.). New York, USA: Natural History Museum.

Clutton-Brock, J., & Wilson, D. E. (2002). Mammals. *Smithsonian handbooks* (1st US ed.). London: DK.

Cordoni, G., & Palagi, E. (2008). Reconciliation in wolves (Canis lupus): New evidence for a comparative perspective. *Ethology, 114*(3), 298–308.

Coren, S. (2005). How dogs think: Understanding the canine mind. New York: Simon and Schuste.

Crockford, S. J., & Iaccovoni, A. (2000). Dogs through time: An archaeological perspective. *Proceedings of the 1st ICAZ Symposium on the History of the Domestic Dog.* BAR international series, Vol. 889. Oxford, England: Archaeopress.

Dalton, K. M., Nacewicz, B. M., Johnstone, T., Schaefer, H. S., Gernsbacher, M. A., Goldsmith, H. H., Alexander, A. L., et al. (2005). Gaze fixation and the neural circuitry of face processing in autism. *Nature Neuroscience, 8*(4), 519–526.

Davis, S. J. M., & Valla, F. R. (1978). Evidence for domestication of dog 12,000 years ago in Natufian of Israel. *Nature, 276*(5688), 608–610.

Diamond, J. (2002). Evolution, consequences and future of plant and animal domestication. *Nature, 418*(6898), 700–707.

Driscoll, C. A., Macdonald, D. W., & O'Brien, S. J. (2009). From wild animals to domestic pets, an evolutionary view of domestication. *Proceedings of the National Academy of Sciences USA, 106*, 19971–9978.

Duchowski, A. T. (2007). *Eye tracking methodology: Theory and practice.* New York: Springer.

Elgier, A. M., Jakovcevic, A., Mustaca, A. E., & Bentosela, M. (2009). Learning and owner-stranger effects on interspecific communication in domestic dogs (*Canis familiaris*). *Behavioural Processes, 81*(1), 44–49.

Felmingham, K. L., Rennie, C., Manor, B., & Bryant, R. A. (2011). Eye tracking and physiological reactivity to threatening stimuli in posttraumatic stress disorder. *Journal of Anxiety Disorders, 25*(5), 668–673.

Fukuzawa, M., Mills, D. S., & Cooper, J. J. (2005). More than just a word: Non-semantic command variables affect obedience in the domestic dog (*Canis familiaris*). *Applied Animal Behaviour Science, 91*(1), 129–141.

Gomez, J. C. (2005). Species comparative studies and cognitive development. *Trends in Cognitive Sciences, 9*(3), 118–125.

Gredebäck, G., Johnson, S., & Von Hofsten, C. (2009). Eye tracking in infancy research. *Developmental Neuropsychology, 35*(1), 1–19.

Hare, B., Brown, M., Williamson, C., & Tomasello, M. (2002). The domestication of social cognition in dogs. *Science, 298*(5598), 1634–1636.

Hare, B., & Tomasello, M. (2005). Human-like social skills in dogs? *Trends in Cognitive Sciences, 9*(9), 439–444.

Hattori, Y., Kano, F., & Tomonaga, M. (2010). Differential sensitivity to conspecific and allospecific cues in chimpanzees and humans: A comparative eye-tracking study. *Biology Letters, 6*(5), 610–613.

Hemmer, H. (1990). *Domestication: The decline of environmental appreciation* (2nd ed.). Cambridge, England: Cambridge University Press.

Hirata, S., Fuwa, K., Sugama, K., Kusunoki, K., & Fujita, S. (2010). Facial perception of conspecifics: Chimpanzees (Pan troglodytes) preferentially attend to proper orientation and open eyes. *Animal Cognition, 13*(5), 679–688.

Holzman, P. S., Proctor, L. R., Levy, D. L., Yasillo, N. J., Meltzer, H. Y., & Hurt, S. W. (1974). Eye-tracking dysfunctions in schizophrenic patients and their relatives. *Archives of General Psychiatry, 31*(2), 143.

Ittyerah, M., & Gaunet, F. (2009). The response of guide dogs and pet dogs (Canis Familiaris) to cues of human referential communication (pointing and gaze). *Animal Cognition, 12*(2), 257–265.

Kaminski, J., Call, J., & Fischer, J. (2004). Word learning in a domestic dog: Evidence for "fast mapping". *Science, 304*(5677), 1682–1683.

Kaminski, J., Neumann, M., Brauer, J., Call, J., & Tomasello, M. (2011). Dogs, Canis familiaris, communicate with humans to request but not to inform. *Animal Behaviour, 82*(4), 651–658

Kano, F., Hirata, S., Call, J., & Tomonaga, M. (2011). The visual strategy specific to humans among hominids: A study using the gap-overlap paradigm. *Vision research, 51*(23–24), 2348–2355.

Kano, F., & Tomonaga, M. (2009). How chimpanzees look at pictures: A comparative eye-tracking study. *Proceedings of the Royal Society of London. Series B: Biological, 276*(1664), 1949–1955.

Kano, F., & Tomonaga, M. (2010). Attention to Emotional Scenes Including Whole-Body Expressions in Chimpanzees (Pan troglodytes). *Journal of Comparative Psychology, 124*(3), 287–294.

Kano, F., & Tomonaga, M. (2013). Head-mounted eye tracking of a chimpanzee under naturalistic conditions. *PLoS ONE, 8*(3), e59785.

Kjaersgaard, A., Pertoldi, C., Loeschcke, V., & Hansen, D. W. (2008). Tracking the gaze of birds. *Journal of Avian Biology, 39*(4), 466–469.

Koler-Matznick, J. (2002). The origin of the dog revisited. *Anthrozoos, 15*(2), 98–118.

Lakatos, G., Soproni, K., Doka, A., & Miklósi, A. (2009). A comparative approach to dogs' (Canis familiaris) and human infants' comprehension of various forms of pointing gestures. *Animal Cognition, 12*(4), 621–631.

Larson, G., Karlsson, E. K., Perri, A., Webster, M. T., Ho, S. Y., Peters, J., et al. (2012). Rethinking dog domestication by integrating genetics, archeology, and biogeography. *Proceedings of the National Academy of Sciences USA, 109*(23), 8878–8883.

Lupo, K. (2011). A dog is for hunting. *Ethnozooarchaeology: The present and past of human–animal relationships*, 4–12.

Machado, C. J., & Nelson, E. E. (2011). Eye-tracking with Nonhuman primates is now more accessible than ever before. *American Journal of Primatology, 73*(6), 562–569.

Mech, L. D. (1970). *The wolf: The ecology and behavior of an endangered species.* (1st ed.). Published for the American Museum of Natural History. Garden City, New York: Natural History Press.

Mech, L. D. (2012). *Wolf.* New York: Random House Digital.

Merola, I., Prato-Previde, E., & Marshall-Pescini, S. (2012). Dogs' social referencing towards owners and strangers. *PLoS ONE, 7*(10), e47653.

Miklósi, A., Pongracz, P., Lakatos, G., Topal, J., & Csanyi, V. (2005). A comparative study of the use of visual communicative signals in interactions between dogs (Canis familiaris) and humans and cats (Felis catus) and humans. *Journal of Comparative Psychology, 119*(2), 179–186.

Moore, C. E., & Dunham, P. J. (1995). *Joint attention: Its origins and role in development.* New York: Lawrence Erlbaum Associates, Inc.

Muro, C., Escobedo, R., Spector, L., & Coppinger, R. P. (2011). Wolf-pack (Canis lupus) hunting strategies emerge from simple rules in computational simulations. *Behavioural Processes, 88*(3), 192–197.

Olsen, S. J. (1985). *Origins of the domestic dog: The fossil record.* Tucson, Arizona: University of Arizona Press.

Packer, C., & Ruttan, L. (1988). The evolution of cooperative hunting. *The American Naturalist, 132*(2), 159–198.

Palagi, E., & Cordoni, G. (2009). Postconflict third-party affiliation in Canis lupus: Do wolves share similarities with the great apes? *Animal Behaviour, 78*(4), 979–986.

Pilley, J. W., & Reid, A. K. (2011). Border collie comprehends object names as verbal referents. *Behavioural Processes, 86*(2), 184–195.

Ramos, D., & Ades, C. (2012). Two-item sentence comprehension by a dog (Canis familiaris). *PLoS ONE, 7*(2), e29689.

Reader, S. M., Laland, K. N. (2002). Social intelligence, innovation, and enhanced brain size in primates. *Proceedings of the National Academy of Sciences USA, 99*(7), 4436–4441.

Ruusila, V., Pesonen, M. (2004). Interspecific cooperation in human (Homo sapiens) hunting: The benefits of a barking dog (Canis familiaris). *Annales Zoologici Fennici, 41*(4), 545–549.

Serpell, J. (1995). *The domestic dog: Its evolution, behaviour, and interactions with people.* Cambridge: Cambridge University Press.

Shepherd, S. V. (2010). Following gaze: Gaze-following behavior as a window into social cognition. *Frontiers in Integrative Neuroscience, 4*(5), 1–13.

Shultz, S., & Dunbar, R. I. (2007). The evolution of the social brain: Anthropoid primates contrast with other vertebrates. *Proceedings of the Royal Society B: Biological Sciences, 274*(1624), 2429–2436.

Smith, L. B., Yu, C., & Pereira, A. F. (2011). Not your mother's view: The dynamics of toddler visual experience. *Developmental Science, 14*(1), 9–17.

Somppi, S., Tornqvist, H., Hanninen, L., Krause, C., & Vainio, O. (2011). Dogs do look at images: Eye tracking in canine cognition research. *Animal Cognition, 11*(1), 167–174.

Soproni, K., Miklósi, A., Topal, J., & Csanyi, V. (2001). Comprehension of human communicative signs in pet dogs (Canis familiaris). *Journal of Comparative Psychology, 115*(2), 122–126.

Soproni, K., Miklósi, A., Topal, J., & Csanyi, V. (2002). Dogs' (Canis familiaris) responsiveness to human pointing gestures. *Journal of Comparative Psychology, 116*(1), 27–34.

Spencer, J. P., Smith, L. B., & Thelen, E. (2001). Tests of a dynamic systems account of the A-not-B error: The influence of prior experience on the spatial memory abilities of two-year-olds. *Child development, 72*(5), 1327–1346.

Sullivan, J. O. (1978). Variability in the wolf, a group hunter. In R. L. Hall & H. S. Sharp (Eds.) *Wolf and man: Evolution in parallel* (pp. 31–40). New York: Academic Press.

Teglas, E., Gergely, A., Kupan, K., Miklósi, A., & Topal, J. (2012). Dogs' gaze following is tuned to human communicative signals. *Current Biology, 22*(3), 209–212.

Tinbergen, N. (1963). On aims and methods of ethology. *Zeitschrift für Tierpsychologie, 20*(4), 410–433.

Vila, C., Savolainen, P., Maldonado, J. E., Amorim, I. R., Rice, J. E., Honeycutt, R. L., et al. (1997). Multiple and ancient origins of the domestic dog. *Science, 276*(5319), 1687–1689.

Weisdorf, J. L. (2005). From foraging to farming: Explaining the Neolithic Revolution. *Journal of Economic Surveys, 19*(4), 561–586.

Williams, F. J., Mills, D. S., & Guo, K. (2011). Development of a head-mounted, eye-tracking system for dogs. *Journal of neuroscience methods, 194*(2), 259–265.

Yarbus, A. L., Haigh, B., & Rigss, L. A. (1967).*Eye movements and vision* (vol 2. vol 5.10). New York: Plenum press.

Zola, S., & Manzanares, C. (2011). A Simple behavioral task combined with Noninvasive infrared eye-tracking for examining the potential impact of viruses and vaccines on memory and other cognitive functions in humans and Nonhuman primates. *Journal of Medical Primatology, 40*(4), 282–283.

Chapter 7
Cognitive Development in Gray Wolves: Development of Object Permanence and Sensorimotor Intelligence with Respect to Domestic Dogs

Sylvain Fiset, Pierre Nadeau-Marchand and Nathaniel J. Hall

Abstract In this chapter, we explore whether domestic dogs and gray wolves share a similar cognitive development with regards to how they represent physical and/or social objects. To reach this objective, we examine two key components of the Piagetian theory of cognitive development in the gray wolf: object permanence and sensorimotor intelligence. We detail how the capacity to search and locate disappearing objects develops in wolves and compare these data with those observed in previous studies with dogs. We then further describe an observational study of sensorimotor intelligence with these wolves. Overall, the results suggest that the development of object permanence is similar in dogs and wolves, both species reaching Stage 5b of object permanence by the age of 11 weeks. In terms of sensorimotor intelligence, Stage 4 was the upper limit of sensorimotor intelligence we observed in wolves. Moreover, up to 6 weeks of age, the behaviors of wolf puppies are directed predominantly towards their conspecifics, and by Week 8, wolves' interest in inanimate object increases significantly. In discussion, we explore the factors affecting the development of object permanence and sensorimotor intelligence in canines.

7.1 Introduction

In the last two decades, the scientific study of cognition in the domestic dog has grown substantially. For example, the Web of Sciences index (Thompson Reuters) reveals that the rate of publications about cognition of domestic dogs (using dogs,

---

S. Fiset (✉) · P. Nadeau-Marchand
Secteur administration et sciences humaines, Université de Moncton Campus d'Edmundston, Edmundston, NB E3V 2S8, Canada
e-mail: sylvain.fiset@umoncton.ca

N. J. Hall
Department of Psychology, University of Florida, Gainesville, FL 32611, USA

behavior and cognition as keywords) has increased at a mean rate of 35 % per year since the year 1991. Without entering into the reasons that can be put forward to explain this strong interest in the study of the domestic dog's mental capacities [as a starting point, readers are invited to consult Bradshaw (2011) and Miklósi (2007)], it is worth mentioning that the domestic dog's cognition is now investigated from different points of view by several disciplines: behavioral biology, behavioral ecology, comparative psychology, ethology, evolutionary anthropology, functional morphology, and veterinary behavior, to name a few. From a psychological perspective, the Piagetian theory of cognitive development, which was initially developed by Piaget (1954) during the course of his observations of his own children, is one of the most fruitful theoretical frameworks used to study animals' cognition (Pepperberg 2002), including dogs' (e.g. see Fiset and Plourde 2013).

Piaget (1954) divided the general development of children's intelligence into four general periods, from birth to adolescence. The first period of cognitive development, namely the sensorimotor period, is of most interest for comparative researchers since it primarily focuses on the organism's sensory perception and motor activities. Moreover, since the Piagetian's tasks used to measure the cognitive development during the sensorimotor period are easily adaptable to the natural behaviors of non-verbal animals, by the end of the 1970s, the Piagetian theory was endorsed by researchers to investigate the development of cognition in animals (for a summary, see Doré and Dumas 1987). More specifically, the sensorimotor period is characterized by the development of various concepts, such as object permanence (OP), space, time, causality and a general capacity, called sensorimotor intelligence (SI). In the present study, in the context of canine's cognition, we concentrated our attention on two key concepts of the sensorimotor period, that is, OP and SI.

OP is defined as the knowledge that social or physical objects still continue to exist when they are no longer present in one's visual field. In canines, like dogs and wolves, OP is essential for survival. For instance, the capacity to mentally represent a disappearing object is useful in predatory situations, where a prey has moved behind an obstacle (e.g. a tree), or in a social context, when different members of a group move around and momentarily disappear from sight. SI, for its part, is characterized by the organization and coordination of different schemas of action in several logical steps. In the Piagetian sensorimotor period, the repetition of a particular behavior in different circumstances results in a common attribute called a schema of action. It is also the organism's cognitive structure that organizes and coordinates its behavior in logical sequences (Doré and Dumas 1987). As experiences happen in one's life, these schemas of action are modulated, allowing the organism to generalize and transpose its behaviors and/or mental processes to new and different situations. In the Piagetian framework, OP and SI are closely interconnected during the sensorimotor period and both cognitive capacities develop at the same rate through a series of six distinct stages.

During the first two stages of OP, children lack interest in disappearing objects. Thus, they do not exhibit actions when objects disappear from their visual field.

They are, however, capable of briefly following them with their head or body. During the third stage, visuomotor coordination is established and children gain the ability to retrieve partially hidden objects. In Stage 4a, they become capable of retrieving hidden objects, but solely if they initiated a search movement toward the hiding location before the final disappearance of the object. Such a movement ceases to be necessary in the subsequent substage (Stage 4b), and children can now find an object they saw disappear at a specific location. It is also in this stage that the A-not-B error emerges. It manifests when children successively search in the last location they saw the object prior to its disappearance.

Stage 5a of OP marks when children stop committing the A-not-B error, and become capable of retrieving visibly hidden objects in several different locations. In Stage 5b, children can retrieve an object they saw disappear successively in several hiding locations. In the sixth and last stage, children gain understanding of invisible displacements. In Stage 6a, they are capable of solving simple invisible displacement problems. In these problems, an object is first hidden inside a transportation device (i.e. a cup or a hand) and this device is moved behind a box or a screen. There, the object is imperceptibly transferred from the transport container to the hiding location. Since the child cannot perceive either the displacement of the object from one location to another or the transfer of the object from the transportation device to the target location, the displacement is considered invisible. The child must mentally infer the displacement to localize the position of the object. In Stage 6b, children master the ability to relocate objects that were successively hidden in different locations using invisible displacements.

The ontogenetic development of OP in canines has, so far, only been investigated by Gagnon and Doré (1994), who conducted a study on the domestic dog. These authors used a cross-sectional study to assess the stages of OP reached by dogs as a function of age. Most specifically, they selected seven groups of dogs: five of which were young puppies of 4–8 weeks old, and the last two groups included dogs of 3 and 9 months-old. In their study, Gagnon and Doré adapted a procedure previously developed by Dumas and Doré (1987, 1989) for domestic cats for dogs. In their study, the dogs' task was to track and find an attractive object (a toy) that moved and disappeared behind a series of opaque screens. When a group failed a particular task, the same task was administered to the next oldest group and so on. Overall, Gagnon and Doré (1994) found that dogs' understanding of OP developed gradually from 4 to 8 weeks of age. Most specifically, their results revealed that the different stages of OP in dogs emerge as follows: Stage 2 (4 weeks), Stage 3 (5 weeks), Stage 4a (6 weeks), Stage 5a (7 weeks) and Stage 5b (8 weeks). As a group, the dogs did not reach the understanding of invisible displacement, but 11-month-old dogs succeeded at problems of Stage 6a, suggesting an understanding of invisible displacement in older dogs, supporting previous works by Gagnon and Doré (1992) in adult dogs (for an alternative interpretation, however, see Collier-Baker et al. 2004; Fiset and LeBlanc 2007).

SI also develops throughout a series of six stages in which schemas are acquired and modified through direct exploration of the world and its objects. In Stage 1, the behaviors are limited to the reflexes (e.g. suckling) of organisms.

The appearance of a child's first habits, known as primary circular reactions, characterizes the second stage. These behaviors revolve around oneself (e.g. putting his thumb in his mouth) and are repeated over and over, as they produce reactions that the child finds interesting. Next, in Stage 3, secondary circular reactions make their appearance. Secondary circular reactions, contrary to primary circular reactions, are repeated actions towards external objects (either physical or social) rather than oneself. Secondary circular reactions are intentionally repeated and are not coincidental. For instance, an infant extends his hand to grab an object close to him in order to play with it.

Stage 4 of SI corresponds to the coordination of two schemas of action, and consolidates the role of intention in children as illustrated by the emergence of imitation. Schemas are no longer repeated to produce long-lasting fortuitous stimulations, but to obtain an intentional result. For instance, a child throws away a first toy in order to grab a second toy, which is later put in his mouth. Stage 4 is succeeded by Stage 5, in which tertiary circular reactions make their appearance. These reactions introduce the notion of behavioral variation in the intentional repetitions of actions on external objects (either physical or social). An example of this period of trial-and-error is when a child produces different sounds (using cries or hitting toys together) to attract the attention of his caregiver. Finally, Stage 6 (early mental representation) corresponds to the invention of new combinations of actions. In Stage 6, children search for ways to pursue a goal, but, contrary to the preceding stages, this process is done mentally, without the need to experiment on the external object beforehand. For instance, a child exposed to different toys selects the one that is the most likely to make the loudest noise when shaken.

To our knowledge, only Frank and Frank (1985) used the Piagetian's framework of SI to interpret the cognitive development of canines. Most specifically, these authors administered a series of puzzle boxes of increasing complexity to 10-week-old dogs (Malamute) and wolf pups, and the animals' task was to perform increasingly complex manipulations to extract a food dish from a box. Frank and Frank's results suggest that wolves demonstrate behaviors of Stage 5, possibly 6, of SI, while dogs only demonstrate responses of Stage 3, maybe 4. However, as rightly pointed out by Frank and Frank (1985), their conclusions about the acquisition of Stage 5 and/or 6 are doubtful since the wolves were probably able to use skills from inferior stages of SI (e.g. Stage 3 and/or 4) to solve the most complex tasks used in their study.

In order to depict the ontogenetic development of cognition in canines, the first objective of the current study was to determine the development of OP in the gray wolf and compare it with the results observed by Gagnon and Doré (1994) in the domestic dog. To reach this objective, similarly to Gagnon and Doré (1994), we used the scale developed by Uzgiris and Hunt (1975) in human children. Actually, our experimental procedures, as well as the dimensions of our material in the tests of OP, were exactly the same as the ones used by Gagnon and Doré (1994). However, given the small number of wolves that we were able to work with, in contrast to Gagnon and Doré (1994) who used a cross-sectional approach, we used a longitudinal approach in which the same animal is tested several times during the course of its development. It is worth noting that longitudinal studies are frequently used by researchers who deal with reduced sample size when investigating the development of OP (for an example in various animal species, see Pollok et al. 2000; Ujfalussy et al. 2012; Zucca et al. 2007).

The present study also aimed to corroborate the conclusions of Frank and Frank (1985) by establishing an overview of the development of SI in the gray wolf. To do so, we adapted the procedures used by Dumas and Doré (1991) in domestic cats and recorded the natural behaviors of the wolves between their fourth and 11th weeks of life. Finally, since Dumas and Doré (1991) observed that domestic cats attain Stage 5b of OP but solely Stage 4 of SI, we wanted to determine whether or not, in canines, the development of SI synchronized with the development of OP.

7.2 Wolf Study

7.2.1 Method

Participants Four gray wolves (three females and one male) from the same litter began this study. The wolves were from Wolf Park, Battle Ground, Indiana (USA), and had been hand-reared for human socialization from the age of 10 days, as described by Klinghammer and Goodmann (1987). Human caretakers were in contact with the pups 24 h a day, from day 10 to day 28. After 4 weeks, the caretakers reduced the contact with the pups to 16 h a day. Intense human socialization was stopped when the wolves were four months old. Afterward, the wolves were still in regular contact with humans for health care, feeding and behavioral studies.

The wolves were all sick between Day 38 and Day 51, and the study was postponed during this critical period of development. Moreover, one wolf, Devra unexpectedly died on Day 56 of an unknown illness. To determine the cause of death (the autopsy later revealed a congenital liver shunt) and make sure it was not contagious, we interrupted the study a second time from Day 56 to Day 61 before resuming with the three remaining wolves. Given that Devra was tested inconsistently in the OP tests before her death, all her data in the tests were removed from the study. Consequently, our conclusions about the development of OP were limited to the three wolves that completed the study. However, Devra's data in the observational phase of the study were kept until her death, and were adjusted accordingly.

Apparatus In the observational phase of SI (see procedure), the wolves' behaviors were recorded via a Sony HDR-CX110 digital video camera fixed on a tripod. To stimulate the behavior of the wolves, several objects of diverse sizes and forms (puppets, towels, cardboard boxes, ropes, etc.) were disposed on the floor of the room or on the ground inside the outdoor enclosure.

In the OP tests (see procedure), three identical painted white wooden boxes (17 cm wide × 19.5 cm high × 11.5 cm deep with a top, a bottom, a front, and two side panels) served to hide the target object. The bottom of each box was filled with lead bars to increase inertia. They were arrayed in a semicircle at a distance of 30 cm from each other and were equidistant (150 cm) from the wolf's position. In order to maintain the wolves' motivation to search for the target object, different objects were used. The objects were either an orange rubber ball (handled by a translucent nylon thread that was tied to it), a cardboard tube, a small puppet or a white towel. In the invisible displacement trials (see procedure), a small wooden box (9 cm wide × 15 cm high × 9 cm deep), without the top and front panels, was also used. The inside of this box (called *displacement device*) was painted black and its outside was painted white. To help its manipulation, a 117 cm vertical plastic stick was fixed on the back of this box. To reduce the possibility that the wolves used olfaction to find the target object, rose water (1/10 diluted in water) was sprayed over the apparatus. This solution is well known for masking olfaction in canines (Gagnon and Doré 1992). Each trial was recorded via a Zi6 Kodak HD digital video camera fixed on a tripod placed behind the animal.

Procedure At the beginning of the study, the wolves were three weeks of age. Two different approaches were used to assess the development of SI and OP. SI was assessed via the recording of wolves' spontaneous behavior and OP was tested via a series of formal tests administered to each wolf.

7.2.2 Behavioral Observation of Sensorimotor Intelligence

The wolves' spontaneous behavior was recorded from Week 4 to Week 11. As did Dumas and Doré (1991), who studied SI in domestic cats, we recorded the wolves' behavior in intervals of 10 min. However, although we had initially planned to record the wolves' individual behavior three times a week, we were not able to follow the schedule as planned. This divergence from the planned recording schedule was mostly due to (i) the great amount of time needed by the human caregivers to nurture the puppies (feeding, etc.), (ii) the sickness of the animals and (iii) the small amount of time the pups were awake and interacted with each other and/or the toys. The recording schedule was therefore modified. For Weeks 4 and 5, since the wolves' behaviors were limited to the nurturing chamber, we recorded the wolves' behavior as a group. From Weeks 6 to 11, when the wolves were moved to the outdoor enclosure, we were able to record their behaviors individually. However, due to various factors (illness, overbooked testing schedule, etc.), it was impossible to record the behavior of each wolf during each week (see Table 7.1).

The analysis of wolves' spontaneous behavior was based on the behavioral categories identified by Dumas and Doré (1991), who coded domestic cats' SI for each stage of cognitive development during the sensorimotor period. We adapted the criteria used by Dumas and Doré (1991) to the natural behaviors of gray

Table 7.1 Duration (in minutes) of the video recordings as a function of weeks and wolves

Week	Type of recording	Dharma	Devra	Tilly	Gordon
4	Group		—50—		
5	Group		—30—		
6	Individual	–	–	10	–
7	Individual	–	–	–	–
8	Individual	10	10	–	–
9	Individual	–	–	–	–
10	Individual	10	–	–	10
11	Individual	–	–	20	10

Note When the duration of recording is marked as higher than 10 min, several 10 min bouts of recording were recorded during this particular week (e.g. 30 = 3 bouts of 10 min)

wolves (see Table 7.2). To be consistent with our approach, circular activities in wolves, a key component of the development of SI, were also coded as described by Dumas and Doré (1991). In short, to be coded as circular (either primary, secondary or tertiary), a behavior had to be repeated a minimum of five times and last 10 s or more during the same action sequence. Given the limited number of bouts we were able to record, the use of these criteria ensured that the behaviors judged as circular were highly repetitive and reflected the natural behavior of wolves. The different behaviors that served to code the development of SI in wolves were as follows: scratching, pawing or kicking, tugging, howling, biting other wolves, wrestling, and gnawing.

7.2.3 Tests of Object Permanence

From Weeks 4–9, the testing took place in the nurturing room. By doing so, we assured that the wolves were familiar with the testing environment (a pilot study performed by Gagnon and Doré (1994) revealed that dog pups' behaviors were perturbed when tested in a new environment). In Weeks 10 and up, the wolves were tested in a quiet and isolated area of the outdoor enclosure.

The acquisition of Stages 2 and 3 of OP was assessed via the administration of a visual pursuit test in the nurturing room. These tests served to evaluate the wolves' perceptual development, most specifically their visuomotor coordination (Dumas and Doré 1989; Gagnon and Doré 1994). In this test, the wolf was not restrained and was free to move of its own will. The experimenter who performed the manipulation (E1) suspended an attractive object (an orange ball) right in front of the wolf. Then, while ensuring that the wolf was looking at the object, E1 slightly moved the object to the right or left of the wolf a few times and then through an arc of 180° (i.e. behind the wolf). If the wolf followed the object up to the point of disappearance in its visual field but failed to look for the object behind itself, Stage

Table 7.2 Criteria used to assess the development of wolves' SI for each stage

Stage 1	*Reflexes* Behavioral expression in pups is due to reflexes that are present at birth (e.g. suckling)
Stage 2	*Primary circular reactions* Reflexes transform to become a pup's primary habits, and manifest under the form of primary circular reactions. These reactions concern the pup's own body, and are triggered by spiking the interest of the pup, who finds it interesting, and then repeats the motion (e.g. scratching, walking)
Stage 3	*Secondary circular reactions* Secondary circular actions are actions that are repeated by the pup and result in an interesting effect with an external object, which may be either physical or social (e.g. tugging a log)
Stage 4	*Coordination of secondary circular reactions* The pup is now able to coordinate two different schemas of action in order to reach a predetermined goal. The pup's actions (e.g. biting, tugging) are directed towards different aspects of the environment (e.g. toys, conspecifics, etc.)
Stage 5	*Tertiary circular reactions* Behavioral variations are now possible. The pup experiments using trial and error, repeating actions that produce interesting effects on social or physical objects (e.g. pulling a branch to get a piece of food or a toy) with constant variations (e.g. with his mouth or his foreleg)
Stage 6	*Early representational thought* The pup is now capable of coordinating mental schemas without the use of direct experimentation on the environment (e.g. putting down a log and taking instead a branch to pull a piece of food toward him)

2 of OP was reached. However, if the wolf was able to follow the object during the entire trajectory and look for the object behind itself, Stage 3 of OP was reached.

Stage 3 and beyond were assessed via a series of formal OP tests. In the OP tests, E1 stood about 50 cm behind the central box and a second experimenter (E2), who restrained the animal by its shoulders, bent on her knees to its left side. At the beginning of each trial, E1 attracted the attention of the wolf by slightly moving the object back and forth about 50 cm in front of the central box. Once the wolf looked at the object, E1 moved the object as described in the tests (see Table 7.3). If the wolf did not pay attention to the manipulation of the object, the trial was rerun. To prevent involuntary cueing, once the object was put down, E1 looked up at E2. Then, E2 released the animal. If the wolf retrieved the object after its first choice, it was reinforced with social rewards (e.g. *Good boy*) and the opportunity to play with the object. However, if the wolf selected a non target box, the object was immediately removed from behind the box and no reward was given.

Six visible displacement tests (1–6) and one invisible displacement test (7) were administered to the wolves. Since our goal was to determine the development of OP, the tests were administered sequentially, from Test 1 to Test 7—that is, all the visible displacement tests were administered first and followed by the invisible displacement test. Each test was composed of 5 trials. To be successful in a test, the wolf had to succeed 4 trials out of 5 (Binomial test, $p = 0.33$, $\alpha = 0.05$). When a wolf succeeded a test, the next test in the sequence was administered as rapidly as possible, usually at a later time in the same day (47 % of the time the next test was administered within the same day). Otherwise, it was administered in the same

7 Cognitive Development in Gray Wolves

Table 7.3 Tests used to assess the development of wolves' OP for each stage

Test	Stage	Description
Test 1	Stage 3	Partial occlusion. E1 partially hid the object behind the target box. The object was always hidden behind the same box (A or C, depending on the wolf)
Test 2	Stage 4a	Single visible displacement–initiation of movement by the animal. E2 released the wolf right before E1 hid the object behind the target box. The object was always hidden behind the same box (A or C) and for each wolf, the target box was the same as the one used in Test 1
Test 3	Stage 4b	Single visible displacement. E2 released the wolf when the object was totally hidden behind the target box. The object was always hidden behind the same box (A or C) and for each wolf, the target box was the same as the one used in Tests 1 and 2
Test 4	Stage 5a	Sequential visible displacement. The object was hidden behind the box located at the opposing end of the row of boxes to the one previously used in Tests 1–3. For example, if the target box in Tests 1–3 was A, in Test 4 it was box C. The object was always hidden behind the same box (A or C). In the first trials of Test 4, if the animal searched at the box used in Tests 1–3, it displayed an A-not-B error
Test 5	Stage 5b	Double visible displacement. E1 first hid the object behind a box. Then, E1 visibly removed the object from the box and moved it behind a second box. Each box (A, B and C) served as first and second box at least once
Test 6	Stage 5b	Triple visible displacement. E1 first hid the object behind a box. Then, E1 visibly removed the object and hid it behind a second box. Next, E1 removed the object from the second box and hid it inside the box not yet visited. If the object was being moved from box A to box C (or from C to A), it always passed in front of box B (never behind). In each trial, every box was visited at least once
Test 7	Stage 6a	Single invisible displacement. At the beginning of a trial, the transportation device was placed at either the right or left end of the row of boxes, its open side facing the wolf. Then, E1 visibly placed the object inside the transportation device and, to hide the object from the wolf's vision, rotated the device on an axis of 180°. Next, E1 moved the device behind one of the three boxes and unnoticeably transferred the object from the device to the target box. After, E1 removed the device from the box and immediately rotated the transportation device to show to the wolf that it was now empty. Finally, E1 brought the device to the other end of the row of boxes from its initial starting location and rotated it on an axis of 180° to hide the fact that it was now empty. The object was always hidden behind the same box (A or C), which was the opposite of the one used in Tests 1–3. For instance, if the target box was A in the three first tests, the object was now hidden in box C

week. The only exception to this rule was Test 4. Since Test 4 served to determine whether or not the wolves commit the A-not-B error, Test 4 was administered immediately after the wolf had succeeded Test 3. If a wolf failed a test (any of them), the same test was rerun later in the same week. However, due to wolves' illness and the death of Devra, it was impossible to follow the schedule as planned. By consequence, on a few occasions, the delay between one test (success or failure) and the next was over 1-week (see Table 7.4).

Table 7.4 Number of correct choices in each test of OP (out of 5) as a function of week, wolf and stage

Week	Wolf	Stage 3	Stage 4		Stage 5			Stage 6
		Test 1	a Test 2	b Test 3	a Test 4	b Test 5	b Test 6	a Test 7
Week 6	Dharma	4	0					
	Gordon	0, 4	1					
	Tilly	4	0					
Week 8	Dharma		4	4	2			
	Gordon		5	–	–			
	Tilly		5	1	–			
Week 9	Dharma		–		2	–		
	Gordon			5	4	1		
	Tilly			5	5	–		
Week 10	Dharma				3	–	–	–
	Gordon				–	–	–	–
	Tilly				–	4	5	2
Week 11	Dharma				5	5	–	–
	Gordon				–	5	1, 5	1
	Tilly				–	–	–	1
Week 12	Dharma						5	4[a]
	Gordon						–	1
	Tilly						–	–

Note 1 The wolves were ill during Week 7, and so tests were not conducted
Note 2 When two numbers are within the same cell, the wolf was tested two times on the same test within the same week. The first number represents failure and the second represents success
[a] Dharma's score on Test 7 was considered a failure (see text)

7.2.4 Video Analysis

Two coders reviewed the videotapes of the development of SI and of the OP tests. One coder (the second author) viewed all the videos, adapted the coding system, and applied it to the behavior of wolves. For validation purpose, the second coder (the first author) also coded the wolves' behavior by viewing a random selection (50 %) of the SI and OP recordings. Both coders agreed on all behaviors.

7.2.5 Results

7.2.5.1 Development of Sensorimotor Intelligence

Due to the inconsistency of our video recordings (see *Procedure*), instead of quantifying how often the schemas of action associated with each stage occurred, we focused our attention on the presence or absence of these schemas of action.

When a behavior from a specific stage was observed for the first time (regardless of the wolf), we considered this specific day as an indication of the earliest occurrence of any behavior associated with this stage of development. As a consequence, this particular stage of development was then coded as reached by the wolves. Moreover, to specify the nature of the objects the wolves interacted with during each stage of development, we examined the frequency of appearance of behaviors with social and physical objects. To do so, given that the number of video recordings per week varied, we adjusted the frequencies of each behavior as a function of the number of 10 min bouts that were recorded in a week for each wolf. This allowed us to standardize the frequency of each behavior per week.

As mentioned earlier, the pups were already 3-weeks old at the start of the study and were fed by human caretakers. By consequence, behaviors that characterized Stage 1 of the sensorimotor period, which are mostly basic reflexes (e.g. suckling, rooting, Galant's reflex), could not be observed. However, on Day 23, when the observation of the SI of wolves formally began, we noticed that crawling was still occasionally among the behaviors of the wolves. We therefore concluded that the wolves acquired Stage 1 before Day 23 but we could not discern the exact age around which this stage was reached.

Similarly, on Day 23, we also observed primary circular reactions, which are behaviors of Stage 2. For instance, walking on four legs is an example of a primary circular reaction in wolves. This locomotor action has a primarily focus on one's body, and was repeated over and over by the animals. On Day 23, all the wolves were experiencing locomotion on four legs, sometimes with a mix of crawling. Rapidly, the wolves gave up crawling to focus entirely on walking. Once again, although we cannot determine exactly when behaviors of Stage 2 first emerged, we can nevertheless conclude that the wolves were already demonstrating behaviors of Stage 2 at the age of 23 days.

On Day 26, we observed secondary circular reactions in the wolves, which characterized Stage 3. These behaviors were directed towards external objects (social or physical) and produced a strong interest in the wolves. Biting another wolf, gnawing, pawing, and scratching were among the most frequent behaviors we observed (see Table 7.2). One characteristic of these behaviors is that they are voluntary: the wolves initiated these behaviors towards specific objects on their own; these behaviors were not simply the result of an accident or the mere proximity of another wolf. An example of a secondary circular reaction observed in the wolves is as follows: *Dharma bit Gordon's rear leg; Gordon ran away; Dharma then pursued Gordon by biting his leg again; Gordon ran away but Dharma sped up and bit Gordon once more.* During Stage 3, most of the wolves' behaviors were directed toward social objects, that is, their conspecifics, rather than towards non-social objects ($X^2_{(1)} = 20.24$, $p < 0.0001$).

Stage 4 of the SI period began on Day 50. The appearance of coordination between the different schemas of actions characterizes this stage. For example, Tilly grabbed a log with her mouth, ran away with it, and then began to chew it. In Stage 4, compared with the behaviors exhibited in Stage 3, the wolves were more inclined to explore physical objects, as illustrated by the fact that their number of interactions with inanimate objects was comparable to the number of social interactions with each other ($X^2_{(1)} = 0.02$, $p = 0.89$). For instance, in Week 10, a large cardboard box was introduced in the outdoor enclosure, and this captured the wolves' attention. Even if all three wolves simultaneously interacted with the box, there was no social interaction between the wolves, as all of their attention was directed towards the physical object. In Stage 4, it was also observed that the exploration of the environment (physical or social) by the wolves was characterized by the predominant use of the mouth: 88 % of the wolves' secondary circular behaviors involved the mouth. This observation highlights the fact that at this stage wolves explore the world by using their mouths much more than their paws ($X^2_{(1)} = 32.42$, $p < 0.01$).

We did not observe any behavior from Stage 5 or Stage 6 in our video recordings, which ended when the wolves were 76 and 79 days old. It was therefore concluded that tertiary circular reactions are either not present in the wolves' behavioral repertoire, or that these behaviors emerge solely after Week 11 of development.

7.2.5.2 Development of Object Permanence

In the visual pursuit tests, on the first and second days of testing (Day 24 and 25), the three wolves failed to follow the object up to its point of disappearance. On the third day of testing, two wolves (Tilly [Day 29] and Dharma [Day 32]) instantly reached Stage 3 of OP: they demonstrated the ability to follow the object during the entire trajectory and searched for the object behind themselves. However, one wolf (Gordon [Day 26]) solely reached Stage 2 of OP during his third day of testing: he was able to follow the object up to its point of disappearance but did not made any attempt to search behind his back. Gordon reached Stage 3 of OP on his fourth day of testing (Day 29). In summary, the wolves reached Stage 2 and Stage 3 of OP by the mean age of 29 and 30 days, respectively. Then, all wolves moved on the formal tests of OP, which started on Day 35.

Table 7.4 presents the individual performance of each wolf as a function of each test in each week. When a wolf failed a test, the same test was rerun the next week. However, on two occasions, the same wolf (Gordon) failed a test and was retested during the same week. As one can see in Fig. 7.1, the performance in the tests among the wolves was relatively homogenous: the wolves reached the same stages of OP at around the same age.

Most specifically, our wolves succeeded the partial occlusion problem (Test 1–Stage 3) at the mean age of 36 days. These first results are consistent with those obtained during Week 5 in the visual pursuit test. Success in the single visible displacement problem with initiation of movement by the wolves (Test 2–Stage 4a) was reached by the mean age of 53 days. Two wolves (Dharma and Gordon) passed Test 3 (Stage 4b) on their first attempt, but one wolf (Tilly) failed it,

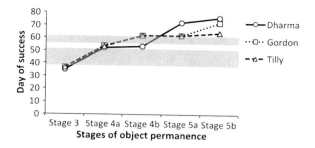

Fig. 7.1 Day of success for each wolf as a function of each stage of development of OP. The *two shaded areas* represent the periods of time during which the study was interrupted due to illness (Day 38–51) or the death of Devra (Day 56–61)

meaning that she may have committed the A-not-B error. However, out of the 4 errors she made on her first attempt in Test 3, Tilly searched four times behind box B and never behind box A (which was her target box in Tests 1–3). So, in spite of her failure on her first attempt in Test 3, we concluded with confidence that Tilly did not make any perseveration errors. Therefore, the wolves did not commit any A-not-B errors and they succeeded on Test 3 by the mean age of 59 days.

The wolves succeeded the sequential visible displacements (Test 4–Stage 5a) by the mean age of 65 days. By succeeding at these problems, the wolves demonstrated their first real understanding of the visible displacement of objects. Double visible displacements (Test 5–Stage 5b) were mastered by the mean age of 67 days, and triple visible displacements (Test 6–Stage 5b) were mastered by the mean age of 71 days. The result of the latter was judged as the date the wolves understandably mastered the visible displacement of objects.

In Test 7 (Stage 6a), two wolves failed the single invisible displacement problem and one wolf (Dharma) succeeded it. However, Dharma's performance in Test 7 was later discarded from the results. Indeed, both coders who reviewed the videotape of this particular test for this wolf agreed that methodological artefacts could explain her success. For instance, the target object occasionally came out of the transport device when the object was being moved to the target box, and, in one trial, the wolf was released by E2 before the end of the manipulation. On the other hand, when the manipulations were performed correctly (Gordon's and Tilly's Test 7), the wolves failed the problem and lost interest in the task, suggesting incomprehension of the invisible displacement problem. We therefore concluded that the wolves did not reach this stage of OP before the age of 12 weeks, that is, when we terminated the study.

7.2.5.3 Comparison with Domestic Dogs

As one of the objectives of the current study was to provide a comparison between dogs and wolves in regards to the development of their understanding of OP, we tentatively compared the performance of wolves in the OP tests with that observed by Gagnon and Doré (1994) in domestic dogs. As a reminder, Gagnon and Doré compared seven groups of dogs of different age, ranging from 4 weeks to

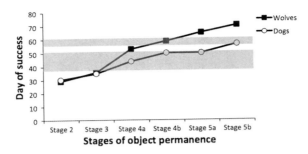

Fig. 7.2 Mean day of success for dogs (original data from Gagnon and Doré 1994) and wolves (current study) as a function of each stage of OP. The *two shaded areas* represent the two periods of time during which the study in wolves was interrupted. Data for Stage 2 were obtained by the visual pursuit tests and those for Stages 3–5b were obtained via the formal OP tests

9 months. In Gagnon and Doré (1994), when a group of a particular age failed to pass a test, the same test was administered to the next oldest group. Based on Gagnon and Doré's data in the visual pursuit tests and formal OP tests, we were able to identify the average age at which domestic dogs reached each stage of OP. Figure 7.2 illustrates the mean day of success of our three wolves (as a group) and of the dogs in Gagnon and Doré's study as a function of each stage of OP. As one can see, dogs' OP appears to develop at a faster pace than the one in wolves. In Stage 2 and 3, there were practically no differences between either species. From Stage 4a to 5b, however, the difference between dogs and wolves was striking: the dogs reached each stage at a much earlier age than the wolves. Depending on the stage, the difference ranged from 9 to 15 days.

However, given that the testing of OP in wolves had to be interrupted twice during the course of the current study, the results in wolves must be asterisked. We can suspect that the wolves would have been able to succeed some of the tests at an earlier age if they had been tested sooner during their development. To provide a better comparison between dogs' and wolves' development of OP, we estimated the mean age to which the wolves would have succeeded if testing had not beed delayed. To do so, we first calculated a differential ratio between the mean day to which dogs and wolves reached Stage 2 and Stage 3 (i.e. before the first interruption of the study). Then, we estimated the mean day of success of the wolves from Stages 3 to 5 by multiplying the mean day of success of dogs by this ratio. The result of this estimation is illustrated in Fig. 7.3. As one can see, if the wolves' testing had not been delayed, the wolves' results would have been quite similar to those of the dogs. However, we are plentifully conscious that this later approach is far from being perfect. Nevertheless, it helps to refine the results observed in wolves and illustrates that both species possibly have a very similar development of OP.

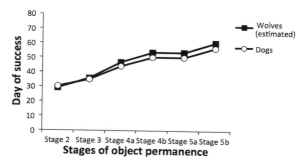

Fig. 7.3 Estimated mean day of success for wolves (to take into consideration the two periods of time during which the current study was interrupted) and the mean day of success for dogs (original data from Gagnon and Doré 1994) as a function of each stage of OP. Data for Stage 2 were obtained by the visual pursuit tests and those for Stages 3–5b were obtained via the formal OP tests

7.2.6 Discussion of Results

This study had two principal objectives. The first was to assess the development of OP in the gray wolf and compare it with the previous results found for the domestic dog (Gagnon and Doré 1994). The second was to observe the development of SI in the gray wolf and determine whether OP and SI develop at the same rate. To simplify the presentation, the results of the development of OP and SI are discussed separately. But first, a few methodological factors should be considered.

First of all, our small sample size reduces the certainty with which we can make generalized conclusions, and our results should be interpreted with caution. The same goes for our conclusions pertaining the exact moment in which wolves reached the different stages of OP, as the study was interrupted twice during a critical period of the wolves' cognitive and physical development. However, it should be noted that the wolves failed some OP tests that were administered after these two periods of interruption, suggesting that had testing not been suspended, they would not have reached that level of OP anyway. In addition, it is without a doubt that the small number of videos we were able to record limits the extent of the richness of the conclusions that can be made about the development of SI in wolves. For instance, we may have missed the appearance of some behaviors that characterized a particular stage. Consequently, the current study should be largely perceived as a pilot study, especially in regards to the timing at which the different stages of OP and SI occur. Nevertheless, we are confident that the conclusions regarding the general development of OP and SI are properly judged and well founded.

7.2.6.1 Development of Object Permanence

Overall, our results suggest that the development of OP in wolves is very similar to the one observed in domestic dogs (Gagnon and Doré 1994). It was observed that by the age of 4 or 5 weeks, both the dog and the wolf are capable of keeping track of an object that moves behind them. Around 5 weeks of age, both species succeed at Stage 3 problems. Thereafter, although it is difficult to establish with certitude the exact moment when wolf pups truly achieve Stage 4 and 5 of OP, the rate of development is mostly alike in both species. Our results also reveals that wolf pups, just like dog pups, do not commit the A-not-B error (Fiset and Plourde 2013; Gagnon and Doré 1992, 1994). In addition, between Weeks 8 and 10, dogs and wolves are both capable of succeeding at triple visible displacement problems. By the age of 12 weeks, however, both species fail invisible displacement problems.

The closeness of the rate at which dogs and wolves develop OP in the first stages can potentially be explained in part by the fact that the visual abilities of both species mature around the same age. In a study conducted by Lord (2012), dog pups and wolf pups were examined to determine the precise moment in which their olfactory, auditory, and visual senses fully develop. Their vision, which is critical in succeeding at OP tests, is only fully mature on Week 6 of life, which is also the time the dog and wolf pups in our study began demonstrating the capacity to truly locate disappearing objects. Prior to this time, neither species was capable of tracking and retrieving objects that had disappeared (Stage 3). This then poses the question of whether this failure was due to inability to see properly during their first weeks of life. Further testing is necessary to explore this idea.

In the current study, the 11-week-old wolves were unable to solve single invisible displacements. On the other hand, Gagnon and Doré (1994) reported that domestic dogs can resolve invisible displacement problems around the end of their first year of life. This introduces the possibility that wolves also reach Stage 6 of OP when they are around one year old. This would then explain the absence of this stage in our results, as our experiment was terminated by the end of Week 11 of life. This hypothesis is contradicted by the results of Fiset and Plourde (2013), who, using a spatial translation task administered to domestic dogs and adult wolves, demonstrated the incapability of either species to resolve invisible displacement problems. However, it should be noted that Fiset and Plourde (2013), in contrast with Gagnon and Doré (1994), did not use the Piagetian invisible-displacement problem. Given that translational problems are perceived as more difficult than those developed by Piaget (Fiset and Plourde 2013), it still remains possible that the adult wolves could solve the latter. However, recent work (Collier-Baker et al. 2004; Fiset and LeBlanc 2007) rejected the conclusion that adult dogs may solve Piagetian invisible displacement problems. These authors found that dogs primarily search as a function of the position of the displacement device and that, to a lesser extent, the presence of an experimenter behind the apparatus increases success in invisible displacement problems. Based on these last observations, we do not believe that adult wolves are capable of solving Piagetian invisible displacement problems. Nevertheless, this hypothesis remains to be confirmed empirically.

7.2.6.2 Development of Sensorimotor Intelligence

In investigating the development of SI in wolves, we can conclude that Stage 1 and 2 are both attained before the age of 23 days, Stage 3 is acquired during the fourth week of life, and Stage 4 is reached during the eighth. However, our observations did not allow us to detect any behaviors of Stage 5 or 6 in wolves by the end of Week 11 of life, when our study was terminated. This last conclusion contrasts with that of Frank and Frank (1985) who suggested that 10-week-old wolves may reach Stage 5, and possibly Stage 6, of SI. How can we explain the discrepancy between our study and the one performed by Frank and Frank? First, it is possible—like Dumas and Doré (1991) argued about domestic cats, which also do not reach Stage 5 of SI—that this is simply due to the complexity of such behaviors, and that observing these behaviors during natural interactions with the environment is very rare. Secondly, our study, similar to the approach used by Piaget (1954) with children, relied exclusively on behavioral observations to determine the stages of SI reached by the wolves. Contrarily, Frank and Frank created an experimental procedure in which young wolves had to perform tasks of increasing complexity to extract a food dish from a box. They then interpreted the behaviors of the wolves following criteria for SI. However, as pointed out earlier, in Frank and Frank, the wolves may have used strategies from Stage 3 to resolve problems that were supposed to be Stage 5, rendering their results difficult to interpret. We are therefore confident that the current study reinforces the conclusion that wolves' SI is limited to Stage 4, which is similar to what was observed by Frank and Frank in domestic dogs, supporting the hypothesis that all canine species present a similar development of SI.

Based on our observations, several general conclusions on the development of SI in wolves can be made. Firstly, the role of secondary circular reactions in young wolves seems to be a dominant factor in their development of SI. Behaviors like biting other wolves or objects, pawing or kicking, scratching the ground, wrestling, and tugging objects, were displayed with great frequency. Primary circular reactions, in contrast, were barely present, only being exhibited during activities like moving around, or gregarious behaviors such as snuggling. Even at a very young age, the pups seemed significantly more oriented towards external objects. This interest in the external world and the pups' tendency towards secondary circular reactions can be explained by the fact that wolves demonstrate an early development of motor coordination (Lord 2012). This early acquisition of locomotor skills, which emerges by the end of the second week of life, allows wolves to explore their environment, most likely allowing them to develop an intrigue for external features. By comparison, locomotor development in dogs emerges by Week 3 of age and is fully functional by Week 4 (Lord 2012). However, since no one has yet investigated the development of SI in dogs, it still remains to determine the exact role of locomotion on the development of SI in canines.

In the current study, it was also noted that the majority of secondary circular reactions, despite their large number, were expressed principally through the wolf pups' mouths. This link between the dominance of behaviors issued through the mouth and secondary circular reactions is characteristic of the third stage of sensorimotor development. In fact, in children, the third stage of the sensorimotor period is defined by the ability to grip objects, which is also first expressed through the mouth. Moreover, in wolves, the third stage of SI is reached before the third stage of OP, which is unlike in children, where the development of these two concepts occurs in a relatively synchronized manner. We hypothesis that the early development of mobility and grip—an ability demonstrated by the use of the mouth in wolves—favors the rapid development of the first three stages of SI, which leads to asynchrony between OP and SI development in wolves. Interestingly, this asynchrony in the wolf is very similar to that observed in the domestic cat (Dumas and Doré 1991), another species that acquires mobility and grip at an early age.

Finally, it was remarked that during their first interactions with the external world (Stage 3), the wolves predominantly oriented their behaviors towards their conspecifics, supporting observations by Lord (2012), who found that the wolf's period of socialization begins at 2 weeks of age. Interestingly, the wolves' constant interactions with each other may explain, in part, their development of OP. Indeed, since wolves are highly mobile and move around, they frequently disappear from each other's sight. From this we can postulate that they gradually learn or acquire the ability to keep track of their conspecifics by developing the mental capacity to remember where they have disappeared in the surrounding environment. Future work should explore the possible link between the development of OP and wolves' tendency to interact with mobile social objects.

7.3 Conclusion

In conclusion, the current study suggests that both domestic dogs and gray wolves share a similar evolutionary past that, over several thousand years, shaped their ontogenetic development of OP (and possibly SI) in a similar way. For instance, both species are gregarious and in the wild chase prey for survival. Since these skills necessitate the ability to keep track of moving objects, this fact potentially explains why both dogs and wolves present a similar development of OP. However, to acquire a more complete understanding on this question, future studies must include other canine species, as well as increase the sample size of species. In addition, the current research provides a first description of the ontogenetic development of SI in wolves by using criteria from Piaget's theory of cognitive development, suggesting that OP and SI in canines develop in an unsynchronized manner. A systematic comparison between dog and wolf development of SI and OP, however, is needed to further understand the reasons underlying this unbalanced relationship, especially in regards to the development of canines' senses.

Acknowledgments The authors wants to thank Clive Wynne, director of research at Wolf Park, Battle Ground, IN, for enthusiastically supporting this research project. They also want to express their gratitude to the Wolf Park's staff for their warm welcome and help during the course of this study. Special thanks are also due to Kathryn Lord and Monique Udell for their assistance in nurturing the wolf pups and collecting data, and to Catherine Fiset, who revised several previous versions of this manuscript as part of her internship sponsored by Shad Valley International. This research was supported by a discovery grant from NSERC (203747-07) and a research grant from Université de Moncton. The experiment received approval from the Comité de protection des animaux from the Faculté des Études supérieures et de la recherche de l'Université de Moncton, which is responsible for the application and enforcement of rules of the Canadian Council on Animal Care.

References

Bradshaw, J. (2011). *Dog sense: How the new science of dog behavior can make you a better friend to your pet*. New York: Basic Books.
Collier-Baker, E., Davis, J. M., & Suddendorf, T. (2004). Do dogs (*Canis familiaris*) understand invisible displacement? *Journal of Comparative Psychology, 118*(4), 421–433.
Doré, F. Y., & Dumas, C. (1987). Psychology of animal cognition: Piagetian studies. *Psychological Bulletin, 102*(2), 219.
Dumas, C., & Doré, F. Y. (1989). Cognitive development in kittens (*Felis catus*): A cross-sectional study of object permanence. *Journal of Comparative Psychology, 103*(2), 191.
Dumas, C., & Doré, F. Y. (1991). Cognitive development in kittens (*Felis catus*): An observational study of object permanence and sensorimotor intelligence. *Journal of Comparative Psychology, 105*(4), 357.
Fiset, S., & LeBlanc, V. (2007). Invisible displacement understanding in domestic dogs (Canis familiaris): The role of visual cues in search behavior. *Animal Cognition, 10*(2), 211–224.
Fiset, S., & Plourde, V. (2013). Object permanence in domestic dogs (*Canis lupus familiaris*) and gray wolves (*Canis lupus*). *Journal of Comparative Psychology, 127*(2), 115–127.
Frank, H., & Frank, M. G. (1985). Comparative manipulation-test performance in ten-week-old wolves (Canis lupus) and Alaskan malamutes (*Canis familiaris*): A Piagetian interpretation. *Journal of Comparative Psychology, 99*(3), 266.
Gagnon, S., & Doré, F. Y. (1992). Search behavior in various breeds of adult dogs (*Canis familiaris*): Object permanence and olfactory cues. *Journal of Comparative Psychology, 106*(1), 58.
Gagnon, S., & Doré, F. Y. (1994). Cross-sectional study of object permanence in domestic puppies (*Canis familiaris*). *Journal of Comparative Psychology, 108*(3), 220.
Klinghammer, E., & Goodmann, P. A. (1987). Socialization and management of wolves in captivity. *Man and wolf: Advances, issues and problems in captive wolf research*. W. Junk, Dordrecht, 31–61.
Lord, K. (2012). A comparison of the sensory development of wolves (*Canis lupus lupus*) and dogs (*Canis lupus familiaris*). In D. Zeh, (Ed.) *Ethology, 119*(2), 110–120.
Miklósi, Á. (2007). *Dog Behaviour, Evolution, and Cognition*. New York: Oxford University Press.
Pepperberg, I. M. (2002). The value of the Piagetian framework for comparative cognitive studies. *Animal Cognition, 5*(3), 177–182.
Piaget, J. (1954). *The construction of reality in the child* (pp. xii–386). New York: Basic Books.
Pollok, B., Prior, H., & Güntürkün, O. (2000). Development of object permanence in food-storing magpies (*Pica pica*). *Journal of Comparative Psychology, 114*(2), 148.
Ujfalussy, D. J., Miklósi, Á., & Bugnyar, T. (2012). Ontogeny of object permanence in a non-storing corvid species, the jackdaw (Corvus monedula). *Animal Cognition, 16*(3), 405–416.

Uzgiris, I. C., & Hunt, J. M. (1975). *Assessment in infancy: Ordinal scales of psychological development. Assessment in infancy: Ordinal scales of psychological development.* Champaign, IL US: University of Illinois Press.

Zucca, P., Milos, N., & Vallortigara, G. (2007). Piagetian object permanence and its development in Eurasian jays (*Garrulus glandarius*). *Animal Cognition, 10*(2), 243–258.

Part III
The Future of Dog Research: Critical Reassessment of Methods and Practice, and Practical Applications

Chapter 8
Measuring the Behaviour of Dogs: An Ethological Approach

Claudia Fugazza and Ádam Miklósi

Abstract What are 'dog cognition' studies actually studying? What role does the dog play in behaviour research? In this essay we consider how to study this species from the ethologist's perspective by providing a critical summary of the various approaches and explaining how these can answer questions on function, evolution, mechanism, and development and by highlighting the potential pitfalls in methodology. It is impossible to claim that one dog is more *dog-like* than others and it is now evident that even the concept of *breed* per se presents some problematic issues from the canid ethologist's perspective. Thus for any sampling it is fundamental to keep in mind what is the research question and to choose the subjects according to what aspects are expected to be relevant. In general the researcher should include a wide range of purebred dogs and mongrels living in human families in a representative sample. When sampling for investigations comparing wolves and dogs, we must bear in mind that dogs present a mosaic pattern of wolf-like traits and cannot be ranked along a strict continuum when assessing their differences from wolves. Therefore for comparative studies and also when the research question regards general dog abilities, a mixed sample of purebred dogs and mongrels is advisable. It is probably even more important to ensure that all animals have had similar past experience, especially with regard to humans. It should be noted that those dog owners that participate on a voluntary basis in a research program are already a specific sub-sample because it is very likely that those owners are particularly interested and take special care of their dogs. Thus it is also likely that these owners want to be part of the experiment and these dogs may actually also 'need' the presence of the owner. If the experiment is designed carefully the presence of the owner should not interfere with the outcome. The lack of a generally accepted ethogram is hindering behaviour research on dogs. The scientific community should aim for developing a categorical list of behaviour units that forms the basis of behavioural observations and experimental work.

C. Fugazza (✉) · Á. Miklósi
Department of Ethology, Eötvös University, Budapest, Hungary
e-mail: claudia.happydog@gmail.com

8.1 Introduction

Dogs are quite remarkable creatures when it comes to their study from the behavioural perspective. Many see them as companions of humans, and our specific, sometimes intimate, relationship with them may make any scientific inquiry necessarily subjective. For others dogs represent a specific case in the evolution of social cognition because dogs may have evolved a potential to share their lives with humans. Finally, there are many practitioners, including professional dog trainers, behaviour counsellors or self-made experts who struggle daily to explain dog behaviour by comparing them to either wolves or children, or who rely on their subjective experience for interpretation. This situation brings about unavoidable disputes and misunderstandings, which are caused mainly by the lack of a commonly agreed foundation on how to study dog behaviour. Along with many others we suggest that this common ground should be deduced from the biological study of animal behaviour, that is, ethology.

The key advantage of this approach is that ethology always regards behaviour as an outcome of four basic processes (function, evolution, mechanism, development) and emphasises the need for a detailed behavioural description of the natural behaviour (Tinbergen 1963; Miklósi 2007). This chapter deals with the last issue in detail.

We assume that humans provide the natural environment for modern dogs even if this anthropogenic environment is quite variable, ranging from a flat on the top of a skyscraper to a village in the savannah. However, in order to keep the topic focused we restrict our discussion to dogs that are regarded as members of human families. Nevertheless many issues raised in this chapter could be useful in studying dogs that do not live in a (very) close relationship with humans. The thoughts presented here are aimed to establish a common ground for behaviour research in dogs that is essential if researchers want to improve the internal and external validity of their data and share their results.

8.2 What is a Dog?

Finding a sample for research that is representative of 'dogs' is not as easy as it might seem at first sight and it is imperative to have a clear idea about the aims of the research and the associated specific questions in order to make correct sampling.

If we consider that a 2 kg Chihuahua and an 80 kg Great Dane belong to the same species, it is immediately evident that dogs are one of the species with the largest variation in their morphological phenotype. Importantly, from the evolutionary point of view, a breed is an artificial population of dogs created by humans through backcrossing and inbreeding to set desired traits. Each single breed derives from a mixture of different dog populations, not from a single ancestor and even if breeders often claim ancient origin for their breed, genetic studies show that many dog breeds have been recreated over the history using individuals from other breeds (e.g., Larson et al. 2012). The subsequent sexual isolation of these populations resulted in a unique genetic pattern for each breed (Parker 2012).

From an ecological point of view, the dog population consists of two different sub-populations that live in very different environments: (1) family dogs, that live in families as pets and whose reproduction is usually controlled by humans and (2) village dogs, that live mainly in the areas of the world where artificial breeding is not widespread (e.g., Africa, Asia) and do not have any ancestor that belongs to any specific breed but are thought to be shaped most recently by natural selection rather than by artificial human selection (Coppinger and Coppinger 2001). Actually, this situation may have changed during the last few 50–80 years in places where purebred dogs abandoned by humans hybridised with the local dogs population (e.g., Australia). Few studies have been conducted on village dogs (but see Pal et al. 1998; Pal 2008, 2010) whereas much of the research on dogs' behaviour has been conducted on family dogs. The conclusions drawn from such studies should not be automatically extended to other sub-populations that evolved in different environments.

8.2.1 Dog Breeds

A breed is defined as an intra-species, semi-closed breeding population that shows relatively uniform phenotypic (morphological) characteristics developed under controlled conditions by human action (Irion et al. 2003) (see also Duffy and Serpell, this volume). Unfortunately this definition does not take into account that there is also a variation over time in some dog breeds because they are subjected to genetic drift and genetic influx from other dog populations (Fondon and Gardner 2004). Furthermore, as the result of artificial human selection the phenotype and the genotype of dog breeds can change within relatively short time.

According to the international breeding rules of the FCI (Fédération Cynologique Internationale), '*puppies from pure-bred dogs of the same breed holding FCI recognised pedigrees are considered to be pedigree puppies and are therefore entitled to be issued FCI recognised pedigrees*' (Federation Cynologique Internationale AISBL International breeding rules of the FCI 1979; http://www.fci.be; p.7). It is immediately evident how pedigrees deal more with kinship than with general phenotype. The official description of a breed is the breed standard. Unfortunately, despite being very precise regarding morphological characteristics, breed standards only very rarely contain the depiction of any behavioural trait and, even when this is mentioned, it is often vague and misleading (e.g., the Bolognese is described by the ENCI—Ente Nazionale Cinofilia Italiana, the Italian national institution affiliated to the FCI—as: '*very serious, apparently not very lively. Creative, docile, attached to its companions until selflessness*'; www.enci.it). Breeders rarely base their breeding decisions on the behaviour of the dogs, and behavioural standards have never been established.

Originally breeds were selected for some specific functions (e.g., retrieving game, pulling sledges etc.) and therefore they were characterised by specific physical and behavioural features for carrying out these particular tasks. Nevertheless, for the traits that were not specifically selected for, which represent a high percentage of the overall phenotype, there is a substantial overlap among the breeds (Coppinger and Coppinger 2001; Overall and Love 2001). It is often estimated that the behavioural variation within a breed is comparable to the one that present among the breeds. Furthermore by the beginning of the 1900s dog breeding became more focused on physical or aesthetic criteria (e.g., selecting traits that were appreciated in the dog shows) rather than on specific working abilities. A well-known example of this changed tendency in breed selection is the German shepherd: those subjects that were originally bred for working purposes look very different to from those that are appreciated today in the shows. The latter ones would probably be unable to perform their original task due to the exaggerations in their morphological conformation.

8.2.2 Mixed Breeds

A large proportion of the dog population does not belong to any pure breed. Apart from the subpopulation of village dogs, mixed breeds (mongrels) are the offspring of either two purebred dogs of different breeds (F1 of two purebred dogs), of other mongrels, or even of two dogs that seemingly belong to a breed but do not own a pedigree. Due to the fact that the official definition of purebreds and mixed breeds does not take into account any evolutionary background, if the researcher aims to study some trait that is hypothesised to be specific for mixed breed dogs, it seems wise to exclude subjects that look like a purebred but have no pedigree and dogs that are F1 crosses of two breeds, and include only dogs that descended from mongrels. Gácsi et al. (2009a) found that such a sample of mongrels performed worse than purebred dogs of cooperative working breeds in utilizing the human pointing gesture to locate food. The authors suggested that while the cooperative breeds have been selected for attending human indications, present day mongrels originate from a population that has been under selection for skills that promote independence (e.g., their reproductive success was not supported by humans). Thus it is likely that mongrels' independent problem-solving abilities prevail over the motivation to be guided by humans.

Mongrels often have a different rearing background than purebreds. The latter are usually bred on purpose by a breeder and thus are kept in a facility or in a house and (usually) experience human contact from very early in development. Mongrels are usually the result of some unwanted or uncontrolled litter and may live the first part of their puppyhood independently from humans. Early experience is known to affect behaviour in many respects (e.g., Scott and Fuller 1965) and this has to be taken into account in order to prevent any unwanted bias in the results. Particularly if one aims to compare purebred dogs and mongrels it is very important to make sure that the subjects in the two groups have the same raising history.

8.2.3 Single-Subject Studies

The reliance on single subjects in studies on behaviour is usually greeted with mixed feelings in ethology, raising the question whether knowledge gained from a single individual is of any value. The usefulness of single-subject studies depends on the question to be asked. If one wants to find out whether some specific skill is present in a species then such studies may have merits. Such investigations should be regarded as pilot experiments that can, nevertheless, narrow down the phenomenon. However, it is also important that these studies rely on carefully designed protocols (see Kazdin 1982 for single cases designs). Such case studies should also help in finding out the conditions in which the skill can be observed and give a lot of hints about potential limitations in performance. Usually such studies are followed by more extensive research on a larger number of subjects.

Such case studies are often done when studying linguistic abilities. Kaminski et al. (2004) reported that Rico, a 9-year old border collie knew the names of more than 200 items. Especially the finding that Rico could rely on learning by exclusion raised the interest in repeating these findings with other dogs. Subsequently, it was shown that two other border collies showed partly similar performance in similar tasks (Kaminski et al. 2009). Later, Pilley (2013) reported that Chaser, another 9-year old Border collie, could understand the syntax of short sentences. Thus it seems that in special cases dogs may rely on complex communicative signals. It is another question however, what such experiments tell us about the function of the dog mind, and its everyday performance. It is likely that such skills as those of Rico and Chaser fall outside the species-typical range and may be regarded as 'by-products' of other traits selected for by evolution.

What single-subject studies do not do is provide explanations regarding the mechanism of the underlying cognitive abilities. In order to answer these kind of questions the number of subjects must be increased. Following this reasoning, on the basis of a single subject study by Topál et al. (2006), which showed that a dog was able to reproduce human-demonstrated actions, we recently investigated the details of the social learning processes involved in this ability—with special attention to the memory supporting imitative skills in dogs—on a larger sample (n = 8) (Fugazza and Miklósi 2013).

8.3 Comparative Studies

Traditionally comparative cognition focused on human-ape comparisons, but nowadays there is growing interest for comparative studies in a range of species, including wolves and dogs (e.g., Hare et al. 2002; Miklósi et al. 2003; Udell et al. 2008). The rise of this interest stems from the increasing scientific agreement on the consideration of dog domestication as an evolutionary process during which the potential for functionally human-like social skills emerged in dogs as a result of the selective pressure created by the anthropogenic environment.

The observation of this functional similarity in humans and dogs raised the idea of convergent evolution of specific traits between these two species (see Topál et al. 2009) but unfortunately there is limited consensus on the issues raised in such comparative studies (e.g., Hare et al. 2009; Udell et al. 2008; Wynne et al. 2008). The dispute's main focus is to identify and possibly separate genetic and environmental contribution to the emergence of specific skills. We argue that in some cases the clarification of methodological issues and the improvement of the experimental design may also help in avoiding unnecessary debates between scientists with different perspectives.

8.3.1 Wolves and Dogs

Through comparative studies it is possible to tackle questions about skill evolution; the main aim of dog-wolf comparisons is to discover what specific features emerged in dogs as a consequence of the evolutionary process of domestication. For example, Miklósi et al. (2003) demonstrated that dogs show a readiness to look at the human face in problematic situations that is not present in wolves, even after intensive socialization. The authors suggested that these enhanced communicative skills emerged through both evolutionary and ontogenetically positive feedback that in turn allowed complex forms of human–dog communication to take place.

Unfortunately the basic rules for comparative research are not easy to follow and it seems that their violation in several studies in wolf-dog comparisons has resulted in contradictory results. When comparing different species it is essential to keep in mind that the performance is not simply the direct output of some underlying cognitive ability because there are other variables that affect behaviour—such as motivation, previous experience, and the experimental conditions created by the researcher by using a particular experimental design. It is therefore extremely important to equalize as much as possible all the known variables that might affect the behaviour in the two or more populations to be compared.

For example, Hare et al. (2002) compared dogs' and wolves' ability to rely on a human pointing gesture to find hidden food in a two-way choice task. They aimed to test the hypothesis that dogs acquired better communicative skills during domestication. Dogs outperformed wolves on this task and this result led the authors to conclude that domestication enhanced dogs' communicative skills. Although this conclusion was later partially supported by other studies (e.g., Miklósi et al. 2003), the method used in that study did not allow the scientists to exclude alternative hypotheses, because the two compared populations differed in a range of variables. For example, as noted by Packard (2003), dogs and wolves differed in their level of socialization toward humans. The circumstances of the test were also different for the two samples. It is also probable that the wolves and dogs differed regarding their previous experience with the objects and procedures used in the experiment (e.g., in the case of pointing experiments, if food is placed in bowls, it is necessary that both wolves and dogs have experience of eating from bowls). Indeed, the performance of the wolves was uniformly low on any version of those tests, suggesting that they did not have the prerequisite experience on the basic requirements of the task (Miklósi et al. 2004). It was later suggested that intensive socialization improves the performance in wolves because they can learn to attend to the human who performs the pointing (Virányi et al. 2008). In addition, Gácsi et al. (2009a) indicated that the skill of relying on the human pointing gesture develops later in wolves. Udell et al. (2008) reported that socialized wolves outperformed dogs in a simpler form of the pointing task and shelter dogs tested outside performed much worse then socialised wolves. Later Pongrácz et al. (2013a) found that family dogs performed equally well on the human momentary distal pointing test when it was run indoors or outdoors, suggesting that different kinds of pointing procedures can have different sensitivity.

The contradictory results on wolves and dog performances were followed by a heated debate regarding to what extent the differences could be attributed to genetic or environmental factors (Hare et al. 2009; Miklósi and Topál 2011; Udell et al. 2010; Udell and Wynne 2011), somewhat reflecting the old debate on 'nature versus nurture': Udell et al. (2010) concluded that ontogeny plays a major role in determining canids' ability to understand the human communicative gestures, whereas other authors, while not denying the importance of ontogeny, attribute the emerging species difference to genetic predispositions acquired through the domestication process (e.g., Gácsi et al. 2009b; Hare et al. 2009). The most honest observation to make here is that this debate could have been avoided by following the principal guidelines for comparative studies. Importantly, the genetic effect can only be investigated if other confounding factors (e.g., rearing conditions, previous experience) are controlled for and equalized in the two samples. Wolves and dogs need to be socialized at comparable levels and have comparable experience with the situations designed by the experimenter. If this rule is not followed, it is impossible to disentangle the factors that contribute to their performance because both genetic predispositions and/or environmental factors (i.e. previous experience and socialization) might affect the results. When comparing wolves and dogs some other inherent differences must be taken into account:

1. *Differences in their maturation rate* Wolves and dogs may show differences in maturation; already during early puppyhood their sensory systems mature differently (Lord 2012). Typically wolves mature sexually approximately a year later than (most) dogs (Packard 2003), and within dogs there is also a high variation in the age at which they reach sexual maturity. It seems therefore wise to select for testing subjects at the age of 2 years but at this age wolves can be less willing to collaborate in experiments, unless they are used to frequent experimental work and intensively socialized.
2. *Differences in socialization* The different level of socialization can be equalized either by extensively socializing members of both species immediately after birth or by 'feralizing' dogs, avoiding human contact with them and keeping both wolves and dogs in semi-wild conditions. However, if some human interaction is necessary for the testing procedures, the first option is preferable.

In this case, the socialization of wolves requires particular care and they have to be kept separated from conspecifics for their first 4–6 months (Klinghammer and Goodman 1987). Different socialization programs are carried out by different research groups. In some cases wolves and dogs are kept in fenced areas and are raised by humans in peer groups and as adults they receive daily human contact and training (e.g., Utrata et al. 2012). In other cases they are raised as family pets, and live in human families (e.g., Miklósi et al. 2003). The effect of these differences in socialisation on the subsequent performance is not known but it is very likely that extensive training (which is not usual for family dogs) affects their behaviour.

3. *Differences in motivation* Wolves and dogs may be motivated differently for food and for social contact (Frank and Frank 1988). Whereas in laboratory animals deprivation is almost routinely used to equalize motivation, this is not feasible with family dogs for obvious ethical and practical reasons. Access to favourite toys seems to be motivating as much as food for some dogs and experience suggests that there is a huge variability in motivation for different kinds of rewards in this species, but unfortunately at present there are little data on the influence of the quality of reward on dogs' motivation and performance (but see Feuerbacher and Wynne 2012; Pongrácz et al. 2013b).

8.3.2 Comparison of Dog Breeds

The same considerations mentioned above also apply for breed comparisons. Unfortunately, when comparing large specific populations of purebred family dogs those conditions are even more difficult to fulfil—particularly regarding the equalization of rearing condition, socialisation, and previous experience (including training). So far this has probably been only achieved by Scott and Fuller (1965); however, those dogs were kept in confinement. It is easy to imagine how difficult can be to equalize the environmental factors for family dogs that live with their owners, especially if we compare it with the work on laboratory animals where the experimenter has complete control over the subjects' life.

Breeds can be compared with regard to breed-specific or non-specific skills and tasks with very different results. It could be expected that comparing retrievers and livestock-guarding dogs in a retrieving task might lead to significant differences in their performances, while in a non-specific task no difference would probably emerge. For example Pongrácz et al. (2005) found no differences in a detour task—a non-specific task—in a sample of ten different breeds. The lack of difference on this task suggests that this trait may have not been selected for or against in these dogs—although, importantly, similarity in performance could be also explained by different genetic and mental mechanisms and the canalising effect of the human environment.

Any comparison of two breeds may reveal specific differences that are independent from the breed selection. It is more advantageous to compare two (or more) groups of breeds in which a differential selection can be supposed (e.g., "hunting dogs" versus "sheepdogs"). This logic was followed by Gácsi et al. (2009b) who compared dog breeds selected to work in visual contact with the human partner with breeds selected for working while visually separated from him. They found that dogs relying on visual contact for cooperation were better in choosing on the basis of the human pointing gesture. Importantly, all dogs were kept as pets without any specific experience, thus part of the difference in performance is probably of genetic nature. Similarly Passalaqua et al. (2011) tested dogs belonging to three different breed groups (basal breeds (e.g., Alaskan malamute), hunting and herding breeds (e.g., border collie), and molossoid/mastiff-type breeds (e.g., boxer)) for their tendency to look at humans in problem-solving situations. They did not find any significant difference looking behaviour in two-month-old puppies kept in pens by their breeders but a significant difference emerged at four and one-half months old and with adult dogs that lived with their owners. Hunting/herding dogs looked more often at the human than did the basal breeds and molossoids.

It should also be noted that there are differences in the dog–human relationship among different countries due to cultural reasons and this might in turn affect the dogs' behaviour. For example, Wan et al. (2009) noted that German shepherds in Hungary are more likely to be kept outdoors while in the U.S. this breed is more likely to be kept as pets, to be taken to dog schools, and is generally rated higher on a scale for confidence and aggression.

The researcher should be aware that claiming specific differences between dog breeds can have significant consequences, not only in terms of a breed's popularity among breeders and owners but also regarding how this might influence legislation about dog breeds. For example, characterising a breed as 'aggressive' may led to arguments of banning these dogs from breeding (Overall and Love 2001).

In summary, it is necessary to ensure that the subjects under study (1) have been exposed to the same physical and social stimulation, (2) live in the same environment, (3) have comparable behavioural constitution and (4) have the same motivation level. If these factors are not controlled for, it becomes virtually impossible to draw conclusions on behavioural differences in terms of genetic or environmental and experience-based influences.

8.3.3 Dogs and Children

From the beginning of dog research there has been a tendency to propose comparative work with children. For example, Scott and Fuller (1965) pointed out that children and dogs could be compared regarding their behavioural development. Both family dogs and human children share the same environment (i.e. the human family) and this makes the comparisons relatively straightforward because similar levels of socialization and experience of the social environment can be assumed. To some extent their social experiences can also be similar. For example, it has been shown that owners' pet-directed speech and infant-directed speech ('motherese') are similar, as are some behaviours of adult humans toward dogs and infants (Hirsch-Pasek and Treiman 1981; Mitchell 2001; Prato-Previde et al. 2006). In addition, dogs and children can be observed and tested in their natural environments (i.e. the house or the laboratory where they come with the parent or owner) (but see Rooney and Bradshaw, this volume, for a different approach to what is the natural environment of the dog).

It should be also obvious that the behaviour (and mental skills) of children and dogs are very different in many respects. They also differ with regard to freedom of movement (see below) and in their range of actions, including fine motor abilities. In addition, infants rapidly start to develop linguistic skills. In spite of this, comparative research can point to functional similarities and convergences in social skills in infants and dogs. Lakatos et al. (2009) compared dogs, two-year-old, and three-year-old children in their ability to use different pointing gestures and found that dogs' performance is similar to the two-year-olds, whereas the older children were able to attend to more complex communicative signals. Thus the development of this ability is probably facilitated in children by the production of similar pointing gestures in different communicative situations, including language learning.

In the preverbal period of development, infants' cognitive abilities are experimentally studied by means of experimental procedures that are often also exploitable for dogs—if their differences in sensory and manual abilities are taken into account. Topál et al. (1998), for example, tested dogs with a modified version of the Strange Situation Test (Ainsworth 1969) that is commonly used by psychologists to test attachment in children. They found that dogs develop an attachment toward the owner that is functionally similar to the attachment present in infants toward their mothers.

8.4 Notes on Experimental Methodology for Testing Family Dogs

Ethologists prefer to study the behaviour of animals in their natural environments. Nevertheless, when researchers aim to answer questions on animals' mental skills it is often necessary to test the subjects in a laboratory setting. In this regard dogs represent a remarkable case: their natural environment being the same as ours, we can test them in the laboratory—that often looks more like a living room—assuming that this is similar to their natural environment. Thus dog behaviour and mental skills can be studied in carefully controlled experiments the conditions of which closely resemble the natural situation. In this way experiments on dogs often resemble those that are designed for children: dogs (like children) can visit the laboratory for the testing occasions instead of being raised and kept in the laboratory and the testing situation should not be considered very different from what most family dogs experience in their everyday life (e.g., visiting different places and environments with the owner). Importantly, in this laboratory setting, the experimenter can control and/or alter the variables that are expected to affect behaviour.

Raising and keeping dogs in the laboratory and testing them with the traditional experimental paradigms for laboratory animals such as rats should be the last resort when studying dog behaviour. Dogs observed and tested under such impoverished environmental conditions can be hardly considered representative of family dogs because, even if some enrichment is provided, the complexity of their social stimulation is very different from what dogs experience when living in a human family. It is therefore our suggestion not to keep dogs in the laboratory but instead to invite owners to bring along their dogs for testing, which usually can be regarded as a game or a recreational activity for both dog and owner.

8.4.1 Owner Effects

The most relevant component of most dogs' natural environment is the presence of the owner(s) with whom they maintain a special relationship. Living in the human society dogs are often exposed to different human individuals and it has been shown that they attend preferentially to their owner compared to a stranger (Mongillo et al. 2010). Furthermore different studies using a modified version of the Strange Situation Test demonstrated that dogs show patterns of attachment behaviour toward their owner (Gácsi et al. 2001; Gácsi et al. 2013; Mariti et al. 2012; Topál et al. 1998). It was also shown that dogs' performance in problem solving situation is affected by their tendency to behave socially dependently (Topál et al. 1997). In contrast to dogs that were kept outside of the house for some working purpose (e.g., guarding dogs), dogs living in a companion relationship with their owners and who were considered as family members interacted with the equipment in a problem-solving situation only after some encouragement by their owners. The tendency of dogs to interact with humans in a social unit has to be considered when designing experiments and raises the problem of testing with or without the owner. Thus it is important for the researchers to understand what the critical issues of testing the dog in presence or in the absence of the owners are.

8.4.1.1 Testing Without the Owner

The absence of the owner might elicit fearful or stress-related behaviours, especially in the laboratory, because most dogs are not used to being left alone at a new location. This stressful situation may interfere with their performance.

In some cases the owner might be replaced by familiar laboratory personal or another human companion of the dog. Although the presence of the person may help to alleviate the dogs' fear, s/he is not an adequate replacement for the owner. For example, Horn et al. (2013) reported that dogs are more interested in watching their owners than another familiar person; this preference reflects the relationship they have with their owners, rather than mere familiarity.

8.4.1.2 Testing in the Presence of the Owner

Testing in the presence of the owner presents its own problems. In novel situations dogs may rely on the active encouragement of the owner (see above) which can be considered as a form of social facilitation. Furthermore, even a passive owner (as mothers of infants) can make the subject comfortable with the environment and thus contribute to improved subject performance. However, in this social situation dogs may be more prone to rely on owner-directed behaviour when facing a problem situation. Thus it is difficult to separate the performance of the dog from the one of the 'team' (dog plus owner).

The presence of the owner can have both direct and indirect effects on the dog's behaviour. The direct effects might arise when the subject is tested in problem solving situations when the owner could be aware of the goal. In this case the owner might be unaware of displaying cues to the dog that increase the performance: the Clever Hans effect (Pfungst 1911). Such situations can be avoided if the owner is not informed of the goal of the task used in the experiment and of what is considered correct performance (blind experiment) or if the owner is physically blindfolded during the test. Although no Clever Hans effect was found in a two-way object choice test on dogs (Hegedüs et al. 2013), other authors (Horowitz et al. 2013; Prato-Previde et al. 2008) have emphasised that ostensive communication has a significant influence on dogs' performances in different behavioural situations and Clever Hans effect can occur in different tasks, including scent detection (Lit et al. 2011; Szetei et al. 2003).

However, an informed owner might behave in a more natural and relaxed way than an unknowing or blindfolded one and this may in turn indirectly affect the behaviour of the dog (e.g., Topál et al. 1997). This indirect effect should not be considered as a bias because, especially in experiments where performance is expected by the dogs in an unfamiliar situation, it is important that they feel comfortable enough to perform as they would in a more natural condition. Furthermore, as negative results are difficult to interpret, it is important to exclude the possibility that the subjects did not perform because they were too stressed in that situation.

Owners often want to be present during the tests because they want to know what happens to their dogs; instructions should ensure that they do not interfere with the procedure in any unwanted way.

Summing up the problem of testing in presence or in absence of the owner: there is more support for having the owner involved in the experiments with family dogs because this represents a more naturalistic situation for the dogs. This method has also precedent in experimental child psychology. At the same time, it is important to keep the behaviour and knowledge of the owner under control.

8.4.2 The Effects of Dog Training and its Use in Research

Many years ago dogs became 'family members' in a quite natural way, through everyday interactions with other members of the group. Although this is still the situation in many places, a growing number of family dogs today are subjected to specific training. The reasons for this change do not concern us here (but see Miklósi 2014); more important is the specific effect of such experience on the behaviour and performance of dogs in the research setting.

Training experiences may affect dogs' behaviour in the testing situation. Marshall-Pescini et al. (2008) observed trained and untrained dogs in a problem solving situation (Fig. 8.1). For each dog an experimenter demonstrated how to open a box and after that the dogs were allowed to open the box themselves. The results showed that trained dogs were more successful and spent also more time interacting with the box while untrained dogs looked more at their owners. In a subsequent paper the same authors Marshall-Pescini et al. (2009) report that although untrained, search-and-rescue, and agility dogs showed similar performance, dogs in the former group looked back more often at the owner during the trial, and this behaviour was most common in agility dogs. Differences also emerged when the dogs faced an unsolvable task—wherein agility dogs again looked more frequently at their owners. Although looking behaviour can be influenced by many factors (e.g., breed), these differences suggest that previous training may also affect the pattern of communication between dog and owner.

Other observations indicate that training can reduce neophobia in dogs, but it can also affect the motivation of the dog in solving a task; that is, these dogs may be more persistent (Marshall-Pescini et al. 2008). In addition, training may improve the motor skills of dogs and also change the hedonistic value of some incentives. For example, trained dogs could be rewarded by receiving an object (ball) for a short play. This means that the level of previous training should be considered as an important external variable that could affect both the performance in a particular experiment and also the outcome of a comparative study. For example, members of one breed are more likely to be trained then other dogs (e.g., hunting dogs versus lap dogs).

In the case of a problem-solving task researchers should also consider that most family dogs actually do not 'work for their living': that is, they are fed by their owners independently of any performance. This means that these dogs are less likely to cooperate in long training and testing sessions.

Training procedures can be also useful to investigate on particular cognitive abilities. For example, training dogs to copy human demonstrated actions on command enabled scientists to study their imitative abilities (Topál et al. 2006; Huber et al. 2009) (see also Huber et al., this volume) and their memory of human actions (Fugazza and Miklósi 2013). Through this specific 'Do as I do' training procedure the dogs learn that they are required to copy and this makes the testing procedure quite straightforward because the researcher is not required to set up a

Fig. 8.1 Family dogs facing a problem solving situation often try to get the owners' attention by looking at the owner. This seems to be an alternative tactic for 'solving' the task, because earlier experience may have taught them that the owner would help in this case (Photo: Claudia Fugazza)

complicated situation in which the dog is expected to be motivated to copy a demonstrator (Fig. 8.2). Importantly, in case of negative results, the researcher can rule out several different explanations, such as lack of the basic requirements to understand the situation and lack of motivation in the testing situation. Specific training methods may be used to estimate dogs' perceptual abilities—particularly olfaction, (Gazit and Terkel 2003; Johnen et al. 2013)—especially because they have practical applications (see also Gadbois and Reeve, this volume).

Unfortunately the field of dog training has received very little attention from researchers and most of the training methods have not been formally validated by a scientific approach (Miklósi 2014). The obvious problem here is that dog training is usually regarded as a 'politically sensitive' area, and researchers do not want to get involved in this. Although there is a desire for up-to-date training methods to live up to welfare requirements, it should also be noted that dog training should be done more in line with the ethology of the species and there is not a single best method for all cases. This would suggest that researchers should compare the efficiency of training methods objectively with regard to specific contexts and dog breeds (e.g., Fugazza and Miklósi submitted).

In conclusion, it is advisable to avoid lengthy training procedures in behaviour observations of family dogs, but in specific cases training some individuals may be required for detecting more subtle problem-solving and mental skills.

8.5 Describing Behaviour

Students of behaviour have long struggled with the definition of behaviour. In a popular book on behaviour methods Bateson and Martin (1986) define behaviour simply as an 'action and reaction of the whole organism'. More recently, one of the authors (A.M.) has suggested extending the definition offered by Levitis et al. (2009) by defining behaviour as an *'internally coordinated action or pose of the*

Fig. 8.2 It may be necessary to train dogs specifically if the researchers want to study a specific mental skill. Fugazza and Miklósi (2013) trained dogs to repeat a demonstrated action on command. This method of imitation training can help researchers understand the abilities (and limitations) of dogs in performing novel actions after watching a demonstration. **a** Owner (or trainer) demonstrates the action; **b** Dogs executes a matching action after the 'Do it' command (Photo: Claudia Fugazza)

whole living organism (individuals or groups) in space and time projected onto its body plan in relation to the environment to internal and/or external stimuli' (Miklósi 2014). The reason for this more extensive definition is that published descriptions of behaviour—or more precisely *behaviour units*—are usually quite poor and selective. They focus only on a small part of the body and often do not include the relationship with the particular environment.

It is important to state that it is often the researchers' fault that behaviour descriptions are vague. In the case of measuring behaviour the tool of the measurement is the human mind, and this 'equipment' has both advantages and limitations. One limitation is how we sense (see and hear) behaviour with our species-specific sensory capacities; by contrast, we are very good in observing 'patterns' and labelling categories. Providing a solid method for describing behaviour is not easy. An overview of the problems and discussion of ways to improve may help to establish a general framework for dog behaviour.

8.5.1 The Human Observer and Degrees of Freedom in Movement

Actions and poses are features of the body. The more complex the body plan is, the more complex actions and poses can be expected. In mechanics the number of possible actions executed by an object can be calculated by taking into account the independent motions that are possible with that particular configuration considering the rigid parts, joints, etc. The same process can be applied to animals, and accordingly one could determine the maximum number of actions and poses of, for instance, a dog body. As the method for determining behaviour is human observation, such an approach to defining behaviour units is totally useless because (1) some of these possible actions may occur very rarely, (2) some may have no particular function and (3) no observer would be able to remember and distinguish so many actions reliably. Thus ethology relies on the opposite method, according to which human observers (1) look for specific actions of behaviour that occur in specific situations, (2) aim to describe behaviour by using the lowest number of units, and (3) focus on specific (local) aspects of the action, disregarding the form and shape of the whole body. This reductionist approach proves to be very useful in making the behaviour amenable for straightforward statistical analysis. At the same time this process involves a major loss in information, and, thus, changes in the way the human observer defines the same behaviour could lead to different results.

For example, Miklósi (2007) has identified at least six different ways of 'describing' (characterising) agonistic behaviour in wolves of dogs. The simplest way is to use a single dimension (no aggression (1)–threat display (5)) (e.g., Svartberg 2005) which almost totally disregards the temporal and spatial complexity of the behaviour. Other more sophisticated descriptive systems may include different aspects of body movements for defining exclusive behavioural categories. For example Hooff and Wensing (1987) define 'low posture approach' by referring to low head position, backward oriented ears, and bent tails and legs. It is clear that the behavioural observations based on these two methods cannot be compared to each other.

The same is true for the more behaviour-oriented descriptions (e.g., Schenkel 1947; Tami and Gallagher 2009) that refer to different aspects of the same behaviour. In the case of assertive behaviour some studies focus on the body posture or tail, while others include gazing and the visibility of the teeth. It is never really clear whether the parts of the body which are not included in the description are left out because of ignorance or because they do not differ from an assumed 'relaxed' state of the dog.

In order to develop a full account of dog behaviour—an ethogram—researchers need to agree on what parts of the dog's body play a role in displaying any behaviour (structured ethogram). Next they should catalogue all possible states of specific body parts (e.g., eye, leg, ear, etc.). Finally, they should provide a description of the behavioural units by referring to the actual shape or form of all body parts identified in the former analysis. This approach would ensure comparability across studies but would not restrict researchers from lumping behaviour units at some later point in the analysis. Such structural analysis of dog behaviour would help not only in the comparison of wolves and dogs (and dog breeds), but also in a more effective understanding of causative factors of behaviour.

8.5.2 Categorisation and Labelling

Once researchers have detailed structural descriptions they may want to define states or categories of behaviour, label them, and build a hierarchical system of behavioural organisation. Many problems arise because the labelling often precedes the detailed behaviour analysis and the labels do not refer to the behaviour but to some hypothetical inner state or to the presumed effect ('function') of the behaviour. For example, one may label a specific behaviour conformation as 'friendly' or 'confident'. At first sight such solutions may be justifiable because these labels are common with regard to human behaviour, and may help both in the recognition and in the categorisation. However, the problems should also be recognised. First, in these cases behaviour is defined on the basis of a hypothetic inner (mental) state. Second, there are no specific or objective criteria and rules for developing such a categorisation system. As a consequence, the categories of dog behaviour are drawn from anthropomorphic categories a priori, before any deeper investigation (e.g., on function) is actually executed. In an ideal case the dog ethogram should not depend on the richness (or poverty) of human vocabulary (e.g., *friendly*, *sociable*, *amicable*, *affectionate*). Obviously, it would be difficult to deny that in certain situations dogs are in a fearful state which is comparable to those ascribed to humans, but it seems to constrain the objective analysis when 'fear' is used as a label for behaviour.

The classic example for using a label of (possible) function is when researchers describe the encounter of two dogs or a dog and a human as 'greeting'. The function of greeting behaviour could be to strengthen or enforce the social relationship between the partners after long separation (e.g., overnight) but in any case this function should be tested specifically before such statement can be made. The behaviour interactions that take place between two encountering wolves, dogs, or a dog and a human may be similar or different and also may serve the same or actually different functions. Greeting behaviour can be more intensive and frequent in the dog–human relationship as compared to the intraspecific interaction among wolves or dogs. Thus 'greeting' in dogs may be controlled by different factors. McGreevy et al. (2012) present a very useful analysis by comparing some behaviour categories in dog–dog, human → dog or dog → human (the base of the arrow indicates the promoter) interactions. They also show that despite the functional compatibility of dog and human behaviour there are many differences on both sides. For example, both humans and dogs can 'pin down' the other but dogs may use their body while humans use their hand.

Similar problems arise when one refers to 'dominant' or 'submissive' behaviours. This labelling indicates that the observer assumes that the respective behaviour is shown only by the 'dominant' or 'submissive' animal, which is not obvious a priori. Even worse is the use of these behaviours as an explanation for dominance hierarchies, especially because most researchers agree that 'dominance' refers to a relationship between two specific individuals (e.g. Bradshaw et al. 2009; McGreevy et al. 2012). Thus 'dominance' should be clearly distinguished from a tendency to displace the other, which is probably a personality character ('assertiveness').

Nevertheless one may hypothesise that one or the other type of behaviour may be displayed differently at the individual level. For example, Fatjó et al. (2007) observed social interactions in a small captive pack of wolves. It might be noted that social behaviour among captive animals may not resemble interactions in natural packs.

This study also found that only a small subset of 'dominant or submissive behaviours' were reliable indicators of hierarchical relationships. They found that the position of the tail could be used to infer the social relationship while this was not the case with 'teeth baring', 'tongue licking' or 'ear position' (Fatjó et al. 2012). Thus there seems to be no reason to categorise these behaviours as being 'dominant' or 'submissive'. It is also interesting to note that these authors refer to 'ambivalence' when the interacting individuals show a mix of 'dominant' and 'submissive' signals. However, it is not clear how the concept of 'ambivalence' helps in explaining the complex nature of interaction: is 'ambivalence' an individual character or a consequence of the interaction? Based on an analogy to humans, ambivalence may indicate the congruent activation of approach and distancing tendencies in the specific situation ('hesitation') and/or characterise an individual who is generally uncertain (slow) in making decisions. A further possibility mentioned by Fatjó et al. (2007) is that 'ambivalent' behaviour acts as a signal and may affect the behaviour of the other.

It is clear that our tendency to anthropomorphise and the fact that these anthropomorphic categories can be sometimes used efficiently in practice (e.g., Wemelsfelder 2007), strongly hinder the development of purely neutral description of dog behaviour. It is important to be aware of the possible pitfalls of relying on the 'human instrument' for observing and measuring behaviour (see Horowitz and Hecht, this volume).

8.5.3 Toward a More Objective Measure of Dog Behaviour

Theoretically, many of the above concerns could be alleviated if (at least part of) the behavioural measures could be made without direct human involvement. Recent developments in technology have led to many new ways of recording and quantifying animal and human behaviour. One solution to the problem is provided by miniaturised sensors which can be put on the animals' bodies and measure the direction of displacement, speed and acceleration in three-dimensional space. These new tools allow for an unprecedented approach to behaviour because they can measure features of the movement that had been not available before (e.g., intensity/velocity of movement), and also offer a more objective categorisation of actions. Although this latter feature has not been used often, recent studies have shown that this so called bio-logging method can be applied to dog behaviour. Dog-attached accelerometers were applied in order to record locomotor activity (Preston et al. 2012), consequential maintenance energy requirements (Wrigglesworth et al. 2011), and gait patterns (Barthélémy et al. 2009).

Gerencsér et al. (2013) developed a new sensor combination (GPS, accelerometer and gyroscope) in order to measure different behaviours in freely moving dogs. This small device can be placed on the dogs' harness without having any effect on the overall movement pattern. At first, the dogs were walked on a predetermined path and then a human coder used six behaviour categories (lay, sit, stand, walk, trot, canter, gallop) to describe the behaviour of the dog. Next, by the means of a learning algorithm the system was taught to recognise the respective behaviours based on the inputs from the sensors. Following this training the system was tested by using new recordings from the same dogs or other dogs. It could recognise these six behaviours on average with 80–90 % fidelity (Gerencsér et al. 2013). As expected the device performed best when the training and testing was done on the sensor data obtained from the same dog, but very reliable detecting rates were also reported if the system was trained and tested on different breeds (Belgian Malinois and Labrador retriever). The performance of the system could also be improved by including more dogs for the training.

Although at present the bio-logging systems are able only to record simple actions, this can already save time and allow the researcher to concentrate on the finer aspects of the behaviour. The miniaturisation of the sensors and the incorporation of other technological innovations could make the recognition of more complex patterns also possible in the not-too-distant future.

Other specific applications are also available which allow researchers to look at behaviour at a finer level than previously. For example, such service is provided by the 'eye tracker' which follows dogs' eye movement in space (Somppi et al. 2011) (see also Rossi, Smedema, Parada, and Allen, this volume). The general assumption is that the looking pattern indicates visual attention including expectancies about future events. In line with this Téglás et al. (2012) found that dogs displayed a higher rate of gaze-following after the human addressed them in a communicative way. Using this method will allow researchers to get closer to the underlying cognitive processes that control behaviour that would be impossible by using only naturalistic observations.

8.5.4 Direction of Future Research on Dog Behaviour

The behavioural study of dogs is a growing field and will continue to grow. Researchers should be prepared for the accumulation of huge piles of data on various aspects of dog behaviour. This information will be most useful if researchers strive to establish common guidelines for the collection of behaviour data. Given the diversity of approaches ranging from psychology to ethology—and including animal welfare and applied aspects of dog behaviour and dog training—this task is not easy. But the ethological method for describing behaviour at the level of movement pattern may provide a common ground.

Sharing data could be an important step in this direction. But any sharing is meaningful only if researchers are clear about what they actually share. This is the case in the study of genetics, where researchers share a sequence of nucleic acids and the units are well-defined chemical compounds (adenine, thymine, glutamine, and cytosine). However, it makes little sense to share pure numbers representing behaviours (in terms of frequencies, durations) when the definition of the behaviour has not been agreed upon. A few years ago Kampis et al. (2010) developed a web-based system that facilitates the exchange of videos among students of animal behaviour (http://www.cmdbase.org). This system has the further advantage that the videos can be edited, linked to pdf files, and complemented by written remarks. A well-edited video sequence allows the viewer to observe the dog's behaviour in context, in addition to understand the experimental method and the particular details of the investigation. Viewing each other's recordings and interpretations could help to reveal an optimal solution for providing a general framework of describing and defining dog behaviour.

Acknowledgements A.M. is grateful for the support of the Hungarian Academy of Sciences (MTA 01 031), and the European Science Foundation (ESF Research Networking Programme titled "The Evolution of Social Cognition: Comparisons and integration across a wide range of human and non-human animal species"). C.F. is grateful for the support of the APDT Foundation and for the fellowship of the Hungarian Scholarship Board.

References

Ainsworth, M. D. S. (1969). Object relations, dependency and attachment: A theoretical review of the infant-mother relationship. *Child Development, 40*, 969–1025.
Barthélémy, I., Barrey, E., Thibaud, J., Uriarte, A., Voit, T., et al. (2009). Gait analysis using accelerometry in dystrophin-deficient dogs. *Neuromuscular Disorders, 19*, 788–796.
Bateson, P. & Martin, P. (1986). *Measuring behaviour*. Cambridge: Cambridge University Press.
Bradshaw, J. W. S., Blackwell, E. J., & Casey, R. A. (2009). Dominance in domestic dogs—useful construct or bad habit? *Journal of Veterinary Behavior, 4*, 135–144.
Coppinger, R. P. & Coppinger, L. (2001). *Dogs: A startling new understanding of canine origin, behavior and evolution*. Chicago: University of Chicago Press.
Fatjó J., Feddersen-Petersen, D., Ruiz de la Torre, J-R., Amat, M., Mets, M., Braus, B, & Manteca, X. (2007). Ambivalent signals during agonistic interactions in a captive wolf pack. *Applied Animal Behavior Science, 105*, 274–283.
Feuerbacher, E., & Wynne, C. D. L. (2012). Relative efficacy of human social interaction and food as reinforcers for domestic dogs and hand-reared wolves. *Journal of the Experimental Analysis of Behavior, 98*, 105–129.
Fondon, J. W. & Garner, H. R. (2004). Molecular origins of rapid and continuous morphological evolution. In *Proceedings of the National Academy of Sciences of the USA, 28*, 18058–18063.
Frank, M. G. & Frank, H. (1988). Food reinforcement versus social reinforcement in timber wolf pups. *Bulletin of the Psychonomic Society, 26*, 467–468.
Fugazza, C., & Miklósi, A. (2013). Deferred imitation and declarative memory in domestic dogs. Animal Cognition (on line) doi:10.1007/s10071-013-0656-5
Fugazza, C., & Miklósi, A. (submitted). Should old dog trainers learn new tricks? The effectiveness of the Do as I Do and Shaping/Clicker training method in dog training
Gácsi, M., Maros, K., Sernkvist, S., Faragó, T., & Miklósi, Á. (2013). Human analogue safe haven effect of the owner: Behavioural and heart rate response to stressful social stimuli in dogs. *PLoS ONE, 8*(3), e58475
Gácsi, M., Győri, B., Virányi, Z., Kubinyi, E., Range F. Belényi, B., et al. (2009a). Explaining dog wolf differences in utilizing human pointing gestures: Selection for synergistic shifts in the development of some social skills. *PLoS ONE, 4*(8), e6584
Gácsi, M., McGreevy P., Kara, E., & Miklósi, A. (2009b). Effects of selection for cooperation and attention in dogs. *Behavioral and Brain Functions, 5*, 31

Gácsi, M., Topál, J., Miklósi, Á., Dóka, A., & Csányi, V. (2001). Attachment behaviour of adult dogs (*Canis familiaris*) living at rescue centres: Forming new bonds. *Journal of Comparative Psychology, 115*, 423–431

Gazit, I., & Terkel, J., (2003). Explosives detection by sniffer dogs following strenuous physical activity. *Applied Animal Behaviour Science, 81*, 149–161

Gerencsér, L., Vásárhelyi, G., Nagy, N. Vicsek, T., & Miklósi, Á. (2013). Identification of behaviour in freely moving dogs (*Canis familiaris*) using inertial sensors. *PLoS*

Hare, B., Brown, M., Williamson, C., & Tomasello, M. (2002). The domestication of social cognition in dogs. *Science, 298*, 1634–1636.

Hare, B., Rosati, A., Kaminski, J., Brauer, J., & Call, J. (2009). The domestication hypothesis for dogs' skills with human communication: A response to Udell et al. (2008) and Wynne et al. (2008). *Animal Behaviour, 47*, e1–e6.

Hegedüs, D., Bálint, A., Miklósi, Á., & Pongrácz, P. (2013). Owners fail to influence the choices of dogs in a two-choice, visual pointing task. *Behaviour, 150*, 427–443.

Hirsch-Pasek, K., & Treiman, R. (1981). Doggerel: Motherese in a new context. *Journal of Child Language, 9*, 229–237.

Horn, L., Range, F., & Huber, L. (2013). Dogs' attention towards humans depends on their relationship, not only on social familiarity. *Animal Cognition, 16*, 435–443.

Hooff, van J. A. R. A. M., & Wensing, J. (1987). Dominance and its behavioral measures in a captive wolf pack. In H. Frank (ed.), *Man and wolf: Advances, issues and problems in captive wolf research*, pp. 219–252. Junk, Dordrecht.

Horowitz, A., Hecht, J., & Dedrick, A. (2013). Smelling more or less: Investigating the olfactory experience of the domestic dog. *Learning and Motivation, 44*, 207–217.

Huber, L., Range, F., Voelkl, B., Szucsich, A., Virányi, Z., Miklósi, A. (2009). The evolution of imitation: what do the capacities of non- human animals tell us about the mechanisms of imitation? *Philosophical Transactions of the Royal Society B, 364*, 2299–2309.

Irion, D. N., Schaffer, A. L., Famula, T. R., Eggleston, M. L., Hughes, S. S., & Pedersen, N. C. (2003). Analysis of genetic variation in 28 dog breed populations with 100 microsatellite markers. *Journal of Heredity, 94*, 81–87.

Johnen, D., Heuwieser, W., & Fischer-Tenhagen, C. (2013). Canine Scent Detection–Fact or Fiction? *Applied Animal Behaviour Science, 148*, 201–208.

Kaminski, J., Call, J., & Fischer, J. (2004). Word learning in the domestic dog: Evidence for 'fastmapping'. *Science, 304*, 1682–1683.

Kaminski, J. Tempelmann, S. Call, J., & Tomasello, M (2009). Domestic dogs comprehend human communication with iconic signs. *Developmental Science, 12*, 831–837.

Kampis, G. Y., Miklósi, Á., Virányi Z., & Gulyás, L. (2010). Video deep tagging and data archiving in the comparative mind database. *Proceedings of Measuring Behavior 2010 (Eindhoven, The Netherlands, August 24–27)*, pp. 185–188.

Kazdin, A. E. (1982). *Single-case research designs*. Oxford: Oxford University Press.

Klinghammer, E. & Goodman, P.A. (1987). Socialization and management of wolves in captivity. In Frank, H., (Ed.), *Man and wolf* (pp. 31–61). Dordrecht: Junk Publishers.

Lakatos, G., Soproni, K., Dóka, A., & Miklósi, Á. (2009). A comparative approach to dogs' (*Canis familiaris*) and human infants' comprehension of various forms of pointing gestures. *Animal Cognition, 12* 621–631.

Larson, G., Karlsson, E. K., Perri, A., Webster, M. T., Ho, S. Y. W., et al. (2012). Rethinking dog domestication by integrating genetics, archeology, and biogeography. *PNAS, 109*, 8878–8883.

Levitis, D. A., Lidicker Jr., W. Z., & Freund, G. (2009). Behvioural biologists do not agree on what constitutes behaviour. *Animal Behaviour, 78*, 103–110.

Lit, L., Schweitzer, J. B. & Oberbauer, A. M. (2011). Handler beliefs affect scent detection dog outcomes. *Animal Cognition, 14*, 387–394.

Lord, K. (2012). A comparison of the sensory development of wolves (*Canis lupus lupus*) and dogs (*Canis lupus familiaris*) *Ethology, 118*, 1–11

Mariti, C., Ricci, E., Carlone, B., Moore, J.L., Sighieri, C., & Gazzano, A. (2012). Dog attachment to man: A comparison between pet and working dogs. *Journal of Veterinary Behavior, 8*, 135–145

Marshall-Pescini, S., Valsecchi, P., Petak, I., Accorsi, P. A., & Prato Previde, E. (2008). Does training make you smarter? The effects of training on dogs' performance in a problem solving task. *Behavioural Processes, 78*, 449–454.

Marshall-Pescini, S., Passalacqua, C., Barnard, S., Valsecchi, P., & Prato-Previde, E. (2009). Agility and search and rescue training differently affects pet dogs' behaviour in sociocognitive tasks. *Behavioural Processes, 81*, 416–422.

McGreevy, P.D., Starling, M., Branson, N. J., Cobb, M. L., & Calnon, D. (2012). An overview of the dog-human dyad and ethograms within it. *Journal of Veterinary Behavior, 7*, 103–117. doi:10.1016/j.jveb.2011.06.001

Miklósi, Á., Kubinyi, E., Topál, J., Gácsi, M., Virányi, Z., & Csányi, V. (2003). A simple reason for a big difference: Wolves do not look back at humans but dogs do. *Current Biology, 13*, 763–766.

Miklósi, Á., Topál, J., & Csányi, V. (2004). Comparative social cognition: What can dogs teach us? *Animal Behaviour, 67*, 995–1004.

Miklósi, A. (2007). *Dog behaviour evolution and cognition.* Oxford: Oxford University Press.

Miklósi, A. (2014). *Dog behaviour evolution and cognition.* Oxford: Oxford University Press

Miklósi, A., & Topál, J. (2011). On the hunt for the gene of perspective taking: Pitfalls in methodology. *Learning and Behaviour, 39*, 310–313.

Mitchell, R. W. (2001). Americans' talk to dogs: Similarities and differences with talk to infants. *Research on Language and Social Interactions, 34*, 183–210.

Mongillo, P., Bono, G., Regolin, L., & Marinelli, L. (2010). Selective attention to humans in companion dogs, Canis familiaris. *Animal Behaviour, 80*, 1057–1063.

Overall, K. L. & Love, M. (2001). Dog bites to humans—dash demography, epidemiology, injury, and risk. *Journal of the American Veterinary Medical Association, 218*, 1923–1934.

Packard, J. M. (2003). Wolf behaviour: Reproductive, social and intelligent. In Mech, D. and Boitani, L., (Eds.), *Wolves: Behavior, Ecology and Conservation* (pp. 35–65). Chicago: University of Chicago Press.

Pal, S. K., Ghosh, B., & Roy, S. (1998). Dispersal behaviour of free-ranging dogs (*Canis familiaris*) in relation to age, sex, season, and dispersal distance. *Applied Animal Behaviour Science, 62*, 123–132.

Pal, S. K. (2010). Play behaviour during early ontogeny in free-ranging dogs (*Canis familiaris*). *Applied Animal Behaviour Science, 126*, 140–153.

Pal, S. K. (2008). Maturation and development of social behaviour during early ontogeny in free-ranging dog puppies in West Bengal, India. *Applied Animal Behaviour Science, 111*, 95–107

Parker, H. G. (2012). Genomic analyses of modern dog breeds. *Mammalian Genome, 23*, 19–27

Passlacqua, C., Marshall-Pescini, S., Barnard, S., Lakatos, G., Valsecchi, P., & Prato Previde, E. (2011). Human-directed gazing behaviour in puppies and adult dogs, Canis lupus familiaris. *Animal Behaviour, 82*, 1043–1050

Pfungst, O. (1911). Clever hans (The horse of Mr. von Osten): A contribution to experimental animal and human psychology (Trans. C.L. Rahn). New York, NY: Henry Holt. (Originally published in German, 1907)

Pilley, J. W., (2013). Border collie comprehends sentences containing a prepositional object, verb, and direct object. *Learning and Motiviation, 44*, 229–240.

Pongrácz, P., Miklósi, Á., Vida, V., & Csányi, V. (2005). The pet-dogs' ability for learning from a human demonstrator in a detour task is independent from the breed and age. *Applied Animal Behaviour Science, 90*, 309–323.

Pongrácz, P., Gácsi, M., Hegedüs, D., Péter, A., & Miklósi, Á. (2013a). Test sensitivity is important for detecting variability in pointing comprehension in canines. *Animal Cognition, 16*, 721–735.

Pongrácz, P., Hegedűs, D., Sanjurjo, B., Kővári, A., & Miklósi, Á. (2013b). "We will work for you"—Social influence may suppress individual food preferences in a communicative situation in dogs. *Learning and Motivation, 44*, 270–281.

Prato-Previde, E., Fallani, G., & Valsecchi, P. (2006). Gender differences in owners interacting with pet dogs: An observational study. *Ethology, 112*, 64–73.

Prato-Previde, E., Marshall-Pescini, S., & Valsecchi, P. (2008). Is your choice my choice? The owners' effect on pet dogs' (*Canis lupus familiaris*) performance in a food choice task. *Animal Cognition, 11*, 167–174

Preston, T., Baltzer, W., & Trost, S. (2012). Accelerometer validity and placement for detection of changes in physical activity in dogs under controlled conditions on a treadmill. *Research in Veterinary Science, 93*, 412–416

Schenkel, R. (1947). Ausdrucks-Studien an Wölfen. *Behaviour*, 81–129

Scott, J. P. & Fuller, J. L. (1965). *Genetics and the social behaviour of the dog*. University of Chicago Press, Chicago

Somppi, S., Törnqvist, H., Hänninen, L., Krause, C., & Vainio, O. (2011). Dogs do look at images: Eye tracking in canine cognition research. *Journal of Veterinary Behaviour: Clincal applications and Research, 6*, 64–65.

Svartberg (2005). A comparison of behaviour and everyday life: Evidence of three consistent boldness-related personality traits in dogs. *Applied Animal Behaviour Science, 91*, 103–128.

Szetei, V., Miklósi, Á., Topál, J, & Csányi, V. (2003). When dogs seem to lose their nose: An investigation on the use of visual and olfactory cues in communicative context between dog and owner. *Applied Animal Behaviour Science, 83*, 141–152.

Tami, G. & Gallagher (2009). Description of the behaviour of domestic dog (Canis familiaris) by experienced and inexperienced people. *Applied Animal Behaviour Science, 120*, 159–169.

Téglás, E., Gergely, A., Kupán, K., Miklósi, Á., & Topál, J. (2012). Dogs' gaze following is tuned to human communicative signals. *Current Biology, 22*, 209–212.

Tinbergen, N. (1963). On aims and methods of ethology. *Zeitschrift fur Tierpsychologie, 20*, 410–433.

Topál, J., Miklósi, Á., & Csányi, V. (1997). Dog-human relationship affects problem solving behaviour in the dog. *Anthrozoös, 10*, 214–224.

Topál, J., Miklósi, Á., & Csányi, V. (1998). Attachment behaviour in dogs: A new application of Ainsworth's (1969) Strange Situation Test. *Journal of Comparative Psychology, 112*, 219–229.

Topál, J., Byrne, R.W., Miklosi, A., & Csányi, V. (2006). Reproducing human actions and action sequences: "Do as I do!" in a dog. *Animal Cognition, 9*, 355–367.

Topál, J., Miklósi, Á., Gácsi, M., Dóka, A., Pongrácz, P., Kubinyi, E., et al. (2009). The dog as a model for understanding human social behavior. *Advances in the Study of Animal Behaviour, 39*, 71–116.

Udell, M. A. R., Dorey, N. R., & Wynne, C. D. L. (2008). Wolves outperform dogs in following human social cues. *Animal Behaviour, 76*, 1767–1773.

Udell, M. A. R., Dorey, N. R., & Wynne, C. D. L. (2010). The performance of stray dogs (*Canis familiaris*) living in a shelter on human-guided object-choice tasks. *Animal Behaviour, 79*, 717–725.

Udell, M. A. R., & Wynne, C. D. L. (2011). Reevaluating canine perspective-taking behavior. *Learning and Behaviour, 39*, 318–323.

Utrata, E., Viranyi, Z., & Range, F. (2012). Quantity discrimination in wolves (*Canis lupus*) Frontiers in Psychology. doi:10.3389/fpsyg.2012.00505

Virányi, Z., Gácsi, M., Kubinyi, E., Topál, J., Belényi, B., Ujfalussy, D., et al. (2008). Comprehension of human pointing gestures in young human-reared wolves (*Canis lupus*) and dogs (*Canis familiaris*). *Animal Cognition, 11*, 373–387.

Wan, M., Kubinyi, E., Miklósi, Á., & Champagne, F. (2009). A cross-cultural comparison of reports by German Shepherd owners in Hungary and the United States of America. *Applied Animal Behaviour Science, 121*, 206–213.

Wemelsfelder, F. (2007). How animals communicate quality of life: The qualitative assessment of animal behaviour. *Animal Welfare, 16*, 25–31.

Wynne, C. D. L., Udell, M. A. R., & Lord, K. A. (2008). Ontogeny's impact on human-dog communication. *Animal Behaviour, 76*, e1–e4.

Wrigglesworth, D., Mort, E., Upton, S., & Miller, A. (2011). Accuracy of the use of triaxial accelerometry for measuring daily activity as a predictor of daily maintenance energy requirement in healthy adult Labrador retrievers. *American Journal of Veterinary Research, 72*, 1151–1155.

Chapter 9
Looking at Dogs: Moving from Anthropocentrism to Canid *Umwelt*

Alexandra Horowitz and Julie Hecht

Abstract As a companion to humans, the domestic dog is naturally interpreted from a human-centered (anthropocentric) perspective. Indeed, dog behavior and actions are often explained by using anthropomorphisms: attributions to the dog that would hold if the actor were human. While sometimes useful, anthropomorphisms also have the potential to be misleading or incorrect. In this chapter we describe work to replace an anthropocentric perspective with a more dog-centered research program. First we detail research systematically testing anthropomorphisms of emotional complexity—the appearance of guilt and jealousy—that are made of dogs, by testing the context of appearance of the "guilty look" and by testing advantageous and disadvantageous inequity aversion. Relatedly, we describe research looking at the contribution of specific dog physical attributes to human preference and anthropomorphizing. Finally, we identify anthropocentric and canid-centric elements of our own and others' research, and suggest ways that research can be more sensitive to the dog's *umwelt*.

9.1 Introduction

The domestic dog is once again a subject of science. In the twentieth century, dogs were most often recognized in behavioral science for their role in Pavlov's development of conditioning theory. By contrast, the new study of dogs,

---

A. Horowitz (✉) · J. Hecht
Dog Cognition Lab, Department of Psychology, Barnard College,
Columbia University, New York, NY 10027, USA
e-mail: ahorowitz@barnard.edu

J. Hecht
Department of Psychology, The Graduate Center, The City University of New York,
New York, NY 10016, USA

"dog cognition" research, views dogs less as a neural system exemplifying a learning process than as a species itself of interest for its skills.

Canis familiaris is unique among psychological and ethological scientific subjects for its ubiquity in Western culture and households. And this ubiquity itself affects how the dog has been investigated: for in this research with dogs, the subjects are most often *owned* dogs, household pets, whose social group is as much humans as it is conspecifics. This fact distinguishes them from almost all other research subjects, in lab, farm, or field; either domesticated or not.

In part for that reason, dogs are a subject about which much information might appear to be known already: far from exotic and unknown, the species is familiar, recognizable, and navigates a human-centered world daily. The primary interlocutors with dogs are dog owners, who manage what is often one or two decades of living together with relatively few mishaps (Sanders 2003). Owners' alleged "understanding" of their dogs is the backdrop for new scientific studies of dogs. While scientists approach the dog *ab initio*, with no assumptions other than species identification and genus affiliation, owners typically make regular assertions about what their dogs, or even all dogs, can understand, see, experience, want, believe, and know.

A scientific approach to animals calls these assertions likely *anthropomorphisms*—that is, claims that a nonhuman animal has attributes characteristic of (and only proven of) human animals. As a matter of method and philosophy, most researchers try to avoid rampant anthropomorphisms. While any particular claim about another species' abilities may indeed be correct, the determination of the claim's correctness is considered an empirical matter, not one that can simply be asserted. To anthropomorphize, then, is to make an unfounded claim. Neither is it explanatory (Wynne 2007).

Nonetheless, insofar as anthropomorphisms represent simply the human perspective on, or way of seeing, non-humans, some degree of anthropomorphizing may be inevitable (Horowitz and Bekoff 2007). Indeed this is just one kind of anthropocentrism that is characteristic of and in some cases definitional of behavioral science. With this in mind much of the research in our lab has been to address and redress anthropocentrism by (a) directly addressing anthropomorphisms made of dogs, (b) determining what prompts specific anthropomorphisms to begin with, and (c) developing methods for a more dog-centered research approach.

This chapter briefly reviews the history of use of anthropomorphisms, in behavioral science and pre-scientifically. We then describe contemporary attitudes about these assertions, as well as their investigated effects on the dog-human dyad. Having introduced the topic conceptually, we review empirical studies largely from our lab (the Horowitz Dog Cognition Lab at Barnard College) exploring the appropriateness of specific attributions of emotions made to dogs. As a complement, we describe research into why we anthropomorphize.

Relatedly, in the final section we review recent dog cognition research with this anthropocentric/*umwelt* lens: both highlighting methodological elements which are anthropocentric in approach, and could be re-considered, and also methods attentive to the *umwelt* of the dog.

9.2 Anthropomorphisms

Anthropomorphisms pre-date contemporary scientific and lay attitudes to animals; representations in Paleolithic art, forty thousand years ago, have been described as anthropomorphic (Mithen 1996). For millennia, weather and even gods have been anthropomorphized. Anthropomorphisms may stem from the discovered usefulness of using what one knows about oneself to predict others' behavior (Serpell 2003). In that way, anthropomorphism might have proved valuable to our human ancestors trying to anticipate the behavior of, for instance, predators by projecting human emotions and motivations onto them. The contemporary scientific attitude toward these attributions begins, interestingly, with the figure otherwise a radical innovator of our view of animals: Charles Darwin.

Characteristically, in his *The expression of emotions in man and animals*, Darwin was untroubled to claim that back-scratches "pleased" cats, to have seen birds faint when "terrified", and to assert that dogs "pretend" to fight during play (Darwin 1872/1979). His contemporary George Romanes followed suit, explaining his attributions by noting: "…we are justified in inferring particular mental states from particular bodily actions" (Romanes 1883). While a behaviorist backlash was immediate, by the mid-twentieth century a more nuanced approach was emerging, noting the value, for instance, in using anthropomorphic language to describe captive primate behavior (Hebb 1946). Similarly, observers have shown high agreement in qualitative assessments of pig expressions, an indication that even "subjective" descriptions may transcend subjects (Wemelsfelder et al. 2000). More recently, a new study of anthropomorphism itself has emerged—both assessing attributions empirically, and aiming to provide a psychological account of anthropomorphic beliefs.

Seminal work has shown that humans more readily anthropomorphize that which bears a real or superficial similarity to themselves, physically as well as behaviorally (Heider and Simmel 1944; Mitchell and Hamm 1996). Subjects more readily ascribe mental states to objects whose movements appear similar to typical human motion (Morewedge et al. 2007). Additionally, people more readily ascribe complex cognitive abilities to other primates as well as companion animal species, like dogs and cats, than to non-primates and non-pets (Eddy et al. 1993).

The relationship between dogs and humans is indeed special: while farm animals are described as "managed" or "handled" (Hemsworth 2003), for many dog owners the relationship is more familial than managerial, with some viewing their charges as "children" or "fur babies" (Greenebaum 2004). Many caretakers ask their companion dogs to participate in society as if they too were human, dressing them in clothing, staging pet "weddings", and maintaining social media accounts in their name. All are unnecessary for the dog's role as a social canid. Modern companion dogs are members of a new class readily viewed through an anthropomorphic lens.

The fluidity and ease with which dogs move alongside humans contributes to the anthropocentric light in which they are considered. In one study in which people were asked to describe a scene between a dog and a person, subjects largely used "psychological" descriptors (Morris et al. 2000). These attributions, particularly that of "mindedness," carry into individuals' ordinary interactions with dogs. Owners' claims of their dogs' understanding and thought-processes are often transposed thoughts from their own heads. For instance, an owner explaining how, on a rainy day, her dogs will "just put one foot outside the door and then go over to where the cookies are kept [so as to say]: 'Well, technically we went out'" is not an unusual statement (Sanders 1993).

9.2.1 Testing of Anthropomorphisms

One line of work in our lab examines anthropomorphisms of animals. First, we ask whether attributions of secondary emotions made to dogs are well founded. Two studies are described below. Both emanate, in part, from evidence that the great majority of dog owners believes that their dogs experience secondary emotions: three-quarters (74 %) say that their dogs experience guilt; and 81 % believe their dogs experience jealousy (Morris et al. 2008). These attributions are of interest, as dogs' long association with and selection by humans suggests that they may display a rudimentary sense of morality (Bekoff 2004); similarly, the history of domestic dogs suggests that the species might have a highly developed sense of what is called "fairness" or even "justice" in primate literature (Brosnan 2013). Second, we describe research looking at the contribution of dog physical appearance to human anthropomorphizing.

9.2.1.1 Guilt

Attributing "guilt" to dogs has long standing: even the great animal observer Konrad Lorenz (1954) wrote of the dog's "bad conscience" on doing a misdeed. The claim is not simply that a dog looks guilty, but that he may feel guilty, having done something "wrong" or which breaks household rules (Sanders 1993). The attribution appears to be based on the recognizable "guilty look" of the dog, especially when this behavior is noticed around the time of an act which has been forbidden (or in the vicinity of evidence of that past act). The "guilty look", according to owner and behaviorist report, includes a combination of any of the following behaviors: avoiding eye contact (Darwin 1872/1979), rolling over, offering a paw (Lorenz 1954), retreating from the owner (Cheney and Seyfarth 2007; Whitely 2006), holding the tail low and head or ears back (McConnell 2006; de Waal 1997), tongue-flicking, freezing, and a low wag.

Thus, to test the connection of these behaviors associated with the guilty look (hereafter, *ABs*) to *actual guilt*—that is, a misdeed—a simple two-by-two design was used, where the appearance of ABs was the dependent variable (Horowitz 2009a). The experimenter recruited fourteen dog-owner dyads (subjects dogs: 6 m,

8 f; mean age 2:6 (range 0:8–9:0); 6 mongrels, 8 purebreds) and performed the experiment in their homes. The owners were told to instruct their dog not to eat a desired treat, and then the owners left the room. The dogs (videotaped throughout the owners' admonishment, absence, and later return) had the opportunity during the owners' absence to eat the forbidden treat.[1] When each owner returned to the room, she was told whether or not her dog "obeyed" the instruction, and to greet the dog if so, or respond with scolding if not. Each dyad participated in four trials, over which two conditions were varied: dog obedience and owner response. In the first condition, the subject either disobeyed (ate the treat) or obeyed (did not eat the treat). In the second condition, the owner was either informed that the dog had or had not eaten the treat, and thus responded to the dog by greeting or scolding, respectively.

Crucially, the report the owner received of the dog's behavior was not always a true report of the dog's behaviour. In two of four trials the owner was misinformed about the dog's behavior: told that the dog had obeyed when in fact the dog had eaten the treat, and told that the dog had disobeyed when he had not.

The subject dogs' behavior was coded from the videotape for number of ABs after owner return in each trial. There was no significant main effect of the dogs' obedience on the number of ABs ($F(1, 13) = 1.59, p = 0.23, r = 0.33$). This indicates that the rate of ABs was similar whether the dogs ate the treat or did not eat the treat: whether each dog was "guilty" or "not guilty" of violating his owner's command. There *was* a significant main effect of the response of the owner on ABs ($F(1, 13) = 29.22, p < .001, r = 0.83$). Scolding the dog led to significantly more ABs than greeting the dog, whether the dog had obeyed the owner's command or was guilty of violating the command. These two results indicate that ABs were a response to owner scolding more than to the dog's own actions. Interestingly, there was a significant interaction effect between the obedience trials and the owner's response ($F(1, 13) = 5.69, p = 0.03, r = 0.55$): scolding when the dog had not eaten the treat led to the most ABs.

From this study we can conclude that the hypothesis that ABs, the "guilty look", increase when a dog disobeys, is not borne out. Instead, *owner scolding behaviour* caused an increase in the guilty look. Thus, when the dogs were scolded but "innocent", the ABs they displayed were not a reflection of dog guilt but of an owner's *perception* of dog guilt. Since many ABs—such as rolling over, tucking the tail between the legs, and pressing the ears back—are submissive behaviors (Darwin 1872/1979), the ostensible "guilty" look may simply be a look of anticipation of, and attempted avoidance of, punishment (Horowitz 2009a).

In an extension of this experiment, research examined the anecdotal claim that dogs display a "guilty look" to non-scolding owners who are ignorant of any misdeed (Hecht et al. 2012). Thus the question is raised whether owners are

---

[1] Disobedience was assured by the treat being offered to the dog by the experimenter after the owner left the room, and obedience, by the treat's immediate removal. In pilot trials, no difference was seen in the dogs' behavior whether they ate the treat because it was provided by the experimenter or of their own accord.

misremembering (or selectively remembering) past incidents or whether it was the misdeed, not the scolding, that prompted dogs' "guilty look." Here, again, in a related paradigm, the rate of ABs did not rise with disobedience (n = 52; z = −1.512, p = 0.131). Additionally, owners could not determine based on dog behavior alone whether their dog had obeyed or not (p = 0.623).

Thus the attribution of an experience of "guilt" to dogs, based on the "guilty look", is unfounded, a pure anthropomorphism. Of course, these results do not indicate that domestic dogs *do not experience* guilt, only that the "guilty look" is not indication of it. Instead, humans misinterpret a learned or instinctive submissive response as one indicating much more awareness than we have evidence of. Importantly, this mis-attribution could be harmful to dogs if their owners have expectations that the dogs *do* understand rules, correct behavior, and so forth, and believe that dogs either willfully or neglectfully violate these rules.

9.2.1.2 Fairness

According to Morris et al. (2008), "jealousy" is the most-often attributed secondary emotion to dogs. Dogs express jealousy, owners report, when they observe more attention being given to someone else than to themselves. In other words, the attribution of "jealousy" is founded on the observation of the dog's noticing an *unfair distribution of attention*. Our lab next tackled whether in fact, a dog has a sense of "fairness". Certainly related phenomena, such as cooperation, also critical to moral experiences (Brosnan 2006; de Waal 1997), are likely to have been part of the evolutionary history of dogs. To go by the behavior of their relatives today, dogs' progenitors built and maintained intraspecific social relationships in part through cooperative interactions, such as when taking down large prey (Mech 1970; Schenkel 1967). Dogs cooperate with conspecifics—as during play (Bekoff 2004)—and attend to one another's behavior, even displaying evidence of social learning (Pongrácz et al. 2008; Slabbert et al. 1997). The dog-human relationship results, in part, from selection for cooperative behavior—most visibly, when dogs assist the blind or engage in other synchronous activities (Gácsi et al. 2009; Naderi et al. 2001).

Given dogs' sociability, ability to coordinate behavior and cooperate with humans and conspecifics, and ability to distinguish notable quantity differences (Ward and Smuts 2007; West and Young 2002), it is reasonable to inquire whether dogs are sensitive to unequal reward distribution.

An initial study by Range et al. (2009) investigated "inequity aversion" in dogs, a model used in studies of fairness with non-human primates. The primate research found that subjects stop participating in cooperative problem-solving tasks after observing a conspecific receive a better reward for the same effort expenditure (Brosnan and de Waal 2003). In this study of *disadvantageous* inequity aversion with dogs—wherein a subject receives less of some reward than another individual—two dogs sat next to one another and were asked to "give a paw" for a reward. When the subject dog was not rewarded for performing the requested behavior, the subject stopped performing more quickly if the other dog continued to be rewarded than when the subject was alone (Range et al. 2009).

A sense of fairness, however involves more than changing one's behavior after being treated unfairly (disadvantageous inequity); it also involves changing one's behavior when others are treated unfairly (*advantageous* inequity).

While advantageous inequity is rare, even in non-human primates, our lab's investigation of inequity aversion in companion dogs included both these aspects (Horowitz 2012). The method was extrapolated from studies of justice in humans (Pritchard et al. 1972). In each trial, subject dogs (n = 38) received the same amount of reward, and the amount given to the control dog varied. Subject and control dogs approached and were familiarized with two trainers who offered different amounts of a reward: one provided both dogs with equal food quantities for performing the behavior "sit," and the second rewarded the dogs unequally, either over-rewarding (3 pieces of food) or under-rewarding (no pieces of food) the control dog. Subject dogs then chose which trainer to approach by themselves, a "fair" trainer or one of the "unfair" trainers.

In the under-rewarding trial, when the trainer gave the control dog less (no) food, subjects were equally likely to choose the fair (16; 48.5 %) and the unfair (17; 51.5 %) trainer ($c^2(1) = 0.03, p = 0.86$). But in the over-rewarding trial, where the trainer gave the control dog more food, subjects (25; 78.1 %) chose the unfair trainer over the fair ($c^2(1) = 10.13, p = 0.001$), a counter-intuitive result if dogs are concerned with fairness. Ultimately, dogs were less influenced by how they, or another dog, were treated—seeming to ignore the ethics of both advantageous and disadvantageous inequity aversion—and were instead more concerned with which trainer was doling out more food overall.

The research to date suggests that dogs do not consistently show sensitivity to disadvantageous inequitable situations and do not consistently attend to advantageous inequity. The concepts of *fairness*, or even *justice*, make more functional sense in description of cultures with societal norms and explicit rules, such as human culture. Dogs may be better described as *socially aware opportunists*. The question of dogs' experience of jealousy is outstanding, but this research indicates that the correlated behavior may simply be the combination of an attention to where resources are, and a frustration or excitement about attempting to secure them.

9.2.2 Physical Prompts to Anthropomorphisms

Not only the behaviors but also the anatomical—physical—features of dogs which prompt anthropomorphisms can be examined. More broadly, this kind of investigation is part of research on the relationship between physical traits of non-human animals and humans' preferences and attributions based upon them.

Humans value—and are often attracted to—objects which are similar to ourselves: the most favored—and well-protected—species have a decidedly human-like physical appearance and share a phylogenetic similarity to humans

Fig. 9.1 Stimuli with eyes smaller (*left*) and larger (*right*)

(DeKay and McClelland 1996; Kellert 1996). People are specifically interested in the way dogs look (Weiss et al. 2012), and dogs possess many of the elements which could make human-like attributions easy: large round eyes; distinct irises and visible sclera; discernible and flexible facial features; and the ability to approximate a smile, move limbs independently, and even cover or scratch one's face (Horowitz and Bekoff 2007). Value judgments and attributions to dogs are heavily influenced by dog appearance, which is not surprising given the recently proposed theory of anthropomorphic selection: "selection in favor of physical and behavioral traits that facilitate the attribution of human mental states to nonhumans" (Serpell 2003). Breed standards connect physical appearance and character attributions: for instance, the breed standard for the Great Pyrenees describes its expression as "elegant, intelligent and contemplative" (American Kennel Club 2013).

In our own lab we have investigated which specific dog physical qualities attract human attention (Hecht and Horowitz 2013). By presenting study participants with two nearly identical dog images, differing only in that one physical characteristic had been manipulated, we were able to determine people's avowed preference for particular physical attributes. Participants (n = 124) viewed the images for up to 15 s and selected their preferred image. Subjects showed a preference for some features of the "infant schema" (Lorenz 1950/1971), namely large eyes ($z = 3.5929, p < 0.001$) (Fig. 9.1) and a larger space between the eyes ($z = 4.986, p < 0.001$). They also showed a preference for the human-like attributes tested, colored irises ($z = 12.7583, p < 0.001$) and a distinct "smile" ($z = 5.4993, p < 0.001$) (Fig. 9.2). In other research, dog coat color and ear shape have been found to be associated with particular personality attributions (Fratkin and Baker 2013). For instance, an unfamiliar yellow-coated dog received higher ratings of agreeableness, conscientiousness and emotional stability than the same dog with a black coat. These studies demonstrate that as with dog behaviors, the contribution of elements of dog physical appearance to attributions made of them can be investigated empirically.

Fig. 9.2 Stimuli with no smile (*left*) and slight smile (*right*)

9.3 Anthropocentrism and Canid-Centrism

By contrast with owners, researchers of dogs, skilled in studying animal behavior, are generally quite sensitive to the possibility of anthropomorphizing their subjects. Study designs attempt to avoid making undue attributions. Some anthropocentric perspective is natural though, and quite common. Taking a step back from some now-familiar dog-cognition research can aid in seeing how insidious anthropocentrism can be.

For instance, by far the most common category of experiment with dogs examines their response to various human cues—such as facial expressions (Buttelmann and Tomasello 2013)—and especially behaviors—such as pointing (e.g., Soproni et al. 2002). While these studies appropriately recognize the salience, or significance, of humans in the dog's world, there are some limitations to their recognition of what the dog may perceive of the humans. The most straightforward limitation is in modality: is a visual cue likely to be the most salient cue for a species whose primary modality is olfactory? While the vision of the average dog allows him to see a pointing hand, might not the opening of the olfactory lode that is the human armpit be as or more informative? Questions also remain about exactly *what* the pointing gesture might mean for the dog. Is what we think of as a straightforward "point" an *informative* gesture to the dog, as commonly assumed, or a *command* (explored by Scheider et al. 2013)? The dog's interpretation might not match the human's label.

The above studies exemplify the kinds of missteps that can be made: by not considering the dog's *sensory abilities*, and by *labeling* a behavior (of human or dog) in a human way, instead of attempting to interpret dog behavior in a way sensitive to the dog's social and cognitive abilities. Disregarding, or ignorance of the different sensory abilities and constraints of non-human animals has been surprisingly prevalent in animal science. For instance, for many years researchers assumed that aposematic coloration of some animals allowed predators to see the color—but as it turns out, many of these animals' predators are effectively color blind, and may be seeing color contrast only (Rivas and Burghardt 2002). Below we focus on elements of two studies which make *assumptions about sensory abilities*: about dog vision and olfaction. Then we describe components of three studies which *prematurely label* dog behavior with human-relevant terms, disadvantageous to clear interpretation of data. Finally, a series of studies on one topic shows how over time, research has evolved, improving sensitivity to the subject and improving reliability of results.

9.3.1 Assumptions About Sensory Abilities

Because their vision is reasonably good, the particulars of dogs' visual system is often neglected in research studies. For instance, interesting work showing that dogs are more likely to steal forbidden food in "the dark" than in a "lit" room dismisses the idea that dog vision is different than human vision out of hand (Kaminski et al. 2013). But research on canid behavior in natural environments provides good evidence that these species see well—better than humans do—in the dark: many are crepuscular or nocturnal foragers or hunters, and are most active in these hours (Sillero-Zubiri et al. 2004). If their behavior is not entirely due to seeing, which seems likely, it is at least done in hours which are not lit: as a result, "the dark" is not an uninformative environment for dogs. The anatomy of the dog retina makes it plain how dogs see differently at night than humans do: dogs have a tapetum lucidum, as well as a preponderance of rod cells, photoreceptor cells which fire at low light (Miller and Murphy 1995).

Certainly dog breeds vary in visual capacity (e.g. McGreevy et al. 2004). No breed information was given for the subjects of this experiment—but breed differences may very well explain why some subjects chose to steal food more often when the food was illuminated and the person was not, or when the person was illuminated and the food was not (Kaminski et al. 2013).

Notably, even research which tackles common anthropomorphisms, such is the claim that dogs act "heroically" to save owners who are drowning, in a fire, or otherwise in danger (Macpherson and Roberts 2006), is not immune to anthropocentrism. The overweening attribution that the researchers address is not that dogs have acted in a way so as to save their owners—they may indeed have—but that dogs act *with understanding of the situation and intent to assist*. The study explicitly tested this claim by setting up artificial emergency situations, such as an owner, accompanied her dog, conspiring with the experimenters to fake a heart attack or pretend to be trapped under an (actually lightweight) fallen bookcase. No dog acted to help his owner by notifying a bystander (Macpherson and Roberts 2006). While notable for taking on an anthropomorphism, even this study relies on a critical sensory assumption: that dogs cannot distinguish, by smell or other means, a pretend heart attack from a real one—itself a claim which should be tested, not presumed, given the dog's impressive olfactory ability.

9.3.2 Premature Labeling

Comparative psychology research aims to discover the similarities, or dissimilarities, between the psychology of humans and non-human animals. Thus an argument might be made that, by starting with human cognitive capacity, any comparative psychological approach is inherently anthropocentric. However, not all such studies need be anthropomorphic: most research is careful to assess the behaviors of animals and humans in their respective contexts. For instance, while point- and gaze-following by children is part of a normally developing theory of mind, dogs' abilities to follow a point or gaze is not typically considered evidence that they are developing a human-like theory of mind.

Less careful has been research studying empathy, another human ability of interest to comparative psychologists. For instance, in one study interested in dogs' empathetic abilities, researchers measured subject dogs' responses to an experimenter or owner pretending to cry (Custance and Mayer 2012). While exposure to a (feigning) crying person may be a reasonable context for exploring empathy-like responses in human children, there is no a priori reason to believe that dogs would even be sensitive to *actual* crying, as the species does not cry. More apt, and less anthropomorphic, would be gauging the dogs' responses to an intraspecific "cry." Relevant vocalizations dogs produce include whimpers, whines, screams, and yelps (Tembrock 1976); the "isolation" bark is another candidate emotion-rich vocalization (Yin 2002). Furthermore, while human empathy may sometimes be demonstrated by a person's response to conspecific crying, "responding to crying" does not equal "existence of empathy." Neither does it for dogs.

At times, language from a human context is applied to the analysis of dog behavior in such a way that results rest on problematic data, and may interfere with an understanding of the behavior of the species. For instance, research on social play which uses terms of human games—"wins" and "loss"—to characterize kinds of dog behaviors, inevitably finds that some dogs "win" games and some "lose" games (Bauer and Smuts 2007). However, substantial research on social play of mammals, including, specifically, play of dogs, does not indicate that play bouts are resolved with "winners" or "losers," as in a competitive sports match. On the contrary, in ethological work, bouts are not defined by results but by specified play behaviors (Burghardt 2005; Fagen 1981; Rooney et al. 2000), and marked, in dogs, by the regular use of play signals (Bekoff 1972; Bekoff 1974; Fox 1978; Horowitz 2009b) (see also Bekoff, this volume). Conclusory assessments of a category of behavior are sometimes cultivated by a limited ethogram (the set of behaviors noted of relevance to the researchers). Such assessments can be avoided.

Finally, interesting research (discussed above) has demonstrated that dogs are averse to inequity, when they are the recipients of the inequity (Range et al. 2009).

In particular, dogs stopped "giving the paw" or "shaking", when no longer rewarded for the behavior, most quickly when a social partner continued to be rewarded. It is surely relevant that "giving the paw" is considered a "submissive" behavior by dogs (Lorenz 1954), and thus the experimental scenario is tinged with the dynamic of the social setting alone, regardless of the change in the dog's fortunes.

These anthropocentric tendencies can be overcome by appeal to the existing literature of dog ethology. At times improvement comes through running multiple iterations of an experiment. The methods used in the study of yawning behavior in dogs is one research topic which has improved through replication. The first studies used video recordings of owner yawns in presentation to dog subjects (Joly-Mascheroni et al. 2008; Harr et al. 2009) added recordings of dog yawns to the mix. Such an addition presumably helped account for the fact that while dogs can see video images, it is likely not the only salient sensory element of a presentation to the subjects. Silva et al. (2012) attempted to isolate the auditory component of yawn from the visual component, and also contrasted familiar and unfamiliar persons as stimuli for subject dogs. Here the importance of the yawner to the dog was acknowledged. At this point, the role of the yawn as stress modulator, as described by trainers (Rugaas 2006), was not yet discussed. Recently, though, two papers have investigated the correlation of contagious yawning and arousal levels (Buttner and Strasser 2014; Romero et al. 2013). Our understanding of dog yawning has substantively improved based on methodological refinements attending to *what the behavior is like for the dog*.

9.3.3 Umwelt

Happily, many dog studies do take a canid-centered view. Instead of assuming their subjects experience and perceive the world, and the experimental setting, just as humans do, this research attends to the difference between canid and human *umwelt*, the subjective or "self-world" of each individual and species (von Uexküll 1934/1957). Von Uexküll proposed that only through attending to each animal's perceptual ability and noting what is salient for the animal, could one begin to draw a non-biased picture of the animal's world. Differences in *umwelt* exist for individuals as well as across-species: "The best way to find out that no two human *Umwelten* are the same," he suggested, "is to have yourself led through unknown territory by someone familiar with it. Your guide unerringly follows a path that you cannot see."

Between humans and dogs, one profound difference is sensory: while humans are visual creatures, dogs are olfactory (see also Gadbois and Reeve, this volume). Dog noses house hundreds of millions more olfactory cells than humans' do, and their corresponding brain regions are much more developed relative to their visual areas than in humans (Lindsay 2000). Respiration and smelling occur via different flow paths within the nose (Craven et al. 2010), and side-nostril exhalation diminishes odor habituation (Settles et al. 2003). As trainers of drug-, narcotics-, and explosives-detection dogs know, their sense of smell is clearly acute (e.g. Gazit and Terkel 2003; McCulloch et al. 2006).

Thus the dog *umwelt*, insofar as it can be approached, will include, at minimum, a perceptual world different than humans'. A handful of instances of research programs which are sensitive to their subject's *umwelt* follow. They serve as demonstration of the ways this sensitivity is manifest, and how complex it can be to take a non-human species' perspective into account in design and execution of experiments.

Acknowledging the relevance of olfaction for these subjects is perhaps the most obvious way to become more dog-centered in experimental design. It is also, however, non-obvious how to control and systematically vary olfactory stimuli in an ordinary experimental or naturalistic setting. Thus it is rarely pursued in dog behavior and cognition studies.[2] Exceptions exist: most have been driven by concerns that olfactory knowledge might explain the subject dogs' performance in experimental trials. Thus, the olfactory cues are controlled in an additional trial iteration to pursue that concern. For instance, early research in this field found that dogs' performance on invisible displacement tasks was not compromised when olfactory cues were masked (Gagnon and Doré 1992)—evidence that subjects were solving the tasks visually, as they had been designed. In addition, when exploring the response of dogs to a dog-shaped robot toy (the AIBO), researchers considered whether having a dog scent (from material lain in a puppy's sleeping box) on the toy would change the subjects' behavior (Kubinyi et al. 2004), although they did not consider the degradation of the odor over time. Relatedly, one study explicitly examined whether social responses to a child's dummy would be changed when its clothes were impregnated with the odor of a familiar or unfamiliar child (Millot 1994). Our own lab recently looked at whether dogs' visual discrimination of more or less food on plates (and selection of the former) (Prato-Previde et al. 2008) was matched by their olfactory discrimination of these quantities (Horowitz et al. 2013). It was not: dogs did not reliably choose a larger quantity of desired food when given olfactory cues alone. On the other hand, the subjects did attend more to the "larger" quantity (covered) plate, when presented to them. Thus, in this kind of experimental trial, measurement of subjects' *attention* conflicted with data of their plate *selection* (Horowitz et al. 2013). In all cases, the role of olfactory cues in explaining dogs' behavior was part of the studies' designs.

Finally, one study addressed whether olfactory cues were relevant in dogs' success at finding hidden food seemingly through following human pointing gestures (Szetei et al. 2003). While the authors found that dogs could follow either olfactory and visual cues, they also found subjects more willing to follow a human

---

[2] Putting aside, of course, the many studies of dogs as tracking dogs or explosives-, drug-, or disease-detection dogs. These research programs are highly relevant for an understanding not just of these working dogs, but also of the entire species. However, these programs do not arise from the cognitive and behavioral fields with which this chapter is concerned.

point when it conflicted with present olfactory information. This study is to be commended not only for attention to the dogs' perceptual experience of an experimental scenario, but also for revealing more information about the hierarchy of cues a dog follows. The "constraints of the social context"—i.e. a human-dog interaction—may trump the odor information available to dogs (Szetei et al. 2003). This phenomenon is a kind of behavioral version of Heisenberg's Uncertainty Principle: the very *presence* of an owner may change the dog's behavior. Dogs' problem-solving capacity may be reduced when the owner is present (Topál et al. 1997); and Clever-Hans-reminiscent cuing and errors can always occur whether an owner (or experimenter) is blinded to the experiment or not (Hauser et al. 2011).

Similarly, research examining dogs' behavior toward new and old objects, otherwise identical, highlights a characteristic of dogs that is highly relevant in dog cognition studies: the species' neophilia (Kaulfuß and Mills 2008). Other studies have examined whether dogs have a side bias (Miklósi 2007) or are subject to a recency effect (Horowitz et al. 2013; Tapp et al. 2003), either of which could partially explain subjects' behavior on a two-sided forced-choice test. Similarly, the effect of gender—especially over repeated presentations—is beginning to get some attention (Duranton et al. 2013). Individual differences have also been examined: Leonardi et al. (2012) developed a protocol testing a dog's ability to delay gratification, with training, via trading lower-quality food in mouth for an anticipated higher-quality food item. The authors rightly noted the importance of both individual variation and olfactory dominance. Prior to training, each individual's "hierarchy" of food preferences was noted. This was used in determining what was "low value" or "high value" for each subject. (Such is now also done in other paradigms—e.g. Kundey et al. 2010.) Similarly, given the difficulty in determining which food smelled the strongest for dogs, the authors also tested quantitative versions (more quantity) of a successive-exchange task, not just qualitative (higher-value food).

The above studies by no means serve as an exhaustive list of dog-centered research designs; however they give a good sampling of the manifestations of the approach in current dog cognition literature.

9.4 Conclusion

As common, and perhaps inevitable, as some anthropocentric perspective is in studying non-humans, including dogs, our research hints at two alternatives. First, through empirical investigation of anthropomorphisms made of dogs, it is clear that these attributions can be tested—and dismissed, if appropriate. Second, through an appreciation of the *umwelt* of the species, projects can be designed which are more dog-centered in approach. Often, observational and simple experimental practices can lend insight into the dog's *umwelt*, which can then be used in design of more *umwelt*-sensitive research. For instance, a fascinating question for non-human animal researchers is what kind of self-awareness, or understanding of self, animals have.

In our lab we are currently designing a protocol to study self-recognition in dogs by modifying the "mirror self-recognition" task as developed and refined (Gallup 1970; Reiss and Marino 2001) for dogs. While a mirror, which allows for self-examination, is a sensible apparatus in a task given to visually-dominant creatures, an *olfactory mirror* would be more appropriate for a dog (Horowitz 2009c). By testing what olfactory acuity an untrained dog has (Horowitz et al. 2013) and re-defining what object may play the role of "mirror" in a canid's life, this interesting question can be asking appropriately of dogs. It is fascinating, and important, to re-consider this most familiar species in a canid-centered, unfamiliar light.

Acknowledgments Many thanks to the members of the Horowitz Dog Cognition Lab who helped to run the experiments mentioned herein, and to Animal Haven, a non-profit shelter in New York City that generously donated use of their facility for running trials. We are indebted to the dog owners who participate in our experiments, and doubly so to their dogs.

References

American Kennel Club. (2013). *AKC meet the dogs: Great Pyrenees*. Retrieved February 15, 2013, http://www.akc.org/breeds/index.cfm

Bauer, E. B., & Smuts, B. B. (2007). Cooperation and competition during dyadic play in domestic dogs, *Canis familiaris*. *Animal Behaviour, 73*, 489–499.

Bekoff, M. (1972). The development of social interaction, play, and meta-communication in mammals: An ethological perspective. *Quarterly Review of Biology, 47*, 412–434.

Bekoff, M. (1974). Social play in coyotes, wolves, and dogs. *BioScience, 24*, 225–230.

Bekoff, M. (2004). Wild justice and fair play: Cooperation, forgiveness, and morality in animals. *Biology and Philosophy, 19*, 489–520.

Brosnan, S. F. (2006). Nonhuman species' reactions to inequity and their implications for fairness. *Social Justice Research, 19*, 153–185.

Brosnan, S. F. (2013) Justice- and fairness-related behaviors in nonhuman primates. *Proceedings of the National Academy of Sciences, 110*, 10416–10423.

Brosnan, S. F., & de Waal, F. B. M. (2003). Monkeys reject unequal pay. *Nature, 425*, 297–299.

Burghardt, G. M. (2005). *The genesis of animal play: Testing the limits*. Cambridge, MA: MIT Press.

Buttelmann, D. & Tomasello, M. (2013). Can domestic dogs (*Canis familiaris*) use referential emotional expressions to locate hidden food? *Animal Cognition, 16*, 137–145.

Buttner, A. P., & Strasser, R. (2014). Contagious yawning, social cognition, and arousal: An investigation of the processes underlying shelter dogs' responses to human yawns. *Animal Cognition, 17*, 95–104.

Cheney, D. L., & Seyfarth, R. M. (2007). *Baboon metaphysics: The evolution of a social mind*. Chicago: University of Chicago Press

Craven, B. A., Paterson, E. G., & Settles, G. S. (2010). The fluid dynamics of canine olfaction: Unique nasal airflow patterns as an explanation of macrosmia. *Journal of the Royal Society Interface, 7*, 933–943.

Custance, D., & Mayer, J. (2012) Empathic-like responding by domestic dogs (*Canis familiaris*) to distress in humans: An exploratory study. *Animal Cognition, 15*, 851–859.

Darwin, C. (1872/1979). *The expression of emotions in man and animals*. New York: St. Martin's Press.

DeKay, M. L., & McClelland, G. H. (1996). Probability and utility of endangered species preservation programs. *Journal of Experimental Psychology Applied, 2*, 60–83.

de Waal, F. (1997). *Good natured: The origins of right and wrong in humans and other animals.* Cambridge, MA: Harvard University Press.

Duranton, C., Rödel, H. G., Bedossa, T., & Belkhir, S. (2013). Inverse sex differences on performance of domestic dogs (*Canis familiaris*) in a repeated problem solving task. Poster presented at the International Ethological Conference, Newcastle, U.K.

Eddy, T. J., Gallup, G. G., Jr., & Povinelli, D. J. (1993). Attribution of cognitive states to animals: Anthropomorphism in comparative perspective. *Journal of Social Issues, 49,* 87–101.

Fagen, R. (1981). *Animal play behavior.* Oxford: Oxford University Press.

Fox, M. W. (1978). *The dog: Its domestication and behavior.* New York: Garland STPM Press.

Fratkin, J. L., & Baker, S. C. (2013). The role of coat color and ear shape on the perception of personality in dogs. *Anthrozoös, 26,* 125–133.

Gácsi, M., McGreevy, P., Kara, E., & Miklósi, Á. (2009). Effects of selection for cooperation and attention in dogs. *Behavioral and Brain Functions, 5,* 31.

Gagnon, S., & Doré, F. Y. (1992). Search behavior in various breeds of adult dogs (*Canis familiaris*): Object permanence and olfactory cues. *Journal of Comparative Psychology, 106,* 58.

Gallup, G. G. Jr. (1970). Chimpanzees: Self-recognition. *Science, 167,* 86–87.

Gazit, I., & Terkel, J. (2003). Domination of olfaction over vision in explosives detection by dogs. *Applied Animal Behaviour Science, 82,* 65–73.

Greenebaum, J. (2004). It's a dog's life: Elevating status from pet to "fur baby" at Yappy Hour. *Society and Animals, 12,* 117–135.

Harr, A., Gilbert, V., & Phillips, K. (2009). Do dogs (*Canis familiaris*) show contagious yawning? *Animal Cognition, 12,* 833–837.

Hauser, M. D., Comins, J. A., Pytka, L. M., Cahill, D. P., & Velez-Calderon, S. (2011). What experimental experience affects dogs' comprehension of human communicative actions? *Behavioural Processes, 86,* 7–20.

Hebb, D. O. (1946). Emotion in man and animal: An analysis of the intuitive process of recognition. *Psychological Review, 53,* 88–106.

Hecht, J., & Horowitz, A. (2013). Physical prompts to anthropomorphism of the domestic dog (*Canis familiaris*). *Journal of Veterinary Behavior: Clinical Applications and Research, 8,* e30.

Hecht, J., Miklósi, Á., & Gácsi, M. (2012). Behavioral assessment and owner perceptions of behaviors associated with guilt in dogs. *Applied Animal Behaviour Science, 139,* 134–142.

Heider, F., & Simmel, M. (1944). An experimental study of apparent behavior. *The American Journal of Psychology, 57,* 243–259.

Hemsworth, P. H. (2003). Human-animal interactions in livestock production. *Applied Animal Behaviour Science, 81,* 185–198.

Horowitz, A. (2009a). Disambiguating the "guilty look": Salient prompts to a familiar dog behavior. *Behavioural Processes, 81,* 447–452.

Horowitz, A. (2009b). Attention to attention in domestic dog (*Canis familiaris*) dyadic play. *Animal Cognition, 12,* 107–118.

Horowitz, A. (2009c). *Inside of a dog.* New York: Scribner.

Horowitz, A. (2012). Fair is fine, but more is better: Limits to inequity aversion in the domestic dog. *Social Justice Research, 25,* 195–212.

Horowitz, A. C., & Bekoff, M. (2007). Naturalizing anthropomorphism: Behavioral prompts to our humanizing of animals. *Anthrozoös, 20*(1), 23–35.

Horowitz, A., Hecht, J., & Dedrick, A. (2013). Smelling more or less: Investigating the olfactory experience of the domestic dog. *Learning and Motivation, 44,* 207–217.

Joly-Mascheroni, R., Senju, A., & Shepherd, A. (2008). Dogs catch human yawns. *Biology Letters, 4,* 446–448.

Kaminski, J., Pitsch, A., & Tomasello, M. (2013). Dogs steal in the dark. *Animal Cognition, 16,* 385–394.

Kaulfuß, P., & Mills, D. S. (2008). Neophilia in domestic dogs (*Canis familiaris*) and its implication for studies of dog cognition. *Animal Cognition, 11,* 553–556.

Kellert, S. R. (1996). *The value of life: Biological diversity and human society.* Washington, DC: Island Press.

Kubinyi, E., Miklósi, Á., Kaplan, F., Gácsi, M., Topál, J., & Csányi, V. (2004). Social behaviour of dogs encountering AIBO, an animal-like robot in a neutral and in a feeding situation. *Behavioural Processes, 65,* 231–239.

Kundey, S. M. A., De Los Reyes, A., Taglang, C., Allen, R., Molina, S., Royer, E.et al. (2010). Domesticated dogs (*Canis familiaris*) react to what others can and cannot hear. *Applied Animal Behaviour Science, 126,* 45–50.

Leonardi, R. J., Vick, S-J., & Dufour, V. (2012). Waiting for more: The performance of domestic dogs (*Canis familiaris*) on exchange tasks. *Animal Cognition, 15,* 107–120.

Lindsay, S. R. (2000). *Handbook of applied dog behavior and training* (Vol 1). Ames, Iowa: Blackwell Publishing.

Lorenz, K. (1950/1971): Part and parcel in animal and human societies. In K. Lorenz (Ed.), *Studies in animal and human behaviour* (Vol. 2, pp. 115–195). London: Methuen

Lorenz, K. (1954). *Man meets dog.* London: Methuen.

Macpherson, K., & Roberts, W. A. (2006). Do dogs (*Canis familiaris*) seek help in an emergency? *Journal of Comparative Psychology, 120,* 113–119.

McConnell, P. B. (2006). *For the love of a dog: Understanding emotion in you and your best friend.* New York: Ballantine.

McCulloch, M., Jezierski, T., Broffman, M., Hubbard, A., Turner, K., & Janecki, T. (2006). Diagnostic accuracy of canine scent detection in early- and late-stage lung and breast cancers. *Integrative Cancer Therapies, 5,* 30–39.

McGreevy, P., Grassia, T. D., & Harmanb, A. M. (2004). A strong correlation exists between the distribution of retinal ganglion cells and nose length in the dog. *Brain, Behavior and Evolution, 63,* 13–22.

Mech, D. (1970). *The wolf: The ecology and behaviour of an endangered species.* Garden City, NY: Natural History Press.

Miklósi, Á. (2007). *Dog behaviour, evolution, and cognition.* New York: Oxford University Press.

Miller, P. E., & Murphey, C. J. (1995). Vision in dogs. *Journal of the American Veterinary Medical Association, 207,* 1623–1634.

Millot, J. L. (1994). Olfactory and visual cues in the interaction systems between dogs and children. *Behavioural Processes, 33,* 177–188.

Mitchell, R .W., & Hamm, M. (1996). The interpretation of animal psychology: Anthropomorphism or behavior reading? *Behaviour, 134,* 173–204.

Mithen, S. (1996). *The prehistory of the mind: The cognitive origins of art, religion and science.* London: Thames and Hudson Ltd.

Morewedge, C. K., Preston, J., & Wegner, D. M. (2007). Timescale bias in the attribution of mind. *Journal of Personality and Social Psychology, 93,* 1–11.

Morris, P. H., Doe, C., & Godsell, E. (2008). Secondary emotions in non-primate species? Behavioural reports and subjective claims by animal owners. *Cognition and Emotion, 22,* 3–20.

Morris, P., Fidler, M., & Costall, A. (2000). Beyond anecdotes: An empirical study of "anthropomorphism." *Society and Animals, 8,* 151–165.

Naderi, S., Miklósi, Á., Dóka, A., & Csányi, V. (2001). Co-operative interactions between blind persons and their dogs. *Applied Animal Behaviour Science, 74,* 59–80.

Pongrácz, P., Vida, V., Bánhegyi, P., & Miklósi, Á. (2008). How does dominance rank status affect individual and social learning performance in the dog (*Canis familiaris*)? *Animal Cognition, 11,* 75–82.

Prato-Previde, E., Marshall-Pescini, S., & Valsecchi, P. (2008). Is your choice my choice? The owners' effect on pet dogs' (*Canis lupus familiaris*) performance in a food choice task. *Animal Cognition, 11,* 167–174.

Pritchard, D., Dunnette, M. D., & Jorgenson, D. O. (1972). Effects of perceptions of equity and inequity on worker performance and satisfaction. *Journal of Applied Psychology, 56,* 75–94.

Range, F., Horn, L., Viranyi, Z. & Huber, L. (2009). The absence of reward induces inequity aversion in dogs. *Proceedings of the National Academy of Sciences, 106*, 340–345.

Reiss, D., & Marino, L. (2001). Mirror self-recognition in the bottlenose dolphin: A case of cognitive convergence. *Proceedings of the National Academy of Science, 98*, 5937–3942.

Rivas, J. A., & Burghardt, G. M. (2002). Crotalomorphism: A metaphor to understand anthropomorphism by omission. In M. Bekoff, A. Colin, & G. M. Burghardt (Eds.), *The cognitive animal: Empirical and theoretical perspectives on animal cognition* (pp. 9–17). Cambridge, MA: MIT Press.

Romanes, G. J. (1883). *Mental evolution in animals*. London: Keegan Paul, Trench & Co.

Romero, T., Konno, A., & Hasegawa, T. (2013). Familiarity bias and physiological responses in contagious yawning by dogs support link to empathy. *PLoS ONE 8*(8), e71365.

Rooney, N. J., Bradshaw, J. W. S., & Robinson, I. R. (2000). A comparison of dog–dog and dog–human play behaviour. *Applied Animal Behaviour Science, 66*, 235–248.

Rugaas, T. (2006) *On talking terms with dogs: Calming signals*. Wenatchee, WA: Dogwise.

Sanders, C. R. (1993). Understanding dogs: Caretakers' attributes of mindedness in canine-human relationships. *Journal of Contemporary Ethnography, 22*, 205–226.

Sanders, C. R. (2003). Actions speak louder than words: Close relationships between humans and nonhuman animals. *Symbolic Interaction, 26*, 405–426.

Scheider, L., Kaminski, J., Call, J., & Tomasello, M. (2013). Do domestic dogs interpret pointing as a command? *Animal Cognition, 16*, 361–372.

Schenkel, R. (1967). Submission: Its features and function in the wolf and dog. *American Zoologist, 7*, 319–329.

Serpell, J. A. (2003). Anthropomorphism and anthropomorphic selection—beyond the "cute response." *Society & Animals, 11*, 83–100.

Settles, G. S., Kester, D. A., & Dodson-Dreibelbis, L. J. (2003). The external aerodynamics of canine olfaction. In F. G. Barth, J. A. C. Humphrey, & T. W. Secomb (Eds.), *Sensors and sensing in biology and engineering* (pp. 323–355). New York: SpringerWein.

Sillero-Zubiri, C., Hoffmann, M., & Macdonald, D. D. W. (2004) *Canids: Foxes, wolves, jackals and dogs: Status survey and conservation action plan*. Gland, CH: IUCN/SSC Canid Specialist Group

Silva, K., Bessa, J., & de Sousa, L. (2012). Auditory contagious yawning in domestic dogs (*Canis familiaris*): First evidence for social modulation. *Animal Cognition, 15*, 721–724.

Slabbert, J. M., Rasa, O., & Anne, E. (1997). Observational learning of an acquired maternal behavior pattern by working dog pups: An alternative training method? *Applied Animal Behaviour Science, 53*, 309–316.

Soproni, K., Miklósi, Á., Topál, J., & Csányi, V. (2002). Dogs' (*Canis familiaris*) responsiveness to human pointing gestures. *Journal of Comparative Psychology, 116*, 27–34.

Szetei, V., Miklósi, Á., Topál, J., & Csányi, V. (2003). When dogs seem to lose their nose: An investigation on the use of visual and olfactory cues in communicative context between dog and owner. *Applied Animal Behaviour Science, 83*, 141–152.

Tapp, P., Siwak, C., Estrada, J., Holowachuk, D., & Migram, N.W. (2003). Effects of age on measures of complex working memory span in the beagle dog (*Canis familiaris*) using two versions of a spatial list learning paradigm. *Learning & Memory, 10*, 148–160.

Tembrock, G. (1976). Canid vocalizations. *Behavioural Processes, 1*, 57–75.

Topál, J., Miklósi, A., & Csányi, V. (1997). Dog–human relationship affects problem solving behaviour in dogs. *Anthrozoös, 10*, 214–224.

von Uexküll, J. (1934/1957). A stroll through the worlds of animals and men. In C. H. Schiller (Ed.), *Instinctive behavior: The development of a modern concept* (pp. 5–80). New York: International Universities Press.

Ward, C., & Smuts, B. B. (2007). Quantity-based judgments in the domestic dog (*Canis lupus familiaris*). *Animal Cognition, 10*, 71–80.

Weiss, E., Miller, K., Mohan-Gibbons, H., & Vela, C. (2012). Why did you choose this pet?: Adopters and pet selection preferences in five animal shelters in the united states. *Animals, 2*, 144–159.

Wemelsfelder, F., Hunter, E. A., Mendl, M. T., & Lawrence, A. B. (2000). The spontaneous qualitative assessment of behavioural expressions in pigs: First explorations of a novel methodology for integrative animal welfare measurement. *Applied Animal Behaviour Science, 67*, 193–215.

West, R. E., & Young, R. J. (2002). Do domestic dogs show any evidence of being able to count? *Animal Cognition, 5*, 183–186.

Whitely, H .E. (2006). *Understanding and training your dog or puppy.* Santa Fe, NM: Sunstone Press.

Wynne, C. D. L. (2007). What are animals? Why anthropomorphism is still not a scientific approach to behavior. *Comparative Cognition & Behavior Reviews, 2*, 125–135.

Yin, S. (2002). A new perspective on barking in dogs. *Journal of Comparative Psychology, 116*, 189–193.

Chapter 10
A Dog's-Eye View of Canine Cognition

Monique A. R. Udell, Kathryn Lord, Erica N. Feuerbacher and Clive D. L. Wynne

Abstract In this chapter we attempt to put the dog back at the heart of dog cognition studies. We identify that the majority of dogs are not first-world pets, dependent on their owners for the fulfillment of all essential needs, and acting as their "best friends." Rather most dogs are scavengers on the periphery of people's lives. These dogs are more likely to avoid human contact than seek it. The sensitivity of pet dogs to human actions and intentions that has been a major focus of recent research is unlikely to be a special adaptation or case of co-evolution, but rather is the expression of basic processes of conditioning as well as social and biological traits that domesticated and wild canids share. In individuals that have been socialized to humans and rendered completely dependent on them these processes lead to high levels of sensitivity to human actions. The fundamental differences between dog and wolf behavior lie at more basic levels: in the processes of socialization, in foraging, and in reproduction. Small but crucial intertwined changes led to an animal that is (1) more promiscuous than any other canid, (2) can reproduce more rapidly, and (3) is a much less effective hunter but (4) more efficient scavenger than other canids. The indirect consequences of these changes include the fact that we have dogs and not wolves resting at our feet. Though it may be a little less flattering to the human species, we believe this perspective on dogs is at least as fascinating and closer to the historical truth than the story that humans created dogs.

M. A. R. Udell
Department of Animal Sciences, Oregon State University, Corvallis, OR 97331, USA

K. Lord
Department of Biology, Gettysburg College, Gettysburg, PA 17325, USA

E. N. Feuerbacher
Department of Psychology, University of Florida, Gainesville, FL 32611, USA

C. D. L. Wynne (✉)
Department of Psychology, Arizona State University, Tempe, AZ 85287, USA
e-mail: clivewynne@gmail.com

10.1 Introduction

Dogs are ubiquitous in human societies. In the United States, nearly half all households include a dog (APPMA 2008). However, most of the world's dogs live in the developing world. Here pet dogs are uncommon, but feral, village, or community dogs are plentiful around humans (Coppinger and Coppinger 2001). It has been estimated that there are upwards of a billion dogs on the planet, with pet dogs in developed countries representing only 17–24 % of the total (Lord et al. 2013).

The nature of the relationships that dogs have with people is a source of fascination to many who share their homes with dogs, but also of considerable significance for the welfare of human communities. Even in the United States, where most dogs live as family pets, dogs are responsible for 12,400 bites a day (Gilchrist et al. 2008). In the third world, dogs are an even greater danger. More than 55,000 people, mainly in Asia and Africa, die each year from rabies, a disease for which dogs are the most important vector (World Health Organisation 2005).

Though the warmth of the human–dog relationship is often commented upon, even in scientific communications, the particularly rich relationship that many people enjoy with their pet dogs is both relatively recent and relatively limited to the first world and wealthier classes in the developing world. The often cited phrase, "Dog is Man's best friend" was coined by the King of Prussia in 1789 (Laveaux and King of Prussia 1789); it didn't enter widespread usage until the early twentieth century.

Despite claims for the novelty of the study of the human–dog relationship (Hare and Woods 2013; Miklosi 2009), the use of dogs as psychological subjects, including the study of dogs' relationship with humans, has a history in science dating back to Pavlov at the very beginning of the twentieth century. Pavlov used dogs in a systematic behavioral research program and discovered what he termed the social reflex (Pavlov 1928), in which the presence of a certain person can increase excitatory activity in the dog (Lynch and Gantt 1968).

W. Horsley Gantt, an American who studied with Pavlov, developed Pavlov's research into the social reflex in dogs on his return to the United States. Gantt and his students studied what he termed the 'effect of person' (Lynch 1987). Gantt found that a dog's heart rate would dramatically increase (tachycardia) when a human entered the room, but would decrease (bradycardia) when petted (Gantt et al. 1966). Additionally, when dogs received unsignaled paw shocks that typically produced tachycardia, petting during shock delivery would substantially decrease this change in heart rate. Because the response to human petting could be readily conditioned to other stimuli, Gantt et al. suggested that petting was an unconditioned stimulus. However, the response also showed specificity to certain people. Those that had a "special relationship" (Gantt et al. 1966, p. 152) with the dog produced larger and more variable responses. Pavlov had also noted that the social reflex was produced by an experimenter who "played with (the dog), fed him and petted him" (Pavlov 1928, p. 368). Feuerbacher and Wynne (2011) presented a review of the extensive history of dogs as subjects in psychological research.

10.1.1 What are Humans for Dogs?

More recent studies have continued to develop our understanding of the functions of human contact for dogs within an operant framework—asking what consequences provided by humans can influence the behavior of dogs. Dogs in Western households experience a range of human interactions, including food delivery, petting, and vocal praise. Fonberg et al. (1981) reported that both food and 20–30 s of petting maintained operant responding in dogs. More recently, Feuerbacher and Wynne (2012) found that food maintained more and faster responding in shelter dogs, pet dogs, and hand-reared captive wolves than a brief social interaction comprised of petting and vocal praise. This was true even though, in the case of the pet dogs, their owners provided the petting and praise under the direction of the lead experimenters. Fukuzawa and Hayashi (2013) also found that food produced shorter latencies to respond early in training compared to petting or vocal praise. Both of these studies indicate that in some cases food is a more effective reinforcer than human interaction.

That food is an important consequence for dogs is also supported by developmental evidence in which hand-fed food-deprived puppies showed more approach and fewer avoidance behaviors towards humans (Elliott and King 1960). Similarly, puppies given a choice between two fake cloth dams, one that provided food and one that did not, preferred the dam that provided food over a 20-day test period (Igel and Calvin 1960). However, when given a choice between a cloth dam that did not produce milk, and a wire one that did, dog pups, like Harlow and Zimmerman's (1959) juvenile monkeys, preferred the non-nutritive cloth mother over the nutritive wire mother (Igel and Calvin 1960).

The potency of food as a reinforcer may be instrumental in producing social interaction between humans and dogs, since in many contexts humans provide access to food for dogs. Nevertheless, Fonberg et al.'s (1981) finding that both food and petting maintained operant responding indicates that human interactions other than food delivery, such as petting, might also function as reinforcers. Petting was sufficient to maintain military dogs' operant responding to the cues, 'sit', 'down', 'come', 'stay', and 'heel' (McIntire and Colley 1967). Furthermore, Feuerbacher and Wynne (in prep.) observed that both shelter and owned dogs remained in proximity to a person providing petting, and showed no signs of satiation to that stimulus. Given a choice between a person providing food and another providing petting, some shelter and some owned dogs (which were being petted by their owner) preferred petting to food when food was readily available. Those dogs that preferred food to petting shifted their preference towards petting as the rate of food delivery was reduced (Feuerbacher and Wynne in prep). This parallels Gantt et al.'s (1966) findings that petting might be an unconditioned stimulus for dogs and Igel and Calvin's research, noted above, in which puppies preferred a non-food producing cloth mother to a food-producing wire mother. Together these results support the idea that petting, or "contact comfort" (Harlow and Zimmerman 1959), is a reinforcer for dogs as it is for primates.

Unlike petting, vocal praise seems to be at best a conditioned reinforcer for dogs. Latencies to respond to basic obedience cues (such as *sit, down, come, stay,* and *heel*) increased when only vocal praise was provided for correct responses (McIntire and Colley 1967), and dogs spent as little time near a human providing vocal praise as when the person provided no interaction at all, even when the human was the dog's owner (Feuerbacher and Wynne in prep). Vocal praise likely has to be explicitly paired with another reinforcer to become a conditioned reinforcer and extinguishes rapidly if not backed up by a primary reinforcer.

Finally, the dog's relationship to the person providing the interaction impacts its behavior towards humans. Owned dogs showed more social approach behavior, as well as more redirected and appeasement behavior, when being petted by a familiar person than someone unfamiliar (Kuhne et al. 2012). Barrera et al. (2010) found that shelter dogs emitted more appeasement behaviors and remained closer to a stranger than did owned dogs, which more often stayed by the door of the enclosure. However, owned dogs remained in proximity to their owner just as much as shelter dogs stayed in proximity to a stranger for petting (Feuerbacher and Wynne in prep.). This suggests that humans might be a source of comfort for socialized dogs in stressful situations such as a shelter or an unfamiliar laboratory. One possible mechanism for this effect may be petting-induced increases in serum levels of hormones associated with pleasure and social bonding such as b-endorphin, oxytocin, prolactin, b-phenylethylamine, and dopamine (Odendaal and Meintjes 2003).

10.1.2 Dogs' 'Human-like' Social Cognition

Over the last two decades there has been a growing interest in the cognitive abilities of dogs, especially with regard to their social cognition. Much of this research has sought to elucidate the acute sensitivity to human actions of many pet dogs (Udell and Wynne 2008; Udell et al. 2010a).

One task that has become axiomatic for the responsiveness of dogs to humans is the human-guided choice task, also known as the pointing task (see Fig. 10.1) (see also Rossi et al. this volume). In this test, an experimenter points at one of two or more locations where food can be found if the dog approaches that location first. Typically, dogs living in human homes excel on this type of problem (Miklósi and Soproni 2006; Udell et al. 2010a). Although points are often made with the experimenter's arm and hand, research shows that many pet dogs can also follow a wide range of other gesture types, including those made with other body parts like a human leg or head, those made from a greater distance, or even after a short delay (Miklósi et al. 1998; Soproni et al. 2001, 2002; Udell et al. 2008b, 2011).

Pet dogs have also proven responsive to human attentional state, at least under some conditions. In situations in which a dog can beg from a human looking at it or from a human with her back turned—simple begging tasks—dogs typically choose the person looking at them (Cooper et al. 2003; Gácsi et al. 2004; Udell et al. 2011). However this ability likely has more to do with dogs' experiences in their home

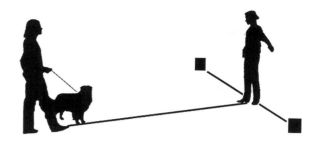

Fig. 10.1 Sketch of the typical layout of a human-guided choice task, also known as a pointing task. The person on the right points at one of two containers on the ground before the dog is released to make its choice

environment where they experience the consequences of begging from people who have different levels of attentiveness than with true perspective-taking, as dogs do not appear to recognize that a human wearing a bucket over her head and eyes is inattentive (Cooper et al. 2003; Udell et al. 2011). Furthermore, only pet dogs recognize that a human reading (with a book covering her face) is inattentive; in our study, dogs living in a shelter and hand-reared wolves which had less opportunity to experience this scenario in their current environment, begged equally from a person who was reading and a person looking at them (Udell et al. 2011).

Dogs also appear to respond to human attentional state in a related problem: the forbidden food task. Here dogs again have to determine whether a human is attentive or inattentive, only this time the goal is not cooperative in nature. In forbidden food tasks, a piece of food is placed on the floor in front of the dog, which is then instructed by the owner or experimenter not to eat it. While dogs may obey, and leave the food untouched when the human is present and attentive, they are increasingly likely to disobey if the human's eyes are closed, her back is turned, she leaves the room, or if a barrier blocks the human's view of the food and the dog's approach (Bräuer et al. 2004; Call et al. 2003).

Dogs have also been reported to use their gaze to guide the attention of humans towards desired objects that are out of reach (Miklósi et al. 2000), to succeed in human-guided detour tasks (Pongracz et al. 2001), and some have even proven capable of emulating human actions (Topál et al. 2006) or responded to large numbers of human words with extensive training. So far the record for words understood rests with Chaser—a Border Collie from South Carolina which has learned names for around 1,200 items (Pilley and Reid 2011).

These findings have led to evolving hypotheses about canine cognition, beginning with the proposal that wolves might be more oriented towards physical (means-end type) cues and thus better at problem solving tasks such as maze and barrier tasks (see Frank et al. 1987, for an overview). Dogs, on the other hand, might be more socially oriented and thus better on training tasks (such as leash training and coming when called) (Frank and Frank 1987). These differences between dogs and wolves were explained primarily as a byproduct of dogs' neotenization (particularly the slowing of social and physical development, making young dogs easier to handle and thus socialize), as well as a long history of human provisioning of dogs, weakening the selection pressure for physical problem solving competence. Later it was proposed by Hare et al. (2002) that dogs could have evolved a human-like form of social cognition during domestication resulting in a higher level of sensitivity to human actions than seen in wolves. Proponents of this hypothesis suggested that domestic dogs were uniquely prepared to respond to human actions, arguing that "...dogs' ability to follow human communicative cues is a skill present in dogs before exposure to humans can have ontogenetically major influences on dogs' behaviour... this skill therefore represents a special adaptation in dogs which is present from early (sic) age" (Riedel et al. 2008, p. 10).

Despite earlier claims that dogs show more human-like social cognition than wolves (Hare et al. 2002, 2005; Miklósi et al. 2003), several recent studies have demonstrated that if wolves are properly socialized to humans and have the opportunity to interact with humans regularly, then they too can succeed on some human-guided cognitive tasks (Gácsi 2009a; Range and Virányi 2011; Udell et al. 2008b, 2011, 2012), in some cases outperforming dogs at the individual level (Udell et al. 2008b). Like dogs, wolves have also proven capable of following more complex point types made with body parts other than the human arm and hand and after a short delay (Udell et al. 2012). It is also important to note, that even though the earlier reports suggested that dogs outperformed wolves and foxes on human-guided tasks, they did not claim that wild-type canids were universally unsuccessful; rather, the wild-type individuals performed above chance on some of the human-guided tasks (or gesture types) under test in every study (Hare et al. 2002, 2005; Miklósi et al. 2003; Virányi et al. 2008).

Various explanations have been offered for why tame wolves may sometimes fail to perform as well as dogs on some versions of a human-guided point-following task. For example, Hare et al. (2002) tested wolves with multiple individuals in the enclosure during testing and from behind a fence, while dogs were tested individually, indoors, with no barrier between them and the experimenter. Udell et al. (2008b) demonstrated that the presence of a fence barrier, like the one that the wolves (but not dogs) had to contend with in Hare et al. (2002), could lead to a similar decrement in performance in domestic dogs when tested under the same conditions. Frank and Frank (1987) had earlier noted that insufficient socialization can lead to poorer performance by wolves on social tasks: "Insofar as socialization to humans might involve sensitization to human behavioral cues, therefore, the incompletely socialized wolf pups may have been operating at a comparative disadvantage in the training situation, much like a nearsighted child trying to learn to read" (p. 35). Frank and Frank (1987) demonstrated this point by rearing a new litter of wolf pups with the socialization procedure established by Klinghammer and Goodman (1987), resulting in wolves with a substantially improved level of responsiveness towards humans. With properly socialized wolves, social interaction with humans could be used as reinforcement, performance on social tasks increased, and handling became easier as these wolves could be called by name.

Whatever the reasons for the failures of some wolves on human-guided tasks, it is just as important to acknowledge that many pet dogs also fail to perform above chance on these tasks. In fact in some studies more than half of the dog subjects did not reliably follow human points (Udell et al. 2008b; Gácsi et al. 2009b). Even those individuals that did follow some gestures or point types often failed to follow others (Udell et al. 2008a, 2010a, b, 2011, 2013), a finding that parallels the results of early studies with wolves. However, the much larger number of dogs tested in these studies can lead to an outcome where a group of dogs can perform above chance on average, even when such an interpretation is inconsistent with the behavior of the majority of the dogs. Given the much larger number of human-socialized pet dogs available for test, it is not surprising that the total number of dogs succeeding on these tasks is greater than wolves. This may be deceptive, however, because the individual success rate for wolves is actually quite high. Furthermore, as previously noted, pet dogs only make up a small percentage of the total domestic dog population and there is little evidence to suggest that the behavior of most dogs world-wide is accurately represented by those living in human homes as pets.

The suggestion that pet dogs perform well on human-guided tasks because of a newly evolved human-like social cognition (Hare et al. 2002) fails to account for the diversity of human-directed behaviors across the broader dog population, as well as successful performances on human-guided tasks not only by wolves, but a wide range of human-socialized non-domesticated species including parrots (Giret et al. 2009), bats (Hall et al. 2011), jackdaws (Von Bayern and Emery 2009), ravens (Schloegl et al. 2008), dolphins (Pack and Herman 2004), elephants (Smet and Byrne 2013), and seals (Scheumann and Call 2004), suggesting that the domestication hypothesis is no longer compatible with current scientific knowledge.

In sum, sometimes dogs may perform better than wolves on human-guided tasks, at other times wolves may perform better than dogs; some individuals from both groups have proven skilled at responding to human gestures, while some individuals from both groups fail. Thus both dogs and wolves have the cognitive capacity for this level of prosocial behavior towards humans, however it is not guaranteed for either subspecies. In the absence of appropriate life experiences, or outside certain contexts, both dogs and wolves may fail. These observations led to the development of a new hypothesis, which takes both phylogeny and ontogeny into account when predicting the social behavior of dogs (and other species) towards humans: the Two Stage Hypothesis (Udell et al. 2010a).

The Two Stage Hypothesis predicts that for canids to perform well on traditional human-guided tasks (like following a human point) both relevant lifetime experiences with humans, including socialization to humans during the critical period for social development, and opportunities to associate human body parts with certain outcomes (such as food being provided by human hands, a human throwing or kicking a ball, etc.) are required.

The Two-Stage hypothesis is not an alternative to an evolutionary approach; instead its predictions focus on the interactions between evolutionary and lifetime factors that contribute to the rich diversity of social behavior in canids (Udell et al. 2010a; Udell and Wynne 2010). While a mechanism for a new heritable form of cognition in dogs is lacking, there are still known heritable biological and developmental traits associated with domestication (like the timing of perceptual and social development—see below) that when combined with the unique environment and experiences of pet dogs would likely be very conducive to success on socio-cognitive tasks. This hypothesis is also more consistent with evidence that environmental and life experience do play a significant role in a dog's response to human actions, and thus provides predictions that can account for the behavior of all dogs, not just pet dogs.

In addition to the fact that many individual pet dogs do not perform above chance on human-guided tasks (Gácsi et al. 2009a, b), there is also evidence that dogs are constantly learning about human actions (Bentosela et al. 2008; Horowitz 2012), gestures (Elgier et al. 2009; Miklósi et al. 1998; Udell et al. 2008a, 2010b, 2013), and attentional state (Udell et al. 2011), and adjust their behavior accordingly. In fact, pet dogs sometimes continue to learn about human gestures over the course of experimental testing, resulting in above chance performance on challenging tasks that may appear spontaneous, but is actually a byproduct of experimental experience. For example, Udell et al. (2013) found that pet dogs which were tested on a series of nine point types (ten trials per point) in order of increasing difficulty performed significantly better on the most difficult point types such as the momentary distal point, where the point is only held in place for a few seconds (returning to a neutral position before the dog is allowed to approach) and the target is located further than 50 cm away from the experimenter's extended arm and hand when compared with naive dogs.

Thus, while it is entirely possible that canines have a predisposition to attend to the actions of their social companions, it appears that individuals (be they dog or wolf) must first learn to recognize that humans are indeed companions worth watching, and then continue to learn about the relationship between human actions and salient outcomes throughout their lives. While in some cases this learning may be a product of experimental setup or explicit training, much of what canines learn about human actions occurs naturally within their home environment without conscious effort by people (Reid 2009; Udell and Wynne 2008). Feral dogs likely also learn about, and respond to, human actions. As obligatory symbiotes even dogs that do not live as pets typically live near and benefit from the presence of humans and human waste (Coppinger and Coppinger 2001; Udell et al. 2010a). However the specific behavioral responses of feral dogs to human actions may not look like the responses one would expect from a pet dog; each would be expected to behave in ways consistent with its environment and life experiences.

10.1.3 So What do we Mean by Dogs?

One important consideration when interpreting research on canine cognition is the exact nature of the population under test: are the dogs succeeding on these tasks a good model for all domestic dogs, or are they unique in significant ways, including lifetime experience or environment, that might provide alternative explanations for their performance than subspecies membership itself? When reading scientific reports about 'dog' cognition, it may seem reasonable to assume that the information presented is representative of the entire subspecies, the domestic dog (*Canis lupus familiaris*). Unfortunately this is often not the case. The great majority of modern research on dog cognition has focused on pet dogs living in human homes (Udell et al. 2010a). Working dogs (search and rescue, guide dogs for the blind, sniffer dogs) have also garnered increasing scientific attention in recent years (e.g., Bensky et al. 2013). However other groups, such as stray dogs or dogs living in shelters, have been vastly underrepresented in canine cognition research, despite the fact that in the US alone roughly 10 % of the domestic dog population lives in shelters (Udell et al. 2010b).

Yet even these three populations combined constitute a minority of dogs worldwide. As mentioned previously, the great majority of dogs (as much as three-quarters of the world dog population) live outside the first world as scavengers. These dogs have never been, and likely never will be, owned by a human, and even those who are owned lead considerably different lives than Westernized pets (Coppinger and Coppinger 2001; Lord et al. 2013). This population has been almost entirely neglected in experimental studies on canine cognition.

Such populations have received some attention from scientists using less intrusive methods, however, and this work has shed light on the general behavior patterns of these populations (e.g., Beck 1973; Berman and Dunbar 1983; Boitani and Ciucci 1995; Daniels 1983; Daniels and Bekoff 1989; Ortolani et al. 2009; Pal 2008; Pal et al. 1998). Some of these studies also illustrate why it has been difficult to include feral dogs in traditional socio-cognitive experiments. For example, Ortolani et al. (2009) looked at the social response of feral village dogs in four Ethiopian villages when approached by a human. They found that the most common response of these dogs was to run away (52 %); another 11 % responded to the approaching human with aggression. Only 4 % of the surveyed population reciprocated the approach in a non-aggressive manner.

Given, first, that pet dogs represent a small minority of the domestic dog population, and, second, that many domestic dogs from other populations (such as free-living and shelter dogs) may show quantifiably different behavior towards humans (including fear and aggression), it seems very unlikely that the social behavior or cognition of pet dogs can be explained as a product or byproduct of domestication alone. What little we know about these free-ranging populations of dogs adds to the ample evidence that domestication does not guarantee excessively social, or even prosocial, behavior in dogs towards humans in the absence of appropriate life experience. In fact, in one of the most comprehensive laboratory studies on the social behavior of dogs to date, Scott and Fuller (1965), demonstrated that domesticated dog puppies raised entirely apart from humans "may later react toward them with extreme fear and hostility" (p. 176). This work was among the first to define a timeline for, and emphasize the importance of, the critical period for social development in dogs: a sensitive period early in a puppy's

		Phylogeny/ Domestication Status	
		Domesticated	Non-domesticated
Ontogeny/ Developmental Experience	Human-socialized	Pet dogs & village dogs raised with early human contact	Hand-reared wolves and other captive canids in research facilities and zoos
	Non-human-socialized	Village and other feral dogs raised outside of early human contact	Wild-living wolves and other wild canids

Fig. 10.2 Examples of canids occupying all four of the logically possible intersections of domestication status and human socialization

life where social interaction has a much greater effect on the development of future social behavior than at any other time.

Domestication has made it easier to tame, or socialize, dogs in comparison to their wild counterparts, primarily due to changes in the timing of development (a topic to which we will return). However it is worth noting that domestication on the one hand and taming or socialization on the other are different processes; the first is a genetic or evolutionary process, the latter a lifetime process (for a review see Udell et al. 2010a). Domestication and socialization/taming can be thought of as occurring on two interacting continua, resulting in four possible canid types illustrated in Fig. 10.2: (1) domesticated and tame; (2) domesticated and not socialized to humans (feral or untame); (3) undomesticated (genetically wild) and tame; (4) undomesticated and not socialized to humans. While in the Western world we most often encounter domesticated dogs that are also tame (category 1) and, less often, undomesticated canines (wolves, coyotes, foxes) that are not socialized to humans (category 4), the other combinations are possible as well. Feral dogs and socialized wolves in research facilities or zoos serve as examples of these two cases respectively.

In other words, dogs may be pets, workers, stray or free living, but all of these individuals are 'domesticated' and belong to the same subspecies. Designations like 'pet' or 'feral' refer to an individual's home environment, lifestyle or current niche, but do not imply significant genetic differences. Domestication, on the other hand, implies genetic change in comparison to wild-type counterparts. While this process may change the probability of tame behavior, it does not determine it (Scott and Fuller 1965; Udell et al. 2010a). Although some dogs (many pets and working dogs) have shown remarkable sensitivity to people, so too have some human socialized wolves. Unsocialized individuals from both groups have been difficult to test, however tame wolves with more intensive socialization have been found to perform better on social tasks than inadequately socialized wolves (Frank et al. 1987).

10.1.4 What is different about the Behavior of Dogs Compared to their Wild Relatives?

Having outlined areas in which the behavior of pet dogs and human-socialized wolves is more alike than some previous authors have claimed, and in full cognizance that dogs and wolves are extremely similar genetically, we nonetheless recognize that there are important behavioral differences between these two species—differences that make dogs widespread pets, whereas socialized wolves are rare. The purpose of this section is to review what we consider the essential behavioral differences between dogs and wolves (see also Fugazza and Miklósi, this volume).

10.1.4.1 Social Imprinting

The first major difference in the behavior of dogs and wolves has already been alluded to. While it is possible to tame a wolf, the process is much more intensive than that required to produce a tame dog.

Dogs require as little as ninety minutes of contact with humans during their 'critical period' of socialization—one of the critical periods of development (see also Fiset et al., this volume)—to form a social attachment (Freedman et al. 1960). This is the minimum requirement and will not result in a highly social pet dog, but it will create a dog that will solicit human attention.

Wolves require twenty-four hours contact a day starting before three weeks of age (Klinghammer and Goodman 1987; Zimen 1987). The standard protocol is to remove pups from the den at about ten days of age. They are then kept in constant contact with humans until they are around four weeks old. At this point they begin to bite their sleeping human companions and thus co-sleeping with humans ends, but the pups still spend all their waking hours in the presence of people. This socialization process continues until the pups are four months old at which point they can live with other captive wolves and will maintain their socialization with humans as long as they continue to get daily human contact (Klinghammer and Goodman 1987).

Despite this intensive process, a well-socialized wolf still behaves very differently from a well-socialized dog. Taming a wolf does not eliminate any of its species-typical behaviors (Gácsi et al. 2005). Tamed wolves still display species-typical hunting and reproductive behaviors, but they will display them in closer proximity to humans (Fentress 1967; Klinghammer and Goodmann 1987). This means it is more likely that they will also display them towards humans. For example, if wolves are not raised with other wolves while being socialized with humans they will direct reproductive behaviors towards humans. During breeding season they may become territorial and compete with human caretakers over perceived potential (human) mates.

Well-socialized wolves do not generalize their socialization to all humans in the same manner as a well socialized dog and are more fearful of novelty in general than socialized dogs (Fentress 1967; Klinghammer and Goodmann 1987; Zimen 1987).

It has long been thought that these differences are due to a change in the timing of the critical period of socialization in dogs and wolves, since this is the time period when social bonds are formed both within an animal's own species and between species.

The critical period for socialization begins with the ability to walk and explore the environment (see Lord 2013, for discussion). Wolves begin to walk and explore at two weeks of age (Frank and Frank 1982; Packard 2003). Dogs do not start to walk and explore until four weeks (Rheingold 1963; Fox 1964). The critical period of socialization closes with the avoidance of novelty, when an animal runs away from, rather than approaching and exploring, novel objects. This threshold has previously been referred to as the onset of fear, but fear actually starts well before the avoidance of novelty. In dogs, fear gradually increases from four weeks, when they show little to no fear of novelty, until around eight weeks when they will run away from a truly novel stimulus (a stimulus having no familiar characteristics). Figure 10.3 shows the development of fear and sensory capacities in dogs and wolves based on Lord (2013)'s study.

It has been hypothesized that wolves reach the end of the critical period for socialization at three weeks (see review in Miklosi 2009). This is based on the fact that it is not possible to socialize a wolf after three weeks of age (Klinghammer and Goodman 1987; Zimen 1987). Wolves also show the first appearance of fear in the form of a startle response to sound at three weeks of age (Zimen 1987). This hypothesis would imply that wolves have a one week critical period of socialization (since socialization begins at two weeks when the wolf pups start to explore), and could thus explain neatly the difference between the ability of dogs and wolves to form interspecies social bonds. However, this hypothesis is based on a confusion concerning the use of the term "onset of fear." The startle response is an altogether different phenomenon from fear and the avoidance of novelty. In fact, dogs, just like wolves, also show their first startle responses to sound at around three weeks of age. And wolves don't show the avoidance of novelty—the true "onset of fear"—until six weeks of age. Thus wolves and dogs both have a four-week critical period for socialization—wolves just go through it two weeks earlier than dogs do.

In itself, the earlier progress of wolves than dogs through the critical period for socialization does not explain the behavioral differences between dogs and wolves. However, Lord (2013) recently found that, despite this two-week difference in the timing of the ability to walk and explore, the dogs and wolves in her study developed the ability to see, hear, and smell at the same time. The consequence of this is that dogs began to explore the world around them at four weeks of age with the senses of sight, hearing, and smell available to them, while wolves began to explore the world at two weeks of age when they had the ability to smell but while functionally blind and deaf (see Fig. 10.3). This change in the interaction between

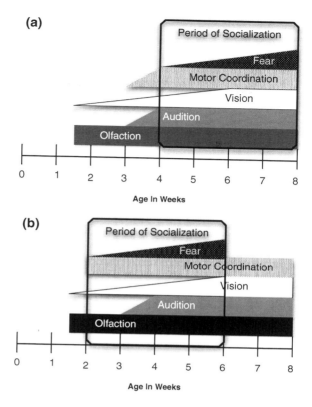

Fig. 10.3 Early sensory development in relationship to critical period for socialization in dogs (*top*) and wolves (*bottom*). The beginning of a bar marks the first appearance of a behavior or sensory capacity, and the full width of a bar indicates mature capacity. From Lord (2013); used by permission

the developing senses and the critical period for socialization means that dogs can generalize familiarity using all of their senses, while wolves must rely primarily on their sense of smell, making more things novel and frightening as adults.

10.1.4.2 Reproductive Behavior

Reproduction is another domain of behavior where wolves and dogs differ importantly. Wolves, and in fact all of the wild members of the genus *Canis*, display complex coordinated parental behaviors. Wolf pups are cared for primarily by their mother for their first three weeks of life (Mech 1970). During this time she remains in the den with them while they rely on her milk for sustenance and her presence for protection from predators. Because of this she cannot spend much time away from them, and the father brings the mother food during this period. Once the pups come out of the den and have enough teeth to chew, the father, mother as well as some pups from previous years, begin to regurgitate food to the pups (Mech et al. 1999). Wolf pups become independent by five to eight months, although they often stay with their parents for years (see Lord et al. 2013 for further discussion).

Dogs, on the other hand, show greatly reduced parental behavior. Pups are still cared for by the mother. They rely on her for milk and protection just like wolves. However, unlike wolves, the mother gets no help from any other dogs during this time. There is no paternal care, let alone help from older siblings. Once pups are weaned at around 10–11 weeks they are independent and receive no further maternal care (see review in Lord et al. 2013).

Lord et al. (2013) reviewed several important differences between dog reproductive behavior and that of the wild canids. Dogs have lost seasonality of reproduction: in other words they do not reproduce solely at a particular time of year (Gipson et al. 1975; Lord et al. 2013). Dogs also reach sexual maturity faster than wolves and can reproduce during their first year of life (Boitani et al. 1995, Ghosh et al. 1984/85). Furthermore, dogs are polygamous, in contrast to wolves, which are generally monogomous (Ghosh et al. 1984/85; Harrington et al. 1982). Thus dogs show no pair bonding and protection of a single mate, but rather have multiple mates in a year.

It is easy to look at these differences in reproductive behavior between dogs and wild canids and assume they are the result of artificial selection by humans or relaxed selection for parental care for pups, as humans intervene to assure pup survival. This hypothesis was proposed by Darwin (1859) among others. It should be kept in mind however that, as noted above, on a global scale, most dogs are not under the direct care of humans. An alternative hypothesis, proposed by Lord et al. (2013), is that reduced parental care in dogs may be an adaptive strategy for a particular niche.

Lord et al. (2013) proposed that these differences in parenting behavior amount to an alternative breeding strategy, one adapted to a life of scavenging instead of hunting. Whereas wolves are constrained to a maximum of one litter a year starting at two years of age, female dogs start reproduction in their first year and can recycle and have another litter as soon as eight months postpartum. The male dog, by not caring for pups or their mother and being unconstrained by seasonality, is free to continue to mate all year long with as many receptive females as he can find. The consequence of this is that whereas wolves put a lot of energy into a few pups, dogs maximize the production of pups. This higher pup production rate enables dogs to maintain or even increase their population with a lower pup survival rate than wolves. It also means that dogs would have a greater capacity than wolves to grow their population after a population crash or when entering a new habitat.

Dogs are only able to adopt this strategy because they no longer have to spend as much energy and ingenuity foraging. Rather than hunting prey, dogs can rely on human refuse, which is more predictably located and available year round. Foraging on garbage is a less complex behavior pattern than hunting and dog pups can forage even before they are entirely weaned. Thus, by the time they are ten weeks old they are perfectly capable of finding their own food (Macdonald and Carr 1995; Pal 2008).

The changes in dog reproductive behaviors and behavioral development that we have noted here are not a consequence of direct human action in the form of people adopting pups on a widespread scale. The role of humans in supporting dogs is indirect. Humans provide a food source, in the form of their garbage, that is easier to find and exploit than live prey. The canids that became dogs adapted to this new niche in several ways. They became more fertile, through earlier onset of reproduction, year-round fecundity, and reduced parental care. Dog parents are able to reduce their investment in each pup because foraging on trash requires less strength and skill than does hunting live prey. This increased fertility enables dogs to more rapidly colonize new niches, and allows more rapid rebound of populations after disease outbreak or other catastrophic population loss.

The changes in behavioral development we note here are likely also adaptations to scavenging on human trash dumps instead of hunting live prey. Whereas wolves go through the critical period for socialization with only olfaction fully functional, dogs have their senses at close to adult levels of functionality during this important phase. This makes it easier for dogs to generalize across sensory dimensions and increases the range of objects to which they can readily be socialized. This is also an adaptation to foraging close to human settlements as it increases the probability that dogs tolerate human proximity.

10.2 Conclusions

We have attempted in this chapter to put the dog back in the center of discussions on dog cognition and behavior (see also Coppinger and Coppinger 2001). Not the dog as viewed by first-world pet owners (among whom we count ourselves), but the dog viewed as a biological object with psychological properties. This is an animal that not only lies at its master's feet as he types, but the dog that scavenges on the fringes of human settlements well away from National Geographic documentaries; the dog that nobody considers their "best friend." This form of dog is still the most common on the planet; it does not come when called, on the contrary, it scurries away when an unfamiliar person appears (Ortolani et al. 2009).

There is little evidence that the particular sensitivity to human actions and intentions that has become a major focus of cognitive studies on dogs in the last two decades is a special adaptation or a case of co-evolution (Hare et al. 2002; Schleidt and Schalter 2003). Indeed, such sensitivity is not widespread among the world's billion-strong dog population, nor is it, as we have summarized above, absent from wolves, in those rare cases where wolves have been effectively socialized to humans. Rather the ability of a few dogs and fewer wolves to respond to human behavior is more likely an expression of basic processes of conditioning operating on animals that have been socialized to and made completely dependent on human beings (Udell et al. 2010a, b), in concert with social and biological traits shared by both wild and domesticated canines.

The real differences between dog and wolf behavior lie at more basic levels: in the process of socialization, in foraging, and in reproduction. The intertwined changes that led to an animal that is (1) more promiscuous than any other member of the genus *Canis*, (2) can reproduce more rapidly, and (3) is a much less effective hunter but (4) more efficient scavenger on human refuse than other members of its genus are small but they have massive downstream effects. These indirect consequences include the fact that we have dogs resting at our feet and not wolves.

We have attempted here to demystify dogs, in the sense that we have sought explanations for their behavior that do not assume special processes or unique relationships, but we submit that a deeper and richer understanding of these fascinating animals will flow from a recognition that they are not human creations, nor co-evolved to be our companions. Rather, dogs are canids that have come to occupy a new niche through natural selection. Though it may be a little less flattering to the human species, we believe this perspective on dogs is at least as fascinating and closer to the historical truth.

References

APPMA. (2008). *Industry statistics and trends.* American Pet Products Manufacturers Association. Retrieved from http://www.appma.org

Barrera, G., Jakovcevic, G., Elgier, A. M., Mustaca, A., et al. (2010). Responses of shelter and pet dogs to an unknown human. *Journal of Veterinary Behaviour, 5*, 339–344.

Beck, A. M. (1973). *The Ecology of Stray Dogs: A Study of Free-ranging Urban Animals.* Lafayette: Purdue University Press.

Bensky, M. K., Gosling, S. D., & Sinn, D. L. (2013). *The world from a dog's view: A comprehensive review of dog cognition research.* In H. J. Brockmann (Ed.), *Advances in the study of behavior*, Vol 45, Amsterdam: Academic Press.

Bentosela, M., Barrera, G., Jakovcevic, A., Elgier, A. M., & Mustaca, A. E. (2008). Effect of reinforcement, reinforcer omission and extinction on a communicative response in domestic dogs (Canis familiaris). *Behavioural Processes, 78*, 464–469.

Berman, M., & Dunbar, I. (1983). The social behaviour of free-ranging suburban dogs. *Applied Animal Ethology, 10*, 5–17.

Boitani, L., & Ciucci, P. (1995). Comparative social ecology of feral dogs and wolves. *Ethology Ecology and Evolution, 7*, 49–72.

Boitani, L., Francisci, F., Ciucci, P., & Andreoli, G. (1995). *Population biology and ecology of feral dogs in central Italy.* In J. Serpell (Ed.) *The domestic dog: Its evolution, behaviour, and interactions with people.* New York: Cambridge University Press.

Bräuer, J., Call, J., & Tomasello, M. (2004). Visual perspective taking in dogs (Canis familiaris) in the presence of barriers. *Applied Animal Behaviour Science, 88*, 299–317.

Call, J., Bräuer, J, Kaminski, J., & Tomasello, M. (2003). Domestic dogs (Canis familiaris) are sensitive to the attentional state of humans. *Journal of Comparative Psychology, 117*, 257–263.

Cooper, J.J., Ashton, C., Bishop, S., West, R., Mills, D.S., & Young, R.J. (2003). Clever hounds: Social cognition in the domestic dog (Canis familiaris). *Applied Animal Behaviour Science, 81*, 229–244.

Coppinger, R., & Coppinger, L. (2001). *Dogs: A startling new understanding of canine origin, behavior and evolution.* New York: Scribner.

Daniels, T. J. (1983). The social organization of free-ranging urban dogs: Non-estrous social behavior. *Applied Animal Ethology, 10*, 341–363.

Daniels, T. J., & Bekoff, M. (1989). Population and social biology of free-ranging dogs, Canis familiaris. *Journal of Mammalogy, 70*, 754.

Darwin, C. (1859). *On the origin of species by means of natural selection.* London: John Murray.

Elgier, A. M., Jakovcevic, A., Mustaca, A. E., & Bentosela, M. (2009). Learning and owner-stranger effects on interspecific communication in domestic dogs (Canis familiaris). *Behavioural Processes, 81*, 44–49.

Elliott, O., & King, J. A. (1960). Effect of early food deprivation upon later consummatory behavior in puppies. *Psychological Reports, 6*, 391–400.

Fentress, J. C. (1967). Observations on the behavioral development of a hand-reared male timber wolf. *American Zoologist, 7*, 339–351.

Feuerbacher, E. N., & Wynne, C. D. L. (2011). A history of dogs as subjects in North American experimental psychological research. *Comparative Cognition and Behaviour Reviews, 6*, 46–71.

Feuerbacher, E. N., & Wynne, C. D. L. (2012). Relative efficacy of human social interaction and food as reinforcers for domestic dogs and hand-reared wolves. *Journal of the Experimental Analysis of Behaviour, 98*, 105–129.

Feuerbacher, E. N. & Wynne, C. D. L. (in prep). Most domestic dogs' (Canis lupus familiaris) prefer food to petting: Population differences and schedule effects in concurrent choice. *Journal of the Experimental Analysis of Behavior.*

Feuerbacher, E. N. & Wynne, C. D. L. (in prep). Domestic dogs' (Canis lupus familiaris) preference for human social interaction in a single-alternative choice.

Fonberg, E., Kostarczyk, E., Prechtl, J. (1981). Training of instrumental responses in dogs socially reinforced by humans. *Pavlovian Journal of Biological Sciences, 16*, 183–193.

Fox, M. W. (1964). The ontogeny of behaviour and neurologic responses in the dog. *Animal Behaviour, 12*, 301–310.

Frank, H., & Frank, M. G. (1982). On the effects of domestication on canine social development and behavior. *Applied Animal Ethology, 8*, 507–525.

Frank, H., & Frank, M. (1987). The University of Michigan canine information-processing project 1979–1981, In H. Frank (Ed.) *Man and wolf: Advances, issues, and problems in captive wolf and wolf-poodle hybrids.* Dordrecht: W. Junk Publishers.

Frank, H., Hasselbach, L. M., & Littleton, D. M. (1987). Socialized vs. unsocialized wolves (Canis lupus) in experimental research. In M. W. Fox & L. D. Mickley (Eds.) *Advances in animal welfare science* (pp. 33–49). New York: Springer.

Freedman, D. G., King, J. A., & Elliot, O. (1960). Critical period in the social development of dogs. *Science, 133*, 1016–1017.

Fukuzawa, M., & Hayashi, N. (2013). Comparison of 3 different reinforcements of learning in dogs (Canis familiaris). *Journal of Veterinary Behavior: Clinical Applications and Research, 8*(4), 221–224.

Gácsi, M., Miklósi, Á., Varga, O., Topal, J., & Csanyi, V. (2004). Are readers of our face readers of our minds? Dogs (Canis familiaris) show situation-dependent recognition of human's attention. *Animal Cognition, 7*, 144–153.

Gácsi, M., Györi, B., Miklósi, A., Virányi, Z., Kubinyi, E., Topál, J., et al. (2005). Species-specific differences and similarities in the behavior of hand-raised dog and wolf pups in social situations with humans. *Developmental Psychobiology, 47*, 111–122.

Gácsi, M., Gyori, B., Viranyi, Z., Kubinyi, E., Range, F., Belenyi, B., et al. (2009a). Explaining dog wolf differences in utilizing human pointing gestures: Selection for synergistic shifts in the development of some social skills. *PLoS ONE, 4*(8), e6584.

Gácsi, M., Kara, E., Belenyi, B., Topal, J., & Miklósi, Á. (2009b). The effect of development and individual differences in pointing comprehension of dogs. *Animal Cognition, 12*, 471–479.

Gantt, W. H., Newton, J. E. O., Royer, F. L., et al. (1966). Effect of person. *Conditioned Reflex, 1*, 146–160.

Gilchrist, J., Sacks, J. J., White, D., & Kresnow, M.-J. (2008). Dog bites: Still a problem? *Injury Prevention, 14*, 296–301.

Gipson, P. S., Gipson, I. K., & Sealander, J. A. (1975). Reproductive biology of wild Canis (Canidae) in Arkansas. *Journal of Mammalogy, 56*, 605–612.

Giret, N., Miklósi, Á., Kreutzer, M., & Bovet, D. (2008). Use of experimenter-given cues by African gray parrots (Psittacus erithacus). *Animal Cognition,12*, 1–10.

Giret, N., Miklósi, Á., Kreutzer, M., & Bovet, D. (2009). Use of experimenter-given cues by African gray parrots (Psittacus erithacus). *Animal Cognition, 12*(1), 1–10.

Ghosh, B., Choudhuri, D. K., & Pal, B. (1984/85). Some aspects of the sexual behaviour of stray dogs, Canis familiaris. *Applied Animal Behaviour Science, 13*, 113–127.

Hall, N. J., Udell, M. A. R., Dorey, N. R., Walsh, A. L., & Wynne, C. D. L (2011). Megachiropteran bats (Pteropus) utilize human referential stimuli to locate hidden food. *Journal of Comparative Psychology, 125*, 341–346.

Hare B., Brown, M., Williamson, C., & Tomasello, M. (2002). The Domestication of Social Cognition in Dogs. *Science, 298*, 1634–1636.

Hare, B., Plyusnina, I., Ignacio, N., Schepina, O., Stepika, A., Wrangham, R., et al. (2005). Social cognitive evolution in captive foxes is a correlated by-product of experimental domestication. *Current Biology, 15*, 226–230.

Hare, B., & Woods, V. (2013). *The genius of dogs*. New York: Dutton.

Harlow, H. F., & Zimmerman, R. R. (1959). Affectional responses in the infant monkey. *Science, 130*, 421–432.

Horowitz, A. (2012). Fair is fine, but more is better: Limits to inequity aversion in the domestic dog. *Social Justice Research, 25*, 195–212.

Harrington, F. H., Paquet, P. C., Ryon, J., & Fentress, J. C. (1982). Monogamy in wolves: A review of the evidence. In F.H. Harrington and P.C. Paquet (Eds.), *Wolves of the world* (pp. 209–222). New Jersey: Noyes Publications.

Igel, G. J., & Calvin, A. D. (1960). The development of affectional responses in infant dogs. *Journal of Comparative and Physiological Psychology, 53*, 302–205.

Klinghammer, E., & Goodmann, P. A. (1987). Socialization and management of wolves in captivity. In H. Frank (Ed.) *Man and wolf: Advances, issues, and problems in captive wolf research*. Dordrecht: W. Junk.

Kuhne, F., Hößler, J. C., & Struwe, R. (2012). Effects of human-dog familiarity on dogs' behavioural responses to petting. *Applied Animal Behaviour Science, 142*, 176–181.

Laveaux, J. C., & King of Prussia, F. (1789). *The life of Frederick the Second, King of Prussia: To which are Added Observations, Authentic Documents, and a Variety of Anecdotes*. London: J. Derbett.

Lord, K. (2013). A comparison of the sensory development of wolves (Canis lupus lupus) and dogs (Canis lupus familiaris). *Ethology, 119*, 110–120.

Lord, K., Feinstein, M., Smith, B., & Coppinger, C. (2013). Variation in reproductive traits of members of the genus Canis with special attention to the domestic dog (Canis familiaris). *Behavioural Processes, 92*, 131–142.

Lynch, J. J., & Gantt, H. W. (1968). The heart rate component of the social reflex in dogs: The conditional effects of petting and person. *Conditioned Reflex, 3*, 69–80.

Lynch, J. J. (1987). W. Horsley Gantt's effect of person. In F.J. McGuigan & T. A. Ban (Eds.) *Critical issues in psychology, psychiatry, and physiology: A memorial to W. Horsley Gantt*, (pp 93–106). Amsterdam: Gordon and Beach.

Macdonald, D. W., & Carr, G. M., (1995). Variation in dog society: Between resource dispersion and social flux. In J. Serpell, (Ed.) *The domestic dog: Its evolution, behaviour, and interactions with people*. Cambridge: Cambridge University Press.

McIntire, R., & Colley, T. A. (1967). Social reinforcement in the dog. *Psychological Reports, 20*, 843–846.

Mech, L. D., Wolf, P. C., & Packard, J. M. (1999). Regurgitative food transfer among wild wolves. *Canadian Journal of Zoology, 77*, 1192–1195.

Mech, L. D. (1970). *The wolf: The ecology and behavior of an endangered species*. Garden City, NY: Natural History Press.

Miklósi, Á., Kubinyi, E., Topál, J., Gácsi, M., Virányi, Z., & Csányi, V. (2003). A simple reason for a big difference: Wolves do not look back at humans, but dogs do. *Current Biology, 13*, 763–766.

Miklósi, Á., Polgárdi, R., Topál, J., & Csányi, V. (1998). Use of experimenter-given cues in dogs. *Animal Cogition, 1*, 113–121.

Miklósi, Á., Polgárdi, R., Topál, J., & Csányi, V. (2000). Intentional behaviour in dog-human communication: An experimental analysis of "showing" behaviour in the dog. *Animal Cognition, 3*, 159–166.

Miklósi, Á., & Soproni, K. (2006). A comparative analysis of animals' understanding of the human pointing gesture. *Animal Cognition, 9*, 81–93.

Miklosi, A. (2009). *Dog behaviour, evolution, and cognition* (1st ed.). New York: Oxford University Press.

Odendaal, J. S .J., & Meintjes, R. A. (2003). Neurophysiological correlates of affiliative behavior between humans and dogs. *Veterinary Journal, 165*, 296–301.

Ortolani, A., Vernooij, H., & Coppinger, R. (2009). Ethiopian village dogs: Behavioural responses to a stranger's approach. *Applied Animal Behaviour Science, 119*, 210–218.

Pack, A. A., & Herman, L. M. (2004). Bottlenosed dolphins (Tursiops truncatus) comprehend the referent of both static and dynamic human gazing and pointing in an object-choice task. *Journal of Comparative Psychology, 118*, 160–171.

Packard, J. M. (2003). Wolf behavior: Reproductive, social, and intelligent. In D. Mech, & L. Boitani (Eds.) *Wolves: Behavior, ecology, and conservation* (pp. 35–65). Chicago: The University of Chicago Press.

Pal, S. K. (2008). Maturation and development of social behaviour during early ontogeny in free-ranging dog puppies in West Bengal, India. *Applied Animal Behaviour Science, 111*, 95–107.

Pal, S. K., Ghosh, B., & Roy, S. (1998.) Agonistic behaviour of free-ranging dogs (Canis familiaris) in relation to season, sex and age. *Applied Animal Behaviour Science, 59*, 331–348.

Pavlov, I. P. (1928). *Lectures on conditioned reflexes* (trans Gantt WH). New York: International.

Pilley, J. W., & Reid, A. K. (2011). Border collie comprehends object names as verbal referents. *Behavioural Processes, 86*, 184–195.

Pongracz, P., Miklosi, A., Kubinyi, E., Gurobi, K., Topall, J., & Csanyi, V. (2001). Social learning in dogs: The effect of a human demonstrator on the performance of dogs in a detour task. *Animal Behaviour, 62*(6), 1109–1117.

Range, F., & Virányi, Z. (2011). Development of gaze following abilities in wolves (Canis lupus). *PloS One, 6*(2), e16888.

Reid, P. J. (2009). Adapting to the human world: Dogs' responsiveness to our social cues. *Behavioural Processes, 80*, 325–333.

Rheingold, H. L. (1963). Maternal behavior in the dog. In H.L. Rheingold (Ed.), *Maternal behavior in mammals* (pp. 169–202). New York: Wiley.

Riedel, J., Schumann, K., Kaminski, J., Call, J., & Tomasello, M. (2008). The early ontogeny of human-dog communication. *Animal Behaviour, 75*, 1003–1014.

Scheumann, M., & Call, J. (2004). The use of experimenter-given cues by South African fur seals (Arctocephalus pusillus). *Animal Cognition, 7*, 224–230.

Schleidt, W., & Schalter, M. (2003). Co-evolution of humans and canids: An alternative view of dog domestication: Homo Homini Lupus? *Evolution & Cognition, 9*, 57–72.

Schloegl, C., Kotrschal, K., & Bugnyar, T. (2007). Do common ravens (Corvus corax) rely on human or conspecific gaze cues to detect hidden food? *Animal Cognition, 11*, 231–241.

Schloegl, C., Kotrschal, K., & Bugnyar, T. (2008). Modifying the object-choice task: Is the way you look important for ravens? *Behavioural Processes, 77*(1), 61–65.

Scott, J. P., & Fuller, J. L. (1965). *Genetics and the social behavior of the dog*. Chicago: University of Chicago Press.

Soproni, K., Miklósi, A., Topál, J., & Csányi, V. (2001). Comprehension of human communicative signs in pet dogs (Canis familiaris). *Journal of Comparative Psychology, 115*, 122–126.

Soproni, K., Miklósi, A., Topál, J., & Csányi, V. (2002). Dogs' (Canis familiaris) responsiveness to human pointing gestures. *Journal of Comparative Psychology, 116*, 27–34.

Smet, A.F., & Byrne, R.W. (2013). African elephants can use human pointing cues to find hidden food. *Current Biology, 23*, 2033-2037.

Topál, J., Gácsi, M., Miklósi, Á., Virányi, Z., Kubinyi, E., & Csányi,V. (2005). Attachment to humans: A comparative study on hand-reared wolves and differently socialized dog puppies. *Animal Behaviour, 70,* 1367–1375.

Topál, J., Byrne, R. W., Miklósi, Á., & Csányi, V. (2006). Reproducing human actions and action sequences: Do as I do! in a dog. *Animal Cognition, 9*(4), 355–367.

Udell, M. A. R, Dorey, N. R., & Wynne, C. D. L. (2008a). Wolves outperform dogs in following human social cues. *Animal Behaviour, 76,* 1767–1773.

Udell, M. A. R, Dorey, N. R., & Wynne, C. D. L. (2010a). What did domestication do to dogs? A new account of dogs' sensitivity to C.D.L actions. *Biological Reviews, 85,* 327–345.

Udell, M. A. R, Dorey, N. R., & Wynne, C. D. L. (2010b). The performance of stray dogs (Canis familiaris) living in a shelter on human-guided object-choice tasks. *Animal Behaviour, 79,* 717–725.

Udell, M. A. R, Dorey, N. R., & Wynne, C. D. L. (2011). Can your dog read your mind? Understanding the causes of canine perspective taking. *Learning & Behavior, 39,* 289–302.

Udell, M. A. R, Giglio, R. F., & Wynne, C. D. L. (2008b). Domestic dogs (Canis familiaris) use human gestures but not nonhuman tokens to find hidden food. *Journal of Comparative Psychology, 122,* 84–93.

Udell, M. A. R, Hall, N. J., Morrison, J., Dorey, N. R., & Wynne, C. D. L. (2013). Point topography and within-session learning are important predictors of pet dogs' (Canis lupus familiaris) performance on human guided tasks. *Revista Argentina de Ciencias del Comportamiento, 5,* 3–20.

Udell, M. A. R, Spencer, J. M., Dorey, N. R., & Wynne, C. D. L. (2012). Human-socialized wolves follow diverse human gestures… and they may not be alone. *International Journal of Comparative Psychology, 25,* 97–117.

Udell, M. A. R, & Wynne, C. D. L. (2008). A review of domestic dogs' (Canis familiaris) human-like behaviors: Or why behavior analysts should stop worrying and love their dogs. *Journal of the Experimental Analysis of Behavior, 89,* 247–261.

Udell, M. A. R, & Wynne, C. D. L (2010). Ontogeny and phylogeny: Both are essential to human-sensitive behaviour in the genus Canis. *Animal Behaviour, 79,* e9–e14.

Virányi, Z., Gácsi, M., Kubinyi, E., Topál, J., Belényi, B., Ujfalussy, et al. (2008). Comprehension of human pointing gestures in young human-reared wolves (Canis lupus) and dogs (Canis familiaris). *Animal Cognition, 11,* 373–387.

Von Bayern, A. M. P., & Emery,. N. J. (2009). Jackdaws respond to human attentional states and communicative cues in different contexts. *Current Biology, 19,* 602–606.

World Health Organisation (2005). WHO expert consultation on rabies: First report. WHO Technical Report Series, 931. World Health Organisation. Geneva Switzerland.

Zimen, E. (1987) Ontogeny of approach and flight behavior toward humans in wolves, poodles and wolf-poodle hybrids. In H. Frank (Ed.), *Man and wolf* (pp. 275–292). Dordrecht: W. Junk Publishers.

Chapter 11
Canine Welfare Science: An Antidote to Sentiment and Myth

Nicola Rooney and John Bradshaw

Abstract Our understanding of the welfare of companion animals is both incomplete and fragmentary. For domestic dogs, most research has focused on animals that do not have stable relationships with people, such as dogs in laboratories and rehoming kennels. The welfare of pet dogs has received limited attention, presumably due to an assumption that owners have their best interests at heart. However, owners' conceptions of their companion's needs can be inconsistent or even contradictory. Dogs are, on the one hand, sentimentalised via anthropomorphic interpretations, but on the other, mythologized as the descendants of savage wolves requiring harsh correction before they will conform to the demands of living alongside people. Canine welfare science attempts to replace such mythos with objective norms that have proved effective when applied to other domesticated species. However, animal welfare science is rarely value-free or unambiguous, since it has variously been defined in terms of physical health, psychological well-being, and the freedom to perform 'natural' behaviour. Here we attempt to strike a balance between each of these approaches while addressing a wide variety of current issues in canine welfare, including: concerns arising from the breeding of pedigree dogs; inappropriate training methods; and the widespread occurrence of behavioural disorders. We finish by describing some barriers to improvement in dog welfare, including owners' anthropomorphisms, the challenges of finding reliable indicators of well-being, and the effects of applying erroneous conceptual frameworks to the dog-owner relationship.

N. Rooney (✉) · J. Bradshaw
Anthrozoology Institute, Animal Welfare and Behaviour Group, School of Veterinary Science, University of Bristol, Langford, Bristol BS40 5DU, UK
e-mail: Nicola.Rooney@bristol.ac.uk

J. Bradshaw
e-mail: J.W.S.Bradshaw@bristol.ac.uk

Fig. 11.1 Our treatment of dogs is mired in contradiction. In a city in China one of the authors recently observed a young puppy marooned amongst six lanes of traffic, clearly terrified, yet human onlookers observed passively and were unmoved to assist. Nearby was a bed-ridden teenager, wheeled into the street to beg for cash, smiling only when caressing her pet dog and upon whose care, some of her precious resources were evidently spent

11.1 Introduction

Given the long relationship between dogs and humans, and the apparent ability of dogs to understand much of what we do, surely if there is one species whose welfare needs we should be able understand and safeguard it will be man's best friend? Of course, in many developing countries dog welfare is less of a priority than it is in the West; in some nations dogs are farmed for food (Podberscek 2009), and in many, large numbers are un-owned, left to stray and roam the streets (e.g. Totton et al. 2011) and thus face welfare problems associated with lack of shelter and veterinary care. However, even where the majority of dogs are owned, vaccinated, and even neutered (PDSA 2011), it is doubtful that humans meet all the needs of their canine companions. Perhaps paradoxically, whilst dogs are widely revered and sentimentalised, and undoubtedly the majority of owners endeavour to do the best for their pet dogs, a lack of scientific knowledge and its incomplete dissemination to the pet owning population, combined with perpetuation of flawed theory, means that dog welfare may often be compromised.

Some owners spend vast sums of money on their canine pets, yet whether these resources are best allocated to enhance the animals' well-being is questionable (see Fig. 11.1). These illogicalities may be due, at least in part, to our long history and apparent co-evolution with dogs (see also Rossi et al. this volume), leading us to assume we know how best to meet their needs, and even in some instances assuming 'what's good for us is good for them.' Perhaps as a side-effect of such anthropomorphisms, the science exploring how best to protect the welfare of dogs lags well behind that of other domesticated species (Rooney in press; Stafford 2007; Yeates 2012). Furthermore, most investigations using dogs have focussed on those in temporary rehoming centres ('shelters') or laboratories, which provide convenient samples and avoid the problems of home-based studies (e.g. lack of standardisation) but are environments that differ considerably from those of the majority of dogs, which are kept as companions in homes with relatively stable inter-and intra-specific relationships (but see Fugazza and Miklosi, this volume).

In this chapter, we first describe what scientific study has suggested may be some of the most critical welfare issues for pet dogs. We summarise research which has been done on each issue, highlighting apparent paradoxes and knowledge gaps, and describing how the adoption of limited views of welfare may have perpetuated these issues historically. We then summarise different approaches to assessing welfare and the challenges of applying these to domestic dogs. We believe that the likelihood that the welfare of millions of dogs is further compromised by unsubstantiated assumptions that they are behaviourally akin to captive wolves, striving for "dominance" (Bradshaw et al. 2009), and consequently in need of coercive training methods which may cause significant suffering (as elaborated below). Finally we discuss how knowledge of the cognitive abilities of dogs is integral to our ability to safeguard their welfare.

11.2 Philosophical Approaches to What Might Constitute 'Well-being' for Domestic Dogs

Historically, animal welfare science has adopted three different approaches to arrive at what constitutes a 'good life' for animals given that they cannot tell us directly (Fraser 2009a). One is the conventional approach of veterinary science, that animal welfare revolves around physical well-being and biological functioning, as indicated by health, growth, productivity and immune function. The second is rooted in neuropsychology, emphasising that it is the animal's subjective experience of the world, its "affective state", that determines its welfare status. Essentially, this is a feelings-based approach: if the animal is in a net positive affective state, it is faring well (e.g. Duncan 1996). The third, based in natural history, identifies well-being as a state in which the animal can perform its natural behaviour, and more generally, express its '*telos*', or 'purpose'.

Whilst it is difficult to argue against any of these approaches, if used in isolation they can sometimes come to conflicting end-points, and this is especially true for dogs. For example, the emphasis on health at all costs has undoubtedly been a contributory factor to the barren conditions in which many kennelled dogs used to be (and in some case still are) kept: such conditions lead to negative feelings both due to physical confinement and from being separated from their human caregivers for long periods of time. Relatedly, by adopting the affective state approach alone, one could rule as benign, some of the abnormalities resulting from pedigree dog breeding, such as congenital deafness, if the dogs affected are not aware of their disability. This stands in direct contrast to the healthy living and natural behaviour approaches, which would both rule such malformations as diminishing welfare.

The 'natural living' approach is especially problematic when applied to domestic dogs. Detractors of this approach sometimes assert that despite being 'natural', it is hardly welfare-friendly for any animal to experience fear of predators, or cold in unheated kennels. This objection is easily remedied by factoring the animals' choices into the equation: if an object/situation is natural to the animal, *and* the animal shows through its behaviour that it has a preference for that object, then it is probably beneficial to the animal's welfare (Fraser 2009b). Note that preference alone is also not sufficient. For example, many dogs choose to consume more food than they require, which if unchecked leads to obesity and thence potentially to physical discomfort and increased risk of cardiovascular and other diseases. However, the greatest barrier to applying the principle of natural living to dog behaviour is defining what is 'natural' for a species which is, cognitively speaking, probably more modified from its wild ancestor than any other. Furthermore, concepts of what constitutes dogs' 'natural' behaviour is still often coloured by misconceptions of what is natural for their best-studied wild counterpart, the American grey wolf (Mech 2008). Moreover, we know little of the behaviour of the wolf that was the direct ancestor of the domestic dog. In addition, we have sparse information of dogs' 'natural' behavioural patterns, for example resting, feeding, and activity (see Rooney in press), beyond the observation that these are easily influenced by humans. The 'natural behaviour' approach has been useful in evaluating the welfare of wild animals kept in zoos, and also for domesticated animals, such as pigs kept in gestation crates, whose behaviour has been usefully compared to that of free-roaming pigs and the ancestral wild boar. But when it comes to companion species, whose behaviour has been artificially selected for many years, one wonders whether 'natural behaviour' is a relevant concept. Furthermore, where this concept has been applied to dogs, it may not have been applied correctly. For example, physical punishment (e.g. pinning a dog to the floor) is often justified as a legitimate component of dog training on the grounds that it mimics the 'natural' behaviour of 'alpha' wolves towards 'subordinate' members of the pack.

The capacity to perform sexual behaviour, however, natural, is often deemed to impact negatively on canine welfare. Animal welfare charities promote neutering as essential for the welfare of individual dogs, despite the procedure's consequence of suppressing 'natural' sexual behaviour. Neutering can improve individual health in a number of ways (Kim et al. 2006), but the welfare gain here is primarily at the level of the population, reducing the number of puppies that cannot find homes, and either die, often after considerable suffering, or are euthanised, depending upon local circumstances. But it is undeniable that a dog that has been desexed has been deprived of the ability to perform one of the most fundamental natural behaviours.

There is a strong argument that good welfare considers all three approaches, and indeed most animal welfare scientists now believe that physical, mental, and 'natural-living' aspects of welfare are interrelated and are all of ethical concern (e.g.

Appleby 1999; Fraser et al. 1997; Hewson 2003). A composite approach, encompassing elements from all three of Fraser (2009a) approaches in different proportions has been applied to dog welfare and identifies five needs: suitable environment; suitable diet; opportunity to exhibit normal behaviour; to be housed with, or apart from, other animals, depending on preference; to be protected from pain, injury, suffering, and disease (DEFRA 2009; Rooney in press). However, the relative value apportioned to each component can still vary according to the assessor or authors' values and as such, welfare science can never be truly objective (Fraser 2003). Any discussion of canine welfare needs to refer to each one of these approaches, and having done so, must attempt to resolve any discrepancies between them. Moreover, the emphasis placed on each approach will need to change as knowledge evolves. On the one hand, veterinary science is a comparatively mature discipline, and it is unlikely that there will be any fundamental shift in what constitutes a healthy dog. On the other, our knowledge of the emotional life of dogs is still in its infancy, and new methods are likely to keep emerging, enabling science to probe emotions far more precisely than is possible today.

11.3 Major Issues in Dog Welfare

Dogs are kept in a wide variety of environments, including laboratories (Hubrecht 1993) and kennel establishments housing large numbers of working dogs for much of their lives (Rooney et al. 2007); short-term housing in rehoming centres (e.g. Hennessy et al. 2002; Hiby et al. 2006; Tuber et al. 1996); kennels within owners' gardens (Kobelt et al. 2007); and within owners' homes with access to roam freely between multiple rooms. Each environment presents its own challenges. Whilst it is often assumed that home-dwelling dogs experience near optimal conditions, and indeed Shore et al. (2006) found that people keeping dogs indoors, compared to in a yard, paid more attention to their social needs, it has also been asserted that modern companion animals experience "a relatively dull life" (Stafford 2007), and it has been shown that owners lack the knowledge necessary to safeguard their dogs' welfare (PDSA 2011).

Recent legislation and social pressure in many cultures has led to dogs being kept predominantly indoors, an environment very different from that in which they were originally domesticated. Below we describe a number of current welfare challenges faced by pet dogs and discuss the extent to which each of the three approaches to welfare discussed above would describe them to be a problem. Since this volume is primarily concerned with cognition and behaviour, and health has tended to dominate much of the literature on dog welfare (Asher et al. 2009; Hiby 2013), we will primarily discuss issues that arise from lack of understanding of, and provision for, dogs' behavioural needs.

However, that is not to say that physical health is not a major issue worldwide. Many street dogs are chronically unhealthy, and Non-Government Organisations invest heavily in mass vaccination and neutering programmes aimed at safeguarding the health of current and future generations (although little attention is paid to affective and behavioural aspects of their welfare). In Western nations, many owners appear to protect the health of their dogs, most dogs are vaccinated against transmissible diseases (PDSA 2011) and many receive prophylactic treatment against parasites (Gates and Nolan 2010). Mass rearing establishments (puppy 'farms' or 'mills') are the obvious exception, where dogs are often bred and reared in large numbers with apparent disregard for disease prevention (Bateson 2010). However, even in the owned population severe health issues persist.

11.3.1 Obesity

Most dogs in developed nations are apparently healthy and well-fed (Houpt et al. 2007), but obesity is now a major issue (Yeates 2012). It has even been suggested as the most important welfare problem of dogs in the post-industrial developed world (Stafford 2007) due in part to the sheer numbers of animals involved: for example, the (Pet Food Manufacturers' Association 2009) state that up to one-third of companion animals in the UK are obese.

Obesity was originally highlighted as a welfare issue due to its effects upon health and longevity, but it is now also known to exacerbate a whole variety of clinical conditions (German et al. 2010) some of which have also shown to be painful (Taylor 2003) and hence to cause psychological distress. In addition, dietary restrictions placed upon animals under treatment for obesity are likely to result in negative feelings of hunger and frustration as the animals will be unaware of their treatment's long term aim or benefit. It is possible that in the process of breeding dogs which are amenable and easily trained using food rewards, we have inadvertently selected individuals which are excessively food-motivated and which feel hungry even when they are physiologically well-nourished, a stereotype anecdotally apportioned to many Labrador Retrievers (among others). If so, it is possible these dogs may suffer psychologically, even when well fed, let alone when they are placed on 'health-enhancing' feed management plans. This situation provides an example of when psychological and health approaches to welfare may be contradictory.

Obesity can also reduce a dog's ability to behave normally, and hence when extreme, would be an issue affecting 'naturalness'. Obesity is a rare example of a welfare issue which all three approaches indicate as a welfare concern; however, the cut-off point at which it would be highlighted as an issue is likely to differ, dependent upon which criteria (health, feelings, or natural behaviour) are paramount.

11.3.2 Pedigree Dog Breeding

Approximately 41 % of dogs in the UK are described by their owners as pedigrees (J. T. Murray unpublished telephone survey). Many such dogs are far from healthy, as has been highlighted both by the popular media (e.g. BBC 2008) and in a range of reports (e.g. Rooney and Sargan 2009; Bateson 2010), reviews, and scientific papers (e.g. Nicholas et al. 2010; Rooney 2009; Steiger et al. 2008; Summers et al. 2010). Breeding dogs primarily for their appearance has led to compromised health and welfare in two different ways, one resulting directly from selection for exaggerated physical features and the other, indirectly resulting in an increased incidence of disease (see also Duffy and Serpell, this volume).

11.3.2.1 Exaggerated Physical Features

Artificial selection has resulted in a wide variety of morphologies in different breeds of dog. Many breeds are anatomically modified in ways which compromise their physical health (Asher et al. 2009; Rooney and Sargan 2010). The English bulldog is a regularly cited example of morphological extremes, resulting in locomotion difficulties, breathing problems, and an inability to mate or give birth without physical and/or surgical interventions (Advocates for Animals 2006). However, there are many other less visually obvious anatomical deformities in other breeds, ranging from overly long backs to heavily wrinkled skin, and flat faces that restrict breathing (Asher et al. 2009).

Systematic studies have started to investigate the effects of these breed modifications. For example, Brachycephalic Obstructive Airway Syndrome (BOAS) has been shown to have detrimental effects on health (Packer et al. 2013), and syringomyelia causes significant pain to Cavalier King Charles Spaniels (Plessas et al. 2012) making it clear that breeding dogs with such traits affects not merely their health and longevity but also psychological well-being. Effects on the dogs' capacity to behave normally are less frequently mentioned (Rooney 2009). Severely reduced limb lengths, for example, may restrict the ability of dwarf dogs to run freely, and breeds with respiratory deformities (e.g. brachycephalic breeds) may be prevented from running without shortness of breath. Their ability to explore and exercise is compromised, and their opportunity for social interactions limited, which likely restricts their socialisation and thus potentially further restricts natural behaviour and diminishes their quality of life.

Numerous breeds are also anatomically modified such that their capacity to signal is drastically reduced (Goodwin et al. 1997): stiff legs prevent adjustments of height; brachycephalic breeds are less able to utilise facial expressions; dogs with very short or curled tails, or with immobile drooping or erect ears lose the function of important signalling structures. Short permanently erect fur leaves dogs unable to raise their hackles, whilst long or dense fur can obscure much body language communication. Short legs and long bodies not only prevent proper locomotion, they may also prevent a dog from play-bowing to invite playful interactions with other dogs. Play behaviour is rewarding (Boissy et al. 2007) and is important for normal social development (Simmel 1979), and high play levels are an indicator of positive welfare (e.g. Jensen et al. 1998). Since play signalling is critical to the initiation and continuation of dog play (Bekoff 1995; Rooney et al. 2001), and can prevent interaction escalating into aggression, this is also a potential welfare concern.

11.3.2.2 Increased Incidence of Inherited Disease

Selective breeding primarily for appearance has also led to dog breeds becoming especially susceptible to a whole suite of disorders (Summers et al. 2010), many of which are acutely painful or chronically debilitating. This is a result of reduced genetic diversity, coupled with ill-advised breeding practices (whereby breeders inadvertently select regions of the genome which happen to contain a disorder as well as the trait they desire), and insufficient selection pressure on health and welfare. The inherited diseases are wide-ranging and include cardiac disease, eye disease, diabetes, glaucoma, and congenital sub-aortic stenosis (Rooney and Sargan 2009, 2010). In a review of fifty UK dog breeds, every breed suffered from at least one of the 312 inherited disorders catalogued, with the nervous system the most commonly affected organ (Summers et al. 2010), pointing to potential effects on psychological well-being that are not yet fully explored.

Selective breeding practices are also likely to impact on behaviour indirectly, since basing breeding choices primarily on physical appearance means attention is diverted away from temperament (McGreevy 2008). Studies reporting breed differences in the incidence of behaviour problems abound, but interpretation of these differences is not always straightforward. Some of these may be a consequence of selective breeding: For example, there is evidence for a genetic predisposition toward aggression in some lines of golden retrievers (Knol et al. 1997; Liinamo et al. 2007). Behaviour described as "dominant-aggressive" varies greatly between differently coloured cocker spaniels (Perez-Guisado et al. 2006; Podberscek and Serpell 1997), and Duffy et al. (2008) found that show-lines of English Springer spaniels were more aggressive to humans and other dogs than were lines bred for work in the field. More generally, Svartberg (2005) compared 13,097 Swedish dogs of 31 breeds, and found that dogs bred for working were less likely to display social and non-social fearfulness, and were more playful and curious, than dogs from show lines (Svartberg 2005). Caution is needed when interpreting such differences, since it is generally impossible to tease apart the consequences of selective breeding from differences in the way breeds (and especially lines within breeds) are reared. However, since most pedigree dogs live the majority of their lives as household pets, and behavioural disorders can both be a consequence of negative affective states and can also result in dogs being surrendered or euthanized, selective breeding remains a significant welfare issue.

11.3.2.3 Why Do Such Effects Persist?

It is difficult to comprehend how breeders and purchasers could condone and even perpetuate such problems. When studying BOAS, Packer et al. (2012) recorded that despite over two-thirds of owners of affected dogs reporting daily breathing difficulties in their dogs during exercise, 58 % stated that their animal did not currently, or had ever had a breathing "problem". This suggests that most owners do not recognise breathing difficulties as a welfare issue for the dog (Packer et al. 2012), perhaps because they rely on a narrow definition of welfare, in which provided they see no obvious signs of pain, then behavioural and mood effects can be ignored. In recent years, breeding associations (e.g. UK Kennel Club) have started both to encourage testing for and breeding to eliminate inherited diseases, and also to alter some breed standards to dissuade breeders from selecting for the most extreme and obviously debilitating conformations. However, since all fifty of the most popular breeds have at least one aspect of their conformation that predisposes them to one or more disorder (Asher et al. 2009) (and there is no reason to assume the remaining 250-plus breeds are not similarly affected), many would argue that more needs to be done in order to prevent suffering (EFRA 2013). Undoubtedly, education of owners as well as breeders is still required.

11.3.3 Lack of Human Company

Dogs have been selectively bred to value and even crave human attention, and the beneficial effects of human contact on dogs have been well demonstrated experimentally (Taylor and Mills 2007). Obedience and other favourable behaviour improved in shelter dogs and in working dogs when they receive increased contact from prisoners (Hennessy et al. 2006) or handlers (Haverbeke et al. 2010; Lefebvre et al. 2007). Positive interactions (stroking, scratching, talking, playing) result in increased beta-endorphin, oxytocin, prolactin, beta-phenylethylamine, and dopamine (Odendaal and Meintjes 2003) and reduced cortisol levels in new arrivals to rehoming kennels (Shiverdecker et al. 2013), and gentle stroking inhibits the increase in cortisol during venipuncture (Hennessy et al. 1998). Longer-term, enhanced human contact not only reduces physiological stress in dogs per se (Coppola et al. 2006) but has also been shown to ameliorate stress responses when dogs are subsequently presented with novel environments (Hennessy et al. 2002) or unfamiliar people (Bergamasco et al. 2010). Contact with people generally has greater success in inhibiting stress in novel environments than does contact with familiar dogs (Tuber et al. 1996); hence, human contact has been suggested to be more important for the well-being of dogs than conspecific contact (Valsecchi et al. 2007; Wells 2004), although this undoubtedly varies with an individual dog's temperament and past experience.

The detriments due to lack of contact with people are well proven for kennelled dogs, but these likely also apply to the majority of owned dogs. One of the two most common reported reasons for owners surrendering a dog to a rehoming centre in the UK is the perception that it requires more attention than the owner is able to provide (Diesel et al. 2010). Increasingly, dogs are kept within households where all members of the family are absent for at least part of the day. Reliable data is lacking, but undoubtedly a large proportion of dogs are routinely left for periods in excess of 4 h per day (e.g. 73 % in Sweden) (Norling and Keeling 2010). So paradoxically whilst we "value the social feedback they give us" and derive great benefits from it we also "leave them alone for lengthy periods" (McGreevy and Bennett 2010), unsurprisingly, many dogs behave in ways that indicate that they find such separation stressful, and this presents one of the biggest challenges of the modern day companion-animal niche.

11.3.4 Separation-Related Behaviour

Separation-related behaviour (usually expressed as destruction, vocalisation and/or toileting) is a common and often overlooked issue. Dog-walkers in southern England reported 24 % of dogs as currently exhibiting, or previously exhibiting, separation-related behaviour (Bradshaw et al. 2002a), and up to 50 % of Border Collies and Labradors filmed in a longitudinal study had shown this behaviour by 18 months of age (Bradshaw et al. 2002b). This problem is not restricted to the UK: 56 % of dogs living in urban Rio de Janeiro were reported to show clinical signs of separation anxiety (Soares et al. 2010).

It may seem contradictory that loving owners accept such signs of apparent distress, but a large proportion seem unaware of their dog's behaviour in their absence, and amongst those that are aware, anthropomorphic attitudes lead many to believe that destructive or eliminative behaviours are motivated by higher-level emotions such as 'revenge' (Borchelt and Voith 1982). Dogs often show no obvious health decrements as a result of their distress, and perhaps for this reason, the severity of separation as a welfare issue has been widely overlooked. Surprisingly, there appear to have been no published studies examining physiological indicators of stress during separation from the owner. However, recently developed cognitive measures (discussed below) have shown that animals which exhibit separation related behaviour are more likely to be experiencing a negative emotional state, not just when they are actually performing the behaviour, but also at other times (Mendl et al. 2010).

Protocols to treat separation-related behaviour can be very effective, producing improvements in 81 % of dogs (Blackwell et al. 2006) and methods to prevent the behaviour developing also show promise (Blackwell et al. 2005). Both can potentially improve a dog's quality of life, as well as reducing the chances of it being surrendered. More general adoption of such techniques, combined with a change in owner expectations such that they no longer routinely leave susceptible dogs alone for long periods of time, has the potential to improve the welfare of large numbers of dogs. However, research is also needed to fully understand the respective roles of past experience, expectation, and owner behaviour in determining why some individuals develop this problem whilst others appear unaffected (Rehn and Keeling 2011).

11.3.5 Fear and Anxiety

Fear and anxiety are both emotional responses to aversive stimuli. Fear may be defined in terms of behavioural responses shown to actual danger (Boissy 1998), whilst anxiety is a state elicited in potentially threatening situations, which can range from simple novelty to situations where some elements of the environment predict a negative outcome (Ennaceur et al. 2006; Massar et al. 2011). These are, critically, distinct emotions whose interaction and development require more detailed description than is possible here (see Casey 2011). Whilst fear and anxiety are adaptive, enabling avoidance of an immediate or anticipated threatening stimulus (Jones and Boissy 2011), they are negative emotional states and if experienced frequently in situations which the dog can neither predict nor control, will induce associated stress responses, with serious implications for welfare (Beerda et al. 1999b).

An experienced observer of dog behaviour only needs to take a walk through a park to notice many dogs displaying frequent signs of fear and anxiety (see Table 11.1 for a partial list). Social fears, both of humans and other dogs, are commonly presented at behavioural clinics, and a recent study confirmed that behavioural signs of fear on exposure to noises are a common, underreported, and significant welfare concern for pet dogs (Blackwell et al. 2013). Almost half of owners interviewed reported that their dog showed at least one behavioural sign typical of fear when exposed to noises, even though only a quarter had reported their dog was 'fearful' in a more general survey. This discrepancy indicates that even when owners recognise behavioural responses, they may not interpret these as associated with an altered subjective state, and hence a welfare concern in their dog.

Not only is psychological well-being affected by perpetual fear and/or anxiety, but so is the dog's behaviour. For example, working dogs' ability to carry out their trained task can be detrimentally affected by fear (e.g. Rooney et al. in prep), and this is also a common reason for guide dog failure (Goddard and Beilharz 1984). Fearful dogs have an impaired ability to learn during an operant task (Blackwell et al. 2010), and those described to have anxiety related disorders take longer to solve a problem solving task (Passalacqua et al. 2013): both may be contributing factors to reduced performance in working dogs.

High levels of fear and anxiety can also be an issue even when considering a health-centred approach, since they can lead to immune-suppression, which in other species has been linked to increased disease risk (Terlouw et al. 1997). Notably, a recent retrospective study of 721 dogs found that how "well-behaved" an owner perceived their dog to be was significantly predictive of greater than average lifespan (Dreschel 2010). This link was not due to owners euthanizing disobedient dogs, nor was it linked to specific diseases. When major factors such as weight and neuter status were controlled for, stranger-directed fear independently and significantly predicted decreased lifespan. Dogs showing extreme non-social fear also tended to have more skin problems and to exhibit higher levels of "separation anxiety". The author suggested that a lifetime of stress may take its

Table 11.1 Behavioural signs utilised as indicators of acute and chronic stress (many of which can be indicative of fear or anxiety (Casey et al. 2011), in a selection of studies of dog welfare

Behavioural sign	Examples of studies using behavioural pattern
Startling	Slabbert and Odendaal (1999)
Low body position; crouching; cowering	Beerda et al. (1997), (1998), (1999a), Doering et al. (2009), Galac and Knol (1997), Schilder and van der Borg (2004)
Low tail position	Doering et al. (2009), Galac and Knol (1997), Schilder and van der Borg (2004)
Yawning	Beerda et al. (1998)
Ears held low or pulled back	Galac and Knol (1997), Schilder and van der Borg (2004)
Trembling/body shaking	Beerda et al. (1998), (1999a), Doering et al. (2009)
Vocalisations e.g. whining, whimpering	Beerda et al. (1999a), Doering et al. (2009), Dreschel and Granger (2005), Shiverdecker et al. (2013)
Yawning	Beerda et al. (1998), Doering et al. (2009)
Licking lips (also referred to as tongue out and tongue flicking)	Beerda et al. (1997), Doering et al. (2009), Schilder and van der Borg (2004)
Changes in locomotory state; increased vigilance, restlessness	Beerda et al. (1999a), Dreschel and Granger (2005)
Self grooming (increased)	Beerda et al. (1999a)
Paw lifting	Hiby et al. (2006), Beerda et al. (1997), (1999a)
Withdrawal/hiding	Doering et al. (2009); Dreschel and Granger (2005); Goddard and Beilharz (1986)
Destructive behaviour e.g. scratching housing	Palestrini et al. (2005)
Panting (not related to heat)	Doering et al. (2009), Hiby et al. (2006), Shiverdecker et al. (2013)
Change in activity	Beerda et al. (1999a), Palestrini et al. (2005)
Coprophagy	Beerda et al. (1999a)
Self mutilation	Prescott et al. (2004)
Repetitive behaviours	Beerda et al. (2000)

toll on a dog's body at a molecular level, causing accelerated aging of cells and earlier death from a number of causes. Although this is single study and the mechanisms behind these effects are unproven, it shows that living with fear and/or anxiety can negatively affect health and lifespan, highlighting the interplay between physical and psychological well-being. Importantly, it also highlights how issues previously ignored due to their irrelevance to one approach to welfare (e.g. health) can gain new importance in the light of new research.

One may wonder why so many dogs are fearful. A dog's ability to cope with environmental stressors is affected by a combination of their individual personality, early rearing environment, and later experiences (Foyer et al. 2013). Fearfulness is one dimension of personality repeatedly seen to be heritable (Goddard and Beilharz 1985, 1986). Hence, by actively selecting predominantly for appearance, selection pressure may have neglected to select against fearfulness, or even inadvertently selected fearful lines. Among environmental factors, gradual, calm, and rewarding introduction to potential fear-inducing stimuli during sensitive periods of learning is well known to reduce the risk of fears and anxiety developing (e.g. Bailey 2008). Thus another important contributing factor to common fear is the number of dogs that are incompletely or inappropriately "socialised" (introduced to a range of stimuli, environments, people, and other animals) during puppyhood. This is exemplified by a study of ex-puppy-farm breeding stock, with limited environmental exposure, which showed significantly higher levels of social and non social fear when compared to a matched sample of companion dogs (McMillan et al. 2011). Furthermore, offspring from such establishments sold in pet shops showed less desirable behaviour than dogs procured from "non-commercial breeders" (McMillan et al. 2013). A third reason is the wide ranging (Tami and Gallagher 2009) and generally poor ability of humans to interpret canine signs, even amongst those living and working with dogs (e.g. Kerswell et al. 2009; Mariti et al. 2012). Hence, although people may be able to interpret dogs' facial expressions (e.g. Bloom and Friedman 2013) many owners mislabel, misinterpret (Tami and Gallagher 2009), and fail to notice other visual signals (Correia et al. 2007) and stress-induced behaviours shown by their own dog (Mariti et al. 2012). Thus even though many dogs that have developed fears can be helped using techniques such as densitization and counter-conditioning (e.g. Levine et al. 2007), a lack of recognition of the signs of fear means that dogs which are referred to qualified clinicians remain a small proportion of those which suffer.

11.3.6 Dog Training

Training methods can be classified according to the emphasis they place on the four types of reinforcement recognised under learning theory: positive reinforcement (e.g. food reward), positive punishment (ranging from harsh vocalisation to the deliberate infliction of acute pain), negative punishment (usually the withholding of an anticipated reward), and negative reinforcement (the cessation of pain or other aversive stimulus) (McGreevy and Boakes 2007). In practice, no training method relies exclusively on just one of these: even "reward-based training" incorporates negative punishment (e.g. withholding human attention) to reduce the performance of undesired behaviour (Ryan 2010). Positive punishment, especially the infliction of pain through hitting, twisting of the ears, choking the windpipe and electric shocks, is intrinsically aversive and hence, according to the 'affective approach', welfare-reducing. It is often justified by its supporters on the grounds that it is more than compensated for by greater benefit to the individual dog's future welfare (in terms of health), because the dog becomes more obedient and hence less likely cause damage to itself, other dogs or livestock, and less likely to be abandoned. There is, however, an increasing body of evidence pointing to the contrary, that dogs trained using methods that incorporate positive punishment tend to be less obedient and/or more aggressive than those trained using "reward", and moreover many such dogs also show signs which may be indicative of fear and anxiety (Arhant et al. 2010; Blackwell et al. 2008; Casey et al. 2013; Herron et al.

2009; Hiby et al. 2004; Hsu and Sun 2010; O'Sullivan et al. 2008; Rooney and Cowan 2011; Tami et al. 2008). Studies that have focused specifically on dogs trained with remotely-activated electric shock collars have found associated evidence of chronic fear to an extent not shown even by dogs trained with other types of positive punishment (Schilder and van der Borg 2004; Schalke et al. 2007) yet such devices remain available to dog owners in many countries, including the USA and parts of the UK (Blackwell et al. 2012). Reward-based training has been criticised on ostensible welfare grounds, as producing "soft" dogs that are disobedient, likely to stray or be injured, and hence at risk of abandonment[1] (McHugh 2009) although there does not appear to be any systematically collected data to support this conclusion, and increasing evidence to the contrary that it is associated with increased obedience (Hiby et al. 2004) and learning ability (Rooney and Cowan 2011).

11.3.7 'Problem' Behaviours

It is traditional to make a distinction between problems that result from seemingly abnormal behaviours—whose relevance to welfare is best understood through the 'affect' approach—and problems that are essentially 'natural behaviour' and arise due to conflict with owners' expectations. The latter (e.g. barking, chasing, and possibly even coprophagy) (Rooney in press) cause a problem when performed within the domestic environment and hence potentially threaten the animal's future welfare through a secondary route by jeopardising the dog-owner relationship (e.g. see Mills and Zulch 2010). However, as our knowledge base increases, this distinction has become increasingly arbitrary, as almost all 'problems', possibly with the exception of neurological disorders, can be explained by 'natural' reactions to unnatural or suboptimal environments. None more so than canine aggression, a highly contentious topic that is too complex to be dealt with here (but see Serpell, this volume).

Undesired or problem behaviour is generally important to dog welfare as it may not only be associated with poor welfare in its own right, it is also often the primary reason given when dogs are surrendered to animal centres, or euthanased. There are few longitudinal studies examining dogs' welfare before and after relinquishment, but since abandonment increases the likelihood of euthanasia, and there are also numerous factors associated with life in a rehoming kennel which can lead to psychological distress (Rooney et al. 2009; Taylor and Mills 2007), it is likely that for many dogs, welfare would be better were the dog to remain in its

---

[1] For example, UK trainer Charlie Clarricoates, who stated "We are seeing dogs now who are spoiled rotten, and never have any discipline, mainly because owners are force-fed incorrect impractical information...This moralistic attitude that you can only train dogs by loving them and being kind is ridiculous. There are some dogs you can't do this with because it doesn't work, even if you have a year with them" (McHugh 2009).

original home. Therefore one of the most effective ways to improve companion dog welfare is to prevent problematic behaviours from developing. This could be aided both by increased owner education and matching of appropriate dogs to owners, and also by mitigating those problems that do develop through professionalized clinical behaviour advice (Wickens 2007).

11.4 Obstacles to Improving Dog Welfare

11.4.1 Misconceptions in Canid Behaviour

The deep-seated and widespread assumption that pain and fear can legitimately be used in the routine training of dogs, whatever its origins in the human psyche, is often justified via a misapplication of the principle that 'natural' behaviour makes a positive contribution to welfare. Misconceived analogies with wolf behaviour are used to justify the infliction of pain on dogs, based on the outmoded precept that wolf society is regulated through the systematic (if controlled) infliction of violence by "dominant" or "alpha" individuals on "subordinates"—the younger or weaker members of the "pack" (e.g. The Monks of New Skete 1978, 1991). This conception of the wolf pack is now considered to be deeply flawed, with aggression actually rare within natural packs, which are cohesive kin-based units, and largely confined to encounters between members of rival packs (Packard 2003; Mech 2008). Moreover, domestication appears to have had such a profound effect on dogs' cognitive and learning abilities that uncritical analogies between dogs and wolves may be generally misleading (Bradshaw 2011). Moreover, dogs clearly alter their social behaviour depending upon whether the other participant is a dog or a person (Rooney et al. 2000). Despite the accumulating evidence that physical punishment does not improve welfare, at least in the hands of the majority of dog owners, training methods based on instilling 'respect' (a euphemism for the infliction of pain and in itself a prime example of an anthropomorphic assumption of higher cognitive processing) are still widely promoted in popular media (Greenebaum 2010). Confusion is also perpetuated by the continued use of the term "dominance aggression" to refer to any aggression towards familiar humans, much of which is motivated by fear or anxiety (Luescher and Reisner 2008).

Other concepts that are widespread in dog "lore" as reflecting the normal behaviour of canids include the supposed benefits of food-depriving a dog on one day each week[2] and the exclusion of 'unnatural' cereals from the dog's diet,[3] despite the evidence that the Canidae are natural omnivores and prefer to eat small regular meals (Axelsson et al. 2013; Morris and Rogers 1989).

---

[2] See, e.g., http://www.rawfoodvets.com/articles/article7.
[3] E.g., http://www.rawfeddogs.org/.

It is not clear how such misconceptions may be eliminated. As Jean Donaldson memorably wrote "Dog training is a divided profession. We are not like plumbers, orthodontists or termite exterminators who, if you put six in a room, will pretty much agree on how to do their jobs. Dog training camps are more like Republicans and Democrats, all agreeing that the job needs to be done but wildly differing on how to do it" (Donaldson 2009). Such disagreements not only have the potential to confuse dog owners, and consequently their dogs, they also promote strongly-held but vastly different opinions as to how best to protect the welfare of dogs.

11.4.2 Anthropomorphism and Anthropocentrism

The welfare of companion animals, especially dogs, is profoundly influenced by the human context in which they live. Unlike farm, laboratory, or wild animals, pet dogs are kept primarily for the companionship they provide: insofar as their owners are concerned, they are often members of a multi species 'family', and may even be perceived in a similar way to children (Blouin 2013, and references therein). Dog owners routinely interpret their dogs' behaviour using a framework based upon human social motivations; indeed, such anthropomorphic perceptions may well be "what ultimately enable people to benefit socially, emotionally and physically from their relationships with companion animals" (Serpell 2005). Furthermore, it seems likely that many owners pay insufficient attention to their dogs' distinctive sensory abilities, anthropocentrically presuming that their pets inhabit the same sensory world that they do (see also Horowitz and Hecht, this volume). We will here consider anthropomorphism, the unjustified attribution of human qualities to animals, and anthropocentrism, the interpretation of reality exclusively in terms of human values and experience, as separate influences on welfare, although in reality they refer to overlapping misinterpretations of dogs' behaviour, starting with misunderstandings of sensory input and continuing with misapprehensions of the motivations underlying the behavioural output. (For a more detailed discussion of these ideas, see Bradshaw and Casey 2007).

11.4.2.1 Anthropomorphism

The "affective states" approach to animal welfare can only be reliable if it adopts a realistic approach to emotional capacity, and therefore anthropomorphism is likely to lead to misleading conclusions. Dogs undoubtedly have the mental capacity for relatively complex social cognition (see Prato-Previde and Marshall-Pescini, this volume), even by comparison with other companion animals, but these can be overestimated or wrongly interpreted, as illustrated in our discussions of social motivations in the Sect. 11.4.1.

Although it was once not the case, it is now commonplace for biologists to accept that all mammals experience a range of basic emotions such as fear, pleasure, and anxiety. Homologies between humans and other mammals are based upon (a) a common set of neural pathways in the brain, generating emotions and relating them to external events (LeDoux 2000), and (b) the difficulty of explaining certain types of spontaneous behaviour, such as play and food-hoarding, unless they are associated with positive emotion, i.e. actions that are apparently "self-rewarding" (Fraser 2009b).

However, tertiary consciousness, arising from the massive expansion of the neocortex that occurred during human evolution, gives dog owners an ability to apply rationality to emotions in a way that dogs appear to be completely incapable of. In Horowitz (2009) seminal experimental demonstration of owners' misattribution of "guilt" to their dogs, the owners were asked to leave their dogs alone with an experimenter who either did or did not allow the dog to eat a forbidden treat. When the owners returned, some were told that their dog had taken the treat, others that it had refused it (in all four possible combinations). Crucially, it was the dogs whose owners had been told, whether rightly or wrongly, that they had taken the treat, that displayed the "guilty" behaviour, and, moreover, that behaviour was more intense in those dogs that were routinely punished for such transgressions. The behaviours that the owners had labelled as "guilty" appeared to be triggered primarily by subtle cues emanating from their owners, rather than upon any recollection of "misbehaviour" (which in the experiment, half of the dogs had not performed). In general, emotions in dogs appear to be tightly connected to temporally contiguous events, meaning that they are probably incapable of experiencing self-conscious emotions such as guilt and pride (Lewis 2002). However, a majority of owners in the UK, at least, express the opinion that their dogs are capable of such complex feelings (Morris et al. 2008), and punish and/or praise their dogs accordingly, for example, for damage done by the dog hours previously (Borchelt and Voith 1982): under these circumstances, what dogs learn probably does not concur with their owners' intentions. This may adversely affect the dog's welfare: for example, punishment meted out by owners returning home to find that their dog has chewed a piece of furniture several hours previously is much more likely to become associated with the owner's demeanour immediately prior to its commencement than the earlier destruction, resulting in subsequent anxiety when the owner returns in the future.

11.4.2.2 Anthropocentrism

Here we use the term 'anthropocentrism' to highlight many owners' proclivity to take it for granted that the world perceived through their own senses is essentially the same as that as perceived by their dog. Although the sensory abilities of dogs and humans overlap considerably (otherwise we would be unable to communicate with one another), there are also striking differences, especially in the chemical senses, both in olfaction (Craven et al. 2010; Quignon et al. 2012; Walker et al. 2006) and in the dog's possession of a second olfactory apparatus, the vomeronasal organ (Adams and Weikamp 1984). While it is self-evident that dogs are far more sensitive to their olfactory surroundings than we are—why else would we wish to train them for scent detection and tracking?—little attention has been paid by animal welfare science to the impact on dogs' well-being of our manipulations of either environmental or social odours (but see Horowitz et al. 2013), despite significant advances in other areas (Rooney et al. 2009). For example, the olfactory ambience in dog kennels must be profoundly altered by cleaning procedures and use of chemical disinfectants, but the impact of these upon dogs' stress levels has not been systematically studied. Even if the effects were understood, there may be institutional resistance to implementation of any necessary changes in routine: for example, dogs' sensitive hearing is widely disregarded in the design of kennels (Coppola et al. 2006). It is now commonplace to consider sensory biases when considering the welfare of livestock (e.g. Grandin 1996) and laboratory animals (e.g. Burn 2008), but for the domestic dog, owners' natural tendency to anthropomorphise means that disproportionate attention is usually paid to physical aspects of the environment, in particular visual features.

In the domestic environment, differences between canine and human sensory abilities may account for some widely-held beliefs about dogs' emotional capacities. Grief, attributed to dogs by about every second owner (Morris et al. 2008), relies on the concept of the finality of death, which does not fully emerge in our own species until the age of four or five and is likely to be too abstract a concept for a dog's cognitive capacities. Anecdotally, however, dogs do appear to search for family members, whether canine or human, who have recently died. Such searching may plausibly be explained by the lingering olfactory "signature" of the missing companion, the dog reverting to normal routines once this has fallen below threshold. The behaviour appears indistinguishable from that observed when a person or dog is simply absent for an extended period.

11.4.3 Challenges in Measuring Dog Welfare

One of the factors hampering our ability to protect and improve dog welfare is the relative infancy of its measurement in this species. Taking each of the elements in turn, measuring health presents few problems: advanced canine veterinary techniques mean that health can reliably easily be assessed, by professionals. However, many owners' apparent inability to recognise pain (Hielm-Bjorkman et al. 2011) or gauge quality of life (PDSA 2011), and the lack of centralised records of morbidity and mortality (Rooney and Sargan 2010) in most countries means that assessment on a population level is well-nigh impossible. Whilst ethology provides us with knowledge and methods by which to record canine behaviour, the difficulty in making valid comparison with 'normal behaviour' in their wild counterparts limits their use. Challenges abound when it comes to assessing feelings.

It is now widely believed that a critical element of welfare assessment is how the animal feels, yet it is impossible to ask the animal directly. Hence scientists working on many species use a range of proxy measures, or 'indicator variables' aimed at assessing feelings indirectly. These primarily fall into two categories, physiological and behavioural, both of which have challenges in all species but perhaps most so in the dog. Physiological measures of activation of the sympathetic nervous system and the hypothalamic pituitary adrenal (HPA) axis are widely used. Those used on dogs include heart rate (e.g. Gillette et al. 2011), heart rate variability (e.g. Hydbring-Sandberg et al. 2004), adrenaline (e.g. Beerda et al. 2000), but most commonly, cortisol (Hiby et al. 2006; Rooney et al. 2007), a hormone produced during periods or stress and arousal which can be measured from plasma, saliva, urine, and faeces.

However, activation of the peripheral stress response and, consequently, elevated cortisol levels, occur with emotions of both positive and negative valence. Hence, for example, the increase in cortisol detected in dogs used for animal-assisted therapy (Haubenhofer and Kirchengast 2007) and in dogs living with children with autism (Burrows et al. 2008) cannot simply be assumed to be detrimental. In addition, the impact of chronic stress on physiological systems and the changes in hormone levels due to diurnal patterns are areas in which we still have incomplete understanding. Therefore studies of dogs increasingly include both physiological and behavioural indicators, so that changes in physiology can be carefully interpreted in conjunction with shifts in behaviour (Beerda et al. 2000).

There is a large range of behaviour patterns postulated to be indicative of acute and chronic stress in dogs (Table 11.1). Their value as 'universal' indicators (applicable to all dogs) has, however, in many cases been inferred from their occurrence in apparently stressful and suboptimal environments (such as rehoming kennels), or by extrapolation from other species, rather than by strict validation with other measures.

Several studies of kennelled dogs have aimed to determine which behaviours are most closely linked to physiological indicators of stress, with varying success. Some detect extreme individual variability even within a single breed and age group (Rooney et al. 2007), or links with specific behaviours such as reduced drinking (Hiby et al. 2006). Other studies have found reliable links between physiological measures and behaviours including paw-lifting, yawning, snout (or lip) licking, lowered body positions, vocalising, panting and increased salivation (Beerda et al. 1997) and repetitive behaviour, increased activity, nosing and increased urination (Beerda et al. 2000).

Such variation is unsurprising. Dog behaviour is affected by time of day and observation method, e.g. observer present or recorded remotely (Gaines et al. 2006). Individual dogs vary greatly in their behavioural responses to stress (Rooney et al. 2007); and recent work suggests the existence of different coping styles, e.g. proactive and passive (Blackwell et al. 2010). These result from differences in underlying temperament and also from differences in the degree of past exposure and previous attempts to adapt to situations that have induced negative affect. This can lead to the animals either interpretting the stimuli differently, or developing different behavioural coping mechanisms. In the case of dogs, such variations are particularly pronounced, as individuals are likely to have experienced a variety of rearing and living environments (especially those in rehoming centres) which affect their later behaviour. In addition, dogs of different breeds have different physical conformations, which limit their behavioural expression. Hence outward expressions of inner affective state, vary considerably, for example between a French bulldog with limited ability to raise and lower its body or ear height, and a husky with very mobile ears and body (Goodwin et al. 1997). Such differences are likely to play a part in concealing links between various putative measures of well-being.

This means that whilst behavioural indicators can be useful measures of individual welfare status, when used to monitor intra-individual changes, or in combination with physiological indicators, there remains much uncertainty regarding which behaviours (if any) can reliably indicate underlying emotional state on a population level. The challenge is illustrated if we consider two specific behaviour patterns: repetitive behaviour and play.

11.4.3.1 Abnormal Repetitive Behaviours

Repetitive behaviours in dogs include circling, pacing, jumping, tail-chasing and wall-bouncing (Hubrecht et al. 1992). They occur in pet dogs (Burn 2011), but are well-studied and common in long-term kennelled dogs: one-quarter of U.S. military dogs (Burghardt 2003) and between 46 % (Hiby 2005) and 93 % (Denham et al. 2013) of the dogs in some UK kennels (although rates vary at differing times of day) (Gaines 2008; Denham et al. 2013). These behaviours can have detrimental health effects, leading to tail damage, sore feet and lameness (Jennings 1991; Gaines 2008), and as such are welfare issues.

Repetitive behaviour in dogs often occurs under conditions thought to cause chronic stress (Beerda et al. 1999a, 2000; Hetts et al. 1992; Hubrecht et al. 1992). They are often described to be 'stereotypes' (invariant and repetitive behavioural patterns, with no obvious function or goal (Mason 1991). On a species population level they may be symptomatic of an underlying welfare decrement, although, individuals which stereotype may have better welfare, at least by some measures, than their non-stereotyping counterparts. Once an animal has developed a stereotypy or repetitive behaviour, carrying out this behaviour can temporarily relieve the animal's feelings of stress (Mason and Latham 2004).

Although numerous studies have used repetitive behaviours as indicators of poor dog welfare, our recent research has suggested that the motivation behind their performance may vary between individuals and particularly high levels are shown in response to husbandry events (Denham et al. 2013; Rooney et al. 2009). Some dogs may have been inadvertently rewarded for showing such behaviours (thus making the behaviours unreliable as indicators of welfare), whilst for other individuals repetitive behaviour may serve as a coping mechanism, thus indicating a long-standing welfare issue. Such observations further highlight the complexity of using behavioural measures in dog populations of mixed origin.

11.4.3.2 Play Behaviour

Whilst historically welfare science has concentrated on measuring poor welfare, increasingly assessments of positive well-being (and the balance of the two) are also being taken into consideration. Potential behavioural indicators include allo-grooming, relaxed postures, and the presence of so-called 'luxury' behaviour such as play. For many species, intraspecific play has been suggested to indicate both the absence of poor welfare and the presence of good welfare (Held and Spinka 2011). In the case of the domestic dog, which also plays readily with humans, high levels of play may be indicative of a successful dog-owner relationship (Rooney and Bradshaw 2003), and likely to induce a positive affective state. However, play may require rather different interpretation in dogs than in other species. Dogs are unusual in that they continue to play at high levels into adulthood, and find interspecific play especially rewarding, as is evidenced by the capacity to train dogs for numerous tasks using only a play reward (Rooney et al. 2004). It is possible that predisposition to play has been artificially selected for during domestication, due to its appeal in a companion, and/or its value as a reward during training. This may not only explain apparent differences in playfulness between breeds, but also raises the question as to whether in this species play is a 'luxury activity', suppressed at times of fitness threat, or whether it has, as a consequence of domestication, become a behavioural need, with inelastic demand. If so, a lack of opportunity to play may pose welfare concerns.

11.4.3.3 New Technologies in Dog Welfare Assessment

Novel physiological indicators of welfare are being developed and show promise for use in dogs including: cortisol extracted from hair (Siniscalchi et al. 2013), acute phase proteins (Casella et al. 2012), and immunological markers (Rammal et al. 2010). It also seems likely that fMRI scanning will soon be able to provide details of a non-anaesthetised dog's brain activity (Berns et al. 2012) and aid welfare assessment.

In light of the uncertainties posed by behavioural and physiological indicators, scientists have recently shown growing interest in the use of cognitive approaches to measure emotional state, for example using the concept of cognitive bias (Paul et al. 2005). People in a negative mood are more likely to attend to and judge ambiguous stimuli or future events as being negative than people in a more positive mood: they show a 'pessimistic' cognitive bias, defined as making a negative judgement about an ambiguous stimulus (Paul et al. 2005, Mendl et al. 2009). Non-linguistic tasks for assessing cognitive bias, developed in rats (Burman et al. 2008), have been adapted for use in many species. They have recently been applied to dogs: measures of 'pessimistic' bias were found to correlate with behaviour indicating distress on separation from human carers in a kennel environment (Mendl et al. 2010), although not between long-and short-term kennelled individuals (Titulaer et al. 2013) or during brief owner absences (Mueller et al. 2012).

These findings support the importance of separation-related behaviour as a major welfare issue in this species, and similar approaches may be valuable to assess other welfare problems.

11.4.3.4 Preference Testing

Dawkins (2004) suggested that many uncertainties in the interpretation of physiological and behavioural indicators can be resolved by 'asking' animals what they want, through the use of preference tests. Since such tests can only indicate the relative value of a resource and not its intrinsic worth, they are usually supplemented with 'economic demand' experiments which, in simple terms, measure how hard an animal will work to gain access to, or conversely to avoid, various environmental resources, and hence elucidate the animal's behavioural priorities. More than 30 years of preference studies on chickens, for example, have armed us with knowledge of resources which are integral to improving their welfare (Nicol et al. 2009). In contrast, we know of no preference or economic-demand tests conducted on dogs, beyond those used to test the relative palatability of different food stuffs and toys (Griffin et al. 1984; Pullen et al. 2010).

Preference testing could help to understand what dogs really want, and thus what resources they should never be deprived of, and hence inform how they should be housed and cared for. For example, techniques could be developed to ascertain the extent to which dogs value human, compared to canine, company (Pullen 2011). This would almost certainly differ dependent upon early experiences, so care would be needed in selecting subjects for testing. Similarly it should be possible to measure the extent to which play and opportunities to run are valued, and whether artificial selection has led to play becoming an inelastic need in dogs as compared to a luxury activity. Such knowledge could be used to develop the most welfare-compatible husbandry routines and housing.

11.4.4 Welfare and Cognition

As our knowledge of the dog's cognitive abilities grows, so it may be necessary to adapt our understanding of dog welfare. Not only is the level of priority we give to an animal's welfare affected by our understanding of their capacities for conscious awareness (Kirkwood and Hubrecht 2001) but also the welfare implications of many aspects of dogs' daily lives are affected by their cognitive abilities. In some aspects, an ability greater than currently presumed may imply a greater capacity to suffer, whilst in other aspects, the reverse is true. For example, numerical competency is likely to affect a dog's ability to predict the day of the week in a home environment, when humans contact is increased for, say, 2 days per week. If dogs become aware of the routine, either by counting days of the week or, more likely, by prediction from simple human-provided cues, their frustration at the restricted contact is likely lessened. Similarly, dogs are frequently exposed to situations in which they witness the distress or pain of conspecific (e.g. a newly neutered dog being placed in an adjacent kennel) or that of humans: the extent to which they are affected by this will depend on their capacity for empathy (see Edgar et al. 2012). Whilst several studies have been interpreted as possible evidence for empathy, such as dogs orientating towards crying owners and strangers (Custance and Mayer 2012) or showing increased cortisol when hearing a baby crying (Yong and Ruffman 2013), others have interpreted human–dog contagious yawning (Madsen and Persson 2013) as low-level imitation (see also Horowitz and Hecht, this volume). Future studies need to avoid pitfalls of anthropomorphism and also distinguish between a valenced emotional response and simple interest or arousal caused by a social stimulus. To demonstrate a truly emotionally empathic response to the plight of another, the response much be positive or negative (Edgar et al. 2012). Understanding how cognitive processes influence empathic responses is fundamental to evaluating the extent to which the welfare of dogs is affected by their social environment.

11.5 Conclusions

Whilst scientists have systematically investigated each aspect of the environment, care, and behaviour of many domesticated species, thereby deriving comprehensive recommendations for optimal conditions and protocol, knowledge remains piecemeal for dogs. We have here highlighted a number of key welfare issues which impact on domestic dogs, and several potential issues for which knowledge is currently lacking. We have framed our discussion in terms of Fraser (2009a) three approaches: health, psychological well-being, and natural behaviour. The past two decades have seen a change from an over-emphasis on a health-based approach to one aiming to achieve a balance between physical health and affective state or psychological well being, and this has highlighted welfare issues previously given little attention. Given the extreme domestication undergone by the domestic dog, especially when considering its cognitive abilities, the 'natural behaviour' approach is superficially less useful for the dog than for many other species. That is, unless its environment of evolutionary adaptation (which must be largely anthropogenic) can be adequately defined and its 'natural' and preferred behaviour patterns (e.g. feeding, exercise and sleeping rhythms) in that environment determined.

Even when all three approaches essentially raise the same welfare issues, in practice the relative importance of each issue still has to be decided, for example when charities plan owner education campaigns. Attempts to prioritise dog welfare issues have been based mainly on consensus stakeholder opinion (e.g. Buckland et al. 2013), but as we have seen, this is likely to incorporate bias based on which dimension of welfare is considered most important, and thus the make-up of the expert group is key (for example what proportion are trained vets). The "Five

Needs" approach adopted in England and Wales as the basis for the Animal Welfare Act (DEFRA 2006) and subsequently the Welfare Code of Practice for Dogs (DEFRA 2009) is essentially pragmatic, categorising good welfare as dependent upon provision of physical environment, diet, opportunities for normal behaviour, social aspects of housing, and protection from pain, suffering, injury and disease. However, balancing the relative importance of each of these still involves some subjectivity. Prioritisation based on scientific investigation is far from complete.

Dissemination and implementation of welfare standards for dogs are also much less straightforward than for other domestic species (particularly production animals), simply because there are millions of humans responsible for the day-to-day care of individual dogs and they are kept in such diverse ways. Routes to implementation should be relatively straightforward for kennelled dogs, partly because the majority of studies of dog welfare have been conducted in such environments, and also because those people responsible for managing the kennels should be easily identified: however, cost implications, or a reluctance to change traditional ways of working, may produce significant barriers to change. Kennelled dogs populations, while providing a readily identifiable target and a valuable research resource, are small in number compared to companion dogs. It is much less easy to quantify the extent of welfare issues among the companion dog population, but the number of fearful dogs we see dragged along the street, the proportion of dogs punished during training, and the number of purebred dogs which struggle to breathe, all provide an indication that there are still significant gaps between what even well-meaning owners would like to provide for their dogs, and the reality of those dogs' lives.

Considering the extent of the welfare problems experienced by companion dogs today, it must be asked whether dogs are really well suited to life as companions in modern-day Western conditions, or, put another way, whether the majority of owners are sufficiently knowledgeable to give their canine companion a good standard of welfare. Undoubtedly many owners can provide their dogs with the opportunities for good health, psychological well-being and, where acceptable, the natural behaviour that they desire, but evidently not all. Hopefully, as canine welfare science advances, and findings both old and new are transmitted more widely to those responsible for the care of dogs, so mankind will finally become "dog's best friend" in the delivery as well as in the intention.

Acknowledgments We thank Dr Sam Gaines (RSPCA) and Dr Corinna Clark for very useful comments on the draft. We also thank our colleagues Dr Rachel Casey and Dr Emily Blackwell for many valuable discussions on canine welfare science.

References

Adams, D. R., & Wiekamp, M. D. (1984). The vasomeronasal organ. *Journal of Anatomy, 138*, 771–787.
Advocates for Animals. (2006). The price of a pedigree: Dog breed standards and breed-related illness. Advocates for Animals Report. http://www.onekind.org/uploads/publications/price-of-a-pedigree.pdf
Appleby, M. (1999). *What should we do about animal welfare?* Oxford: Blackwell Scientific.
Arhant, C., Bubna-Littitz, H., Bartels, A., Futschik, A., & Troxler, J. (2010). Behaviour of smaller and larger dogs: Effects of training methods, inconsistency of owner behaviour and level of engagement in activities with the dog. *Applied Animal Behaviour Science, 123*, 131–142.
Asher, L., Diesel, G., Summers, J. F., McGreevy, P. D., & Collins, L. M. (2009). Inherited defects in pedigree dogs. Part 1: Disorders related to breed standards. *The Veterinary Journal, 182*, 402–411.
Axelsson, E., Ratnakumar, A., Arendt, M. L., Maqbool, K., Webster, M. T., Perloski, M. et al. (2013). The genomic signature of dog domestication reveals adaptation to a starch-rich diet. *Nature, 495*, 360–364.
Bailey, G. (2008). *The perfect puppy.* London: Hamlyn.
Bateson, P. (2010). *Independent inquiry into dog breeding.* London: Dogs Trust.
BBC. (2008). BBC One reveals shocking truth about pedigree dog breeding in UK. Retrieved October 23, 2013 from http://www.bbc.co.uk/pressoffice/pressreleases/stories/2008/08_august/19/dogs
Beerda, B., Schilder, M. B. H., van Hooff, J. A. R. A. M., & de Vries, H. W. (1997). Manifestations of chronic and acute stress in dogs. *Applied Animal Behaviour Science, 52*, 307–319.
Beerda, B., Schilder, M. B. H., van Hooff, J. A. R. A. M., de Vries, H. W., & Mol, J. A. (1998). Behavioural, saliva cortisol and heart rate responses to different types of stimuli in dogs. *Applied Animal Behaviour Science, 58*, 365–381.
Beerda, B., Schilder, M. B. H., van Hooff, J., De Vries, H. W., & Mol, J. A. (1999a). Chronic stress in dogs subjected to social and spatial restriction. I. Behavioral responses. *Physiology & Behavior, 66*, 233–242.
Beerda, B., Schilder, M. B. H., Bernadina, W., van Hooff, J., De Vries, H. W., & Mol, J. A. (1999b). Chronic stress in dogs subjected to social and spatial restriction. II. Hormonal and immunological responses. *Physiology & Behavior, 66*, 243–254.
Beerda, B., Schilder, M. B. H., van Hooff, J., de Vries, H. W., & Mol, J. A. (2000). Behavioural and hormonal indicators of enduring environmental stress in dogs. *Animal Welfare, 9*, 49–62.
Bekoff, M. (1995). Play signals as punctuation—the structure of social play in canids. *Behaviour, 132*, 419–429.
Bergamasco, L., Osella, M. C., Savarino, P., Larosa, G., Ozella, L., & Manassero, M. et al. (2010). Heart rate variability and saliva cortisol assessment in shelter dog: Human–animal interaction effects. *Applied Animal Behaviour Science, 125*, 56–68.
Berns, G. S., Brooks, A. M., & Spivak, M. (2012). Functional MRI in Awake Unrestrained Dogs. *Plos One, 7*, doi: 10.1371/journal.pone.0038027.
Blackwell, E. J., Bodnariu, A., Tyson, J., Bradshaw, J. W. S., & Casey, R. A. (2010). Rapid shaping of behaviour associated with high urinary cortisol in domestic dogs. *Applied Animal Behaviour Science, 124*, 113–120.
Blackwell, E. J., Casey, R. A., & Bradshaw, J. W. S. (2006). Controlled trial of behavioural therapy for separation-related disorders in dogs. *Veterinary Record, 158*, 551–554.
Blackwell, E. J., Bolster, C., Richards, G., Loftus, B., & Casey, R. A. (2012). The use of electronic collars for training domestic dogs: Estimated prevalence, reasons and risk factors for use, and owner perceived success as compared to other training methods. *BMC Veterinary Research, 8*(93), doi: 10.1186/1746-6148-8-93.

Blackwell, E. J., Bradshaw, J. W. S., & Casey, R. A. (2013). Fear responses to noises in domestic dogs: Prevalence, risk factors and co-occurrence with other fear related behaviour. *Applied Animal Behaviour Science, 145*, 15–25.

Blackwell, E. J., Casey, R. A., & Bradshaw, J. W. S. (2005). The prevention of separation-related behaviour problems in dogs re-homed from rescue centres. In D. Mills, E. Levine, G. Landsberg, D. Horwitz, M. Duxbury, P. Mertens et al. (Eds.), *Current Issues and Research in Veterinary Behavioural Medicine. Papers presented at 5th international veterinary behaviour meeting* (pp. 236–238). West Lafayette, Indiana, USA: Purdue University Press.

Blackwell, E. J., Twells, C., Seawright, A., & Casey, R. A. (2008). The relationship between training methods and the occurrence of behavior problems, as reported by owners, in a population of domestic dogs. *Journal of Veterinary Behavior-Clinical Applications and Research, 3*, 207–217.

Bloom, T., & Friedman, H. (2013). Classifying dogs' (Canis familiaris) facial expressions from photographs. *Behavioural Processes, 96*, 1–10.

Blouin, D. D. (2013). Are dogs children, companions, or just animals? Understanding variations in people's orientations toward animals. *Anthrozoös, 26*, 279–294.

Boissy, A. (1998). Fear and fearfulness in determining behavior,. In T. Grandin (Ed.), *Genetics and the behavior of domestic animals* (pp. 67–111). New York: Academic Press

Borchelt, P. L., & Voith. V. L. (1982). Diagnosis and treatment of separation-related behavior problems in dogs. *Veterinary Clinics of North America: Small Animal Practice, 12*, 625–635.

Boissy, A., Manteuffel, G., Jensen. M. B., Moe, R. O., Spruijt, B., Keeling, L. J. et al. (2007). Assessment of positive emotions in animals to improve their welfare. *Physiology & Behavior, 92*, 375–397.

Bradshaw, J. W. S. (2011). *Dog sense: How the new science of dog behavior can make you a better friend to your pet*. New York: Basic Books.

Bradshaw, J. W. S., & Casey, R. A. (2007). Anthropomorphism and anthropocentrism as influences in the quality of life of companion animals. *Animal Welfare, 16*(S), 149–154.

Bradshaw, J. W. S., Blackwell, E. J., & Casey, R. A. (2009). Dominance in domestic dogs: Useful construct or bad habit? *Journal of Veterinary Behaviour, 4*, 135–144.

Bradshaw, J. W. S., Blackwell, E. J., Rooney, N. J., & Casey, R. A. (2002a). Prevalence of separation-related behaviour in dogs in southern England. In J. B. E. Dehasse & E. Biosca Marce (Eds.), *Proceedings of the 8th European Society of Veterinary Clinical Ethology Meeting on Veterinary Behavioural Medicine, Granada, Spain*, October 2, 2002 (pp. 189–193). Paris: Publibook.

Bradshaw, J. W. S., McPherson, J. A., Casey, R. A., & Larter, I. S. (2002b). Aetiology of separation-related behaviour in domestic dogs. *Veterinary Record, 151*, 43–46.

Buckland, E. L., Whiting, M. C., Abeyesinghe, S. M., Asher, L., Corr, S., & Wathes, C. M. (2013). A survey of stakeholders' opinions on the priority issues affecting the welfare of companion dogs in Great Britain. *Animal Welfare, 22*, 239–253.

Burghardt, W. F. (2003). Behavioral considerations in the management of working dogs. *Veterinary Clinics of North America—Small Animal Practice, 33*, 417.

Burman, O. H. P., Parker, R., Paul, E. S., & Mendl, M. (2008). A spatial judgement task to determine background emotional state in laboratory rats, *Rattus norvegicus*. *Animal Behaviour, 76*, 801–809.

Burn, C. C. (2008). What is it like to be a rat? Rat sensory perception and its implications for experimental design and rat welfare. *Applied Animal Behaviour Science, 112*, 1–32.

Burn, C. C. (2011). A vicious cycle: A cross-sectional study of canine tail-chasing and human responses to it, using a free video-sharing website. *PLoS One, 6*(11), e26553. doi: 10.1371/journal.pone.0026553.

Burrows, K. E., Adams, C. L., & Millman, S. T. (2008). Factors affecting behavior and welfare of service dogs for children with autism spectrum disorder. *Journal of Applied Animal Welfare Science, 11*, 42–62.

Casella, S., Fazio, F., Giannetto, C., Giudice, E., & Piccione, G. (2012). Influence of transportation on serum concentrations of acute phase proteins in horse. *Research in Veterinary Science, 93*, 914–917.

Casey, R. A. (2011). Pet personality: Do individual differences influence undesired behaviours? In *Proceedings of the Companion Animal Behaviour Therapy Study Group Meeting (CABTSG)*. Birmingham, UK

Casey, R. A., Clark C. C. A., & Rooney N. J. (2011). How best to achieve a fearless working dog. Report to the Defence Science and Technology Report.

Casey, R. A., Loftus, B., Bolster, C., Richards, G. J., & Blackwell, E. J. (2013). Inter-dog aggression in a UK owner survey: Prevalence, co-occurrence in different contexts and risk factors. *Veterinary Record, 172*, 127.

Craven, B. A., Paterson, E. G., & Settles, G. S. (2010). The fluid dynamics of canine olfaction: Unique nasal airflow patterns as an explanation of macrosmia. *Journal of the Royal Society Interface, 7*, 933–943.

Coppola, C. L., Enns, R. M., & Grandin, T. (2006). Noise in the animal shelter environment: Building design and the effects of daily noise exposure. *Journal of Applied Animal Welfare Science, 9*, 1–7.

Correia, C., Ruiz de la Torre, J. L., Manteca., X., & Fatjo, J. (2007). Accuracy of dog owners to describe and interpret the canine body language during aggressive episodes, 6th IVBM/ECVBM-CA Meeting Riccione.

Custance, D., & Mayer, J. (2012). Empathic-like responding by domestic dogs (Canis familiaris) to distress in humans: An exploratory study. *Animal Cognition, 15*, 851–959.

Dawkins, M. S. (2004). Using behaviour to assess animal welfare. *Animal Welfare, 13*, S3–S7.

DEFRA (Department for Environment Food and Rural Affairs) (2006). Animal Health and Welfare, Animal Welfare Act Secondary Legislation and Codes of Practice. Retrieved October 11 2013, from, http://www.defra.gov.uk/publications/files/pb12460-dutytocare-080312.pdf

DEFRA (Department for Environment Food and Rural Affairs) (2009). Code of Practice for the Welfare of Dogs. London Retrieved October 30 2013, from, https://www.gov.uk/government/uploads/system/uploads/attachment_data/file/69390/pb13333-cop-dogs-091204.pdf

Denhman, H. D. C., Bradshaw, J. W. S., & Rooney, N. J. (2013) Repetitive behaviours in kennelled dogs—stereotypical or not? *Physiology & Behavior*

Diesel, G., Brodbelt, D., & Pfeiffer, D. U. (2010) Characteristics of relinquished dogs and their owners at 14 Rehoming Centers in the United Kingdom. *Journal of Applied Animal Welfare Science, 13*(1), 15–30.

Doering, D., Roscher, A., Scheipl, F., Kuchenhoff, H., & Erhard, M. H. (2009). Fear-related behaviour of dogs in veterinary practice. *Veterinary Journal, 182*, 38–43.

Donaldson, (2009). http://www.urbandawgs.com/divided_profession.html

Dreschel, N. A. (2010). The effects of fear and anxiety on health and lifespan in pet dogs. *Applied Animal Behaviour Science, 125*, 157–162.

Dreschel, N. A., & Granger, D. A., (2005). Physiological and behavioral reactivity to stress in thunderstorm-phobic dogs and their caregivers. *Applied Animal Behaviour Science, 95*, 153–168.

Duffy, D. L., Hsu, Y. Y., & Serpell, J. A. (2008). Breed differences in canine aggression. *Applied Animal Behaviour Science, 114*, 441–460.

Duncan, I. J. H. (1996). Animal welfare defined in terms of feelings. *Acta Agriculturae Scandinavica Section a-Animal Science, 27*, 29–35.

Edgar, J. L., Nicol, C. J., Clark, C. C. A., & Paul, E. S. (2012). Measuring empathic responses in animals. *Applied Animal Behaviour Science, 138*, 182–193.

Environment, Food and Rural Affairs Committee (EFRA) (2013).—Seventh Report Dog Control and Welfare, Parlimentary Copyright. Retrieved November 1, 2013, from http://www.publications.parliament.uk/pa/cm201213/cmselect/cmenvfru/575/57502.htm

Ennaceur, A., Michalikova, S., & Chazot, P. L. (2006). Models of anxiety: Responses of rats to novelty in an open space and an enclosed space. *Behavioural Brain Research, 171*, 26–49.

Foyer, P., Wilsson, E., Wright, D., & Jensen, P. (2013). Early experiences modulate stress coping in a population of German shepherd dogs. *Applied Animal Behaviour Science, 146*, 79–87.

Fraser, D. (2003). Assessing animal welfare at the farm and group level: The interplay of science and values. *Animal Welfare, 12*, 433–443.

Fraser, D. (2009a). Assessing animal welfare: Different philosophies, different scientific approaches. *Zoo Biology, 28*, 507–518.

Fraser, D. (2009b). Animal behaviour, animal welfare and the scientific study of affect. *Applied Animal Behaviour Science, 118*, 108–117.

Fraser, D., Weary, D. M., Pajor, E. A., & Milligan, B. N. (1997). A scientific conception of animal welfare that reflects ethical concerns. *Animal Welfare, 6*, 187–205.

Gaines, S. (2008). *Kennelled dog welfare—effects of housing and husbandry*. PhD Thesis, University of Bristol.

Gaines, S. A., Rooney, N. J., & Bradshaw, J. W. S. (2006). The effects of the presence of an observer, and time of day, on welfare indicators for working police dogs. In M.B. Mendl, J.W.S. Bradshaw, O.H.P. Burman, A. Butterworth, N.J. Harris, S. Held (Eds.), *Proceedings of the 40th International Congress of the International Society for Applied Ethology* (pp.14). Bristol, UK

Galac, S., & Knol, B. W. (1997). Fear-motivated aggression in dogs: Patient characteristics, diagnosis and therapy. *Animal Welfare, 6*, 9–15.

Gates, M. C. & Nolan, T. J. (2010). Factors influencing heartworm, flea, and tick preventative use in patients presenting to a veterinary teaching hospital. *Preventive Veterinary Medicine, 93*, 193–200.

German, A. J., Ryan, V. H., German, A. C., Wood, I. S., & Trayhurn, P. (2010). Obesity, its associated disorders and the role of inflammatory adipokines in companion animals. *Veterinary Journal, 185*, 4–9.

Gillette, R. L., Angle, T. C., Sanders, J. S., & DeGraves, F. J. (2011). An evaluation of the physiological affects of anticipation, activity arousal and recovery in sprinting greyhounds. *Applied Animal Behaviour Science, 130*, 101–106.

Goddard, M. E. & Beilharz, R. G. (1984). A factor analysis of fearfulness in potential guide dogs. *Applied Animal Behaviour Science, 12*, 253–265.

Goddard, M. E., & Beilharz, R. G. (1985). A multivariate analysis of the genetics of fearfulness in potential guide dogs. *Behavior Genetics, 15*, 69–89.

Goddard, M. E., & Beilharz, R. G. (1986). Early prediction of adult behaviour in potential guide dogs. *Applied Animal Behaviour Science, 15*, 247–260.

Goodwin, D., Bradshaw, J. W. S., & Wickens, S. M. (1997). Paedomorphosis affects agonistic visual signals of domestic dogs. *Animal Behaviour, 53*, 297–304.

Grandin, T. (1996). Factors that impede animal movement at slaughter plants. *Journal of the American Veterinary Medical Association, 209*, 757–759.

Greenebaum, J. B. (2010). Training dogs and training humans: Symbolic interaction and dog training. *Anthrozoös, 23*, 129–141.

Griffin, R. W., Scott, G. C., & Cante, C. J. (1984). Food preferences of dogs housed in testing-kennels and in consumers homes—some comparisons. *Neuroscience and Biobehavioral Reviews, 8*, 253–259.

Haubenhofer, D. K., & Kirchengast, S. (2007). Dog handlers' and dogs' emotional and cortisol secretion responses associated with animal-assisted therapy sessions. *Society & Animals, 15*, 127.

Haverbeke, A., Messaoudi, F., Depiereux, C., Stevens, M., Giffroy, J. M., & Diederich, C. (2010). Efficiency of working dogs undergoing a new human familiarization and training program. *Journal of Veterinary Behavior: Clinical Applications and Research, 5*(2), 112–119.

Held, S. D. E., & Spinka, M. (2011). Animal play and animal welfare. *Animal Behaviour, 81*, 891–899.

Hennessy, M. B., Morris, A., & Linden, F. (2006). Evaluation of the effects of a socialization program in a prison on behavior and pituitary-adrenal hormone levels of shelter dogs. *Applied Animal Behaviour Science, 99*, 157–171.

Hennessy, M. B., Williams, M. T., Miller, D. D., Douglas, C. W., & Voith, V. L. (1998). Influence of male and female petters on plasma cortisol and behaviour: Can human interaction reduce the stress of dogs in a public animal shelter? *Applied Animal Behaviour Science, 61*, 63–77.

Hennessy, M. B., Voith, V. L., Young, T. L., Hawke, J. L., Centrone, J., McDowell, A. L. et al. (2002). Exploring human interaction and diet effects on the behavior of dogs in a public animal shelter. *Journal of Applied Animal Welfare Science, 5*, 253–273.

Herron, M. E., Shofer, F. S., & Reisner, I. R. (2009). Survey of the use and outcome of confrontational and non-confrontational training methods in client-owned dogs showing undesired behaviors. *Applied Animal Behaviour Science, 117*, 47–54.

Hetts, S., Derrell Clark, J., Calpin, J. P., Arnold, C. E., & Mateo, J. M. (1992). Influence of housing conditions on beagle behaviour. *Applied Animal Behaviour Science, 34*, 137–155.

Hewson, C. J. (2003). Can we assess welfare? *Canadian Veterinary Journal-Revue Veterinaire Canadienne, 44*, 749–753.

Hiby, E. F. (2005). *The welfare of kennelled domestic dogs*. PhD Thesis, University of Bristol.

Hiby, E. F. (2013). Dog population management. In C. N. L. Macpherson, F. X. Meslin, & A. I. Wandele, (Eds.), *Dogs, zoonoses and public health* (pp. 177–204). Oxford: CABI International.

Hiby, E. F., Rooney, N. J., & Bradshaw, J. W. S. (2004). Dog training methods: Their use, effectiveness and interaction with behaviour and welfare. *Animal Welfare, 13*, 63–69.

Hiby, E. F., Rooney, N. J., & Bradshaw, J. W. S. (2006). Behavioural and physiological responses of dogs entering re-homing kennels. *Physiology & Behavior, 89*, 385–391.

Hielm-Bjorkman, A. K., Kapatkin, A. S., & Rita, H. J. (2011). Reliability and validity of a visual analogue scale used by owners to measure chronic pain attributable to osteoarthritis in their dogs. *American Journal of Veterinary Research, 72*, 601–607.

Horowitz, A. (2009). Disambiguating the "guilty look": Salient prompts to a familiar dog behaviour. *Behavioural Processes, 81*, 447–452.

Horowitz, A., Hecht, J., & Dedrick, A. (2013). Smelling more or less: Investigating the olfactory experience of the domestic dog. *Learning and Motivation, 44*, 207–217.

Houpt, K. A., Goodwin, D., Uchida, Y., Baranyiova, E,. Fatjo, J., & Kakuma, Y. (2007). In Proceedings of a workshop to identify dog welfare issues in the US, Japan, Czech Republic, Spain and the UK. *Applied Animal Behaviour Science, 106*, 221–233.

Hsu, Y. Y., & Sun, L. C. (2010). Factors associated with aggressive responses in pet dogs. *Applied Animal Behaviour Science, 123*, 108–123.

Hubrecht, R. C. (1993). A comparison of social and environmental enrichment methods for laboratory housed dogs. *Applied Animal Behaviour Science, 37*, 345–361.

Hubrecht, R. C., Serpell, J. A., & Poole, T. B. (1992). Correlates of pen size and housing conditions on the behaviour of kennelled dogs. *Applied Animal Behaviour Science, 34*, 365–383

Hydbring-Sandberg, E., von Walter, L., Hoglund, K., Svartberg, K., Swenson, L., & Forkman, B. (2004). Physiological reactions to fear provocation in dogs. *Journal of Endocrinology, 180*, 439–448.

Jennings, P. B. (1991). Veterinary care of the Belgian Malinois working dog. *Military Medicine, 156*, 36–38.

Jensen, M. B., Vestergaard, K. S., Krohn, C. C. (1998). Play behaviour in dairy calves kept in pens: The effect of social contact and space allowance. *Applied Animal Behaviour Science, 56*, 97–108.

Jones, B., & Boissy, A. (2011). Fear and other negative emotions, In M. C. Appleby, B. O. Hughes, J. A. Mench, & A. Olsson (Eds.), *Animal Welfare* (pp. 78–97). Oxford: CABIPublishing.

Kerswell, K. J., Bennett, P., Butler, K. L., & Hemsworth, P. H. (2009). Self-reported comprehension ratings of dog behavior by puppy owners. *Anthrozoös, 22*, 183–193.

Kim, H. H., Yeon, S. C., Houpt, K. A., Lee, H. C., Chang, H. H., & Lee, H. J. (2006). Effects of ovariohysterectomy on reactivity in German Shepherd dogs. *Veterinary Journal, 172*(1), 154–159.

Kirkwood, J. K. & Hubrecht, R. (2001). Animal consciousness, cognition and welfare. *Animal Welfare, 10*, S5–S17.

Knol, B. W. G., Groenewoud, H. J. C., & Ubbink, G. J. (1997). Motivated aggression in Golden Retrievers: No correlation with inbreeding. In *Proceedings of the First International Conference on Veterinary Behavioural Medicine* (p. 112). Birmingham, UK, 1 & 2 April.

Kobelt, A. J., Hemsworth, P. H., Barnett, J. L., Coleman, G. J., & Butler, K. L. (2007). The behaviour of Labrador retrievers in suburban backyards: The relationships between the backyard environment and dog behaviour. *Applied Animal Behaviour Science, 106*, 70–84.

LeDoux, J. E. (2000). Emotion circuits in the brain. *Annual Review of Neuroscience, 23*, 155–184.

Lefebvre, D, Diederich, C., Delcourt, M., & Giffroy, J-M. (2007). The quality of the relation between handler and military dogs influences efficiency and welfare of dogs. *Applied Animal Behaviour Science, 104*, 49–60.

Lewis, M. (2002). Early emotional development. In A. Slater & M. Lewis (Eds.), *Introduction to infant development* (pp. 192–209). Oxford: Oxford University Press.

Levine, E. D., Ramos, D., & Mills, D. S. (2007). A prospective study of two self-help CD based desensitization and counter-conditioning programmes with the use of dog appeasing pheromone for the treatment of firework fears in dogs (Canis familiaris). *Applied Animal Behaviour Science, 105*, 311–329.

Liinamo, A. E., van den Berg, L., Leegwater, P. A. J., Schilder, M. B. H., van Arendonk, J. A. M., & van Oost B. A., (2007). Genetic variation in aggression-related traits in Golden Retriever dogs. *Applied Animal Behaviour Science, 104*, 95–106.

Luescher, A. U., & Reisner, I. R. (2008). Canine aggression toward familiar people: A new look at an old problem. *Veterinary Clinics: Small Animal Practice, 38*, 1107–1130.

Madsen, E. A., & Persson, T. (2013). Contagious yawning in domestic dog puppies (Canis lupus familiaris): The effect of ontogeny and emotional closeness on low-level imitation in dogs. *Animal Cognition, 16*, 233–240.

Mariti, C., Gazzano, A., Moore, J. L., Baragli, P., Chelli, L., & Sighieri, C. (2012). Perception of dogs' stress by their owners. *Journal of Veterinary Behavior: Clinical Applications and Research, 7*, 213–239.

Mason, G. J. (1991). Stereotypies and suffering. *Behavioural Processes, 25*, 103–115.

Mason, G. J., & Latham, N. R. (2004). Can't stop, won't stop: Is stereotypy a reliable animal welfare indicator? *Animal Welfare, 13*, S57–S69.

Massar, S. A. A., Mol, N. M., Kenemans, J. L., & Baas, J. M. P. (2011). Attentional bias in high- and low-anxious individuals: Evidence for threat-induced effects on engagement and disengagement. *Cognition & Emotion, 25*, 805–817.

McGreevy, P. (2008). Comment: We must breed happier, healthier dogs. *New Scientist, 200*, 18.

McGreevy, P. D., & Bennett, P. C. (2010). Challenges and paradoxes in the companion-animal niche. *Animal Welfare, 19*, 11–16.

McGreevy, P., & Boakes, R. A. (2007). *Carrots and sticks: Principles of animal training*. Cambridge: Cambridge University Press.

McHugh, A. (2009). *The gloves are off*. Your Dog, December 2009, 44–47.

McMillan, F. D., Duffy, D. L., & Serpell, J. A. (2011). Mental health of dogs formerly used as 'breeding stock' in commercial breeding establishments. *Applied Animal Behaviour Science, 135*, 86–94.

McMillan, F. D., Serpell, J. A., Duffy, D. L., Masaoud, E., & Dohoo, I. R. (2013). Differences in behavioval characteristics between dogs obtained as puppies from pet stores and those obtained from noncommercial breeders. *Journal of the American Veterinary Medical Association, 242*, 1359–1363.

Mech, D. L. (2008). Whatever happened to the term "Alpha Wolf"? *International Wolf*, Winter 2008, 4–8.

Mendl, M., Burman, O. H. P., Parker, R. M. A., & Paul, E. S. (2009). Cognitive bias as an indicator of animal emotion and welfare: Emerging evidence and underlying mechanisms. *Applied Animal Behaviour Science, 118*, 161–181.

Mendl, M., Brooks, J., Basse, C., Burman, O., Paul, E., & Blackwell, E. et al. (2010). Dogs showing separation-related behaviour exhibit a 'pessimistic' cognitive bias. *Current Biology, 20*, R839–R840.

Mills, D. S., & Zulch, H. (2010). Veterinary medicine and animal behaviour: Barking up the right tree! *Veterinary Journal, 183*, 119–120.

Morris, J. G., & Rogers, Q. R. (1989). Comparative aspects of nutrition and metabolism of dogs and cats. In I.H. Burger, & J. P. W. Rivers (Eds.), *Nutrition of the Dog and Cat, Waltham Symposium* (pp. 35–66). Cambridge: Cambridge University Press.

Morris, P. H., Doe, C., & Godsell, E. (2008). Secondary emotions in non-primate species? Behavioural reports and subjective claims by animal owners. *Cognition & Emotion, 22*, 3–20.

Mueller, C. A., Riemer, S., Rosam, C. M., Schosswender, J., Range, F., & Huber, L. (2012). Brief owner absence does not induce negative judgement bias in pet dogs. *Animal Cognition, 15*, 1031–1035.

Nicholas, F. W., Wade, C. M., & Williamson, P. (2010). Disorders in pedigree dogs: Assembling the evidence. *Veterinary Journal, 183*, 8–9.

Nicol, C. J., Caplen, G., Edgar, J., & Browne, W. J. (2009). Associations between welfare indicators and environmental choice in laying hens. *Animal Behaviour, 78*, 413–424.

Norling, A-Y., & Keeling, L. (2010). Owning a dog and working: A telephone survey of dog owners and employers in Sweden. *Anthrozoös, 23*, 157–171.

O'Sullivan, E. N., Jones, B. R., O'Sullivan, K., & Hanlon, A. J. (2008). Characteristics of 234 dog bite incidents in Ireland during 2004 and 2005. *Veterinary Record, 163*, 37–42

Odendaal, J. S. J., & Meintjes, R. A. (2003). Neurophysiological correlates of affiliative behaviour between humans and dogs. *The Veterinary Journal, 165*, 296–301.

Packard, J. M. (2003). Wolf behavior: Reproductive, social and intelligent. In D. Mech & L. Boitani (Eds.), *Wolves: Behavior, Ecology, and Conservation*, (pp. 35–65). Chicago: Chicago University Press.

Packer, R. M. A., Hendricks, A., & Burn, C. C. (2012). Do dog owners perceive the clinical signs related to conformational inherited disorders as 'normal' for the breed? A potential constraint to improving canine welfare. *Animal Welfare, 21*, 81–93.

Packer, R. M. A., Tivers, M. S., Hendricks, A., & Burn, C. C. (2013). Short muzzle; short of breath? The effect of conformation on the risk of Brachycephalic Obstructive Airway Syndrome (BOAS) in domestic dogs. UFAW Symposium, Barcelona.

Palestrini, C., Previde, E. P., Spiezio, C., & Verga, M. (2005). Heart rate and behavioural responses of dogs in the Ainsworth's Strange Situation: A pilot study. *Applied Animal Behaviour Science, 94*, 75–88.

Passalacqua, C., Marshall-Pescini, S., Merola, I., Palestrini, C., & Previde, E. P. (2013). Different problem-solving strategies in dogs diagnosed with anxiety-related disorders and control dogs in an unsolvable task paradigm. *Applied Animal Behaviour Science, 147*, 139–148.

Paul, E. S., Harding, E. J., & Mendl, M. (2005). Measuring emotional processes in animals: The utility of a cognitive approach. *Neuroscience and Biobehavioral Reviews, 29*, 469–491.

PDSA (2011). PDSA Animal Wellbeing Report 2011. www.pdsa.org.uk/pet-health-advice/pdsa-animal-wellbeing-report. Accessed 23/10/2013

Perez-Guisado, J., Lopez-Rodriguez, R., & Munoz-Serrano, A. (2006). Heritability of dominant-aggressive behaviour in English Cocker Spaniels. *Applied Animal Behaviour Science, 100*, 219–227.

Pet Food Manufacturer's Association. (2009). Pet obesity: The reality in 2009. http://www.pfma.org.uk/_assets/docs/PFMA_WhitePaper%20Final.pdf. Accesssed 23/10/2013

Plessas, I. N., Rusbridge, C., Driver, C. J., Chandler, K. E., Craig, A., & McGonnell, I. M., et al. (2012). Long-term outcome of Cavalier King Charles spaniel dogs with clinical signs associated with Chiari-like malformation and syringomyelia. *Veterinary Record. 171*, doi: 10.1136/vr.100449

Podberscek, A. L. (2009). Good to pet and eat: The keeping and consuming of dogs and cats in South Korea. *Journal of Social Issues, 65*, 615–632.

Podberscek, A. L, & Serell, J. A (1997). Environmental influences on the expression of aggressive behaviour in English Cocker Spaniels. *Applied Animal Behaviour Science, 52*, 7215–7227.

Prescott, M. J., Morton, D. B., Anderson, D., Buckwell, A., Heath, S., & Hubrecht, R. et al. (2004). Refining dog husbandry and care—Eighth report of the BVAAWF/FRAME/RSPCA/ UFAW Joint Working Group on Refinement. *Laboratory Animals, 38*, S1–S94.

Pullen, A. J. (2011). Behavioural indicators of candidate enrichments for kennel housed dogs. PhD thesis, University of Bristol, Chapter 6.

Pullen, A. J., Merrill, R. J. N., & Bradshaw, J. W. S. (2010). Preferences for toy types and presentations in kennel housed dogs. *Applied Animal Behaviour Science, 125*, 151–156.

Quignon, P., Rimbault, M., Robin, S., & Galibert, F. (2012). Genetics of canine olfaction and receptor diversity. *Mammalian Genome, 23*, 132–143.

Rammal, H., Bouayed, J., Falla, J., Boujedaini, N., & Soulimani, R. (2010). The Impact of High Anxiety Level on Cellular and Humoral Immunity in Mice. *Neuroimmunomodulation, 17*, 1–8.

Rehn, T., & Keeling L. J. (2011). The effect of time left alone at home on dog welfare. *Applied Animal Behaviour Science, 129*, 129–135.

Rooney, N. J. (2009). The welfare of pedigree dogs: Cause for concern. *Journal of Veterinary Behavior: Clinical Applications and Research, 4*, 180–186.

Rooney, N. J., & Cowan, S. (2011). Training methods and owner-dog interactions: Links with dog behaviour and learning ability. *Applied Animal Behaviour Science, 132*, 169–177.

Rooney, N. J., & Bradshaw, J. W. S. (2003). Links between play and dominance and attachment dimensions of dog-human relationships. *Journal of Applied Animal Welfare Science, 6*, 67–94.

Rooney, N. J., & Sargan, D. R. (2009). Pedigree dog breeding—An independent review. Report to the RSPCA. http://www.rspca.org.uk/ImageLocator/LocateAsset?asset=document&assetId=1232712491490&mode=prd. Accessed 23/10/2013

Rooney, N. J., & Sargan, D. R. (2010). Welfare concerns associated with pedigree dog breeding in the UK. *Animal Welfare, 19*, 133–140.

Rooney, N. J., Bradshaw, J. W. S., & Almey, H. (2004). Attributes of specialist search dogs: A questionnaire survey of UK dog handlers and trainers. *Journal of Forensic Sciences, 49*, 300–306.

Rooney, N. J., Bradshaw, J. W. S, & Robinson, I. H. (2000). A comparison of dog-dog and dog-human play behaviour. *Applied Animal Behaviour Science, 66*, 235–248.

Rooney, N. J., Bradshaw, J. W. S, & Robinson, I. H. (2001). Do dogs respond to play signals given by humans? *Animal Behaviour, 61*, 715–722.

Rooney, N. J., Gaines, S. A., & Bradshaw, J. W. S. (2007). Behavioural and glucocorticoid responses of dogs (Canis familiaris) to kennelling: Investigating mitigation of stress by prior habituation. *Physiology & Behavior, 92*, 847–854.

Rooney, N., Gaines, S., & Hiby, E. (2009). A practitioner's guide to working dog welfare. *Journal of Veterinary Behavior: Clinical Applications and Research, 4*, 127–134.

Rooney, N. J., Clark, C. C., & Casey, R. A. (in prep). How best to avoid the development of fear and anxiety in working dogs: A review. *Journal Clinical Veterinary Behavior*

Ryan, D. (2010). Who is a Positive Reinforcement trainer? http://www.apbc.org.uk/blog/positive_reinforcement. Accessed 8 January 2014.

Schalke, E., Stichnoth, J., Ott, S., & Jones-Baade, R. (2007). Clinical signs caused by the use of electric training collars on dogs in everyday life situations. *Applied Animal Behaviour Science, 105*, 369–380.

Schilder, M. B. H., & van der Borg, J. A. M. (2004). Training dogs with help of the shock collar: Short and long term behavioural effects. *Applied Animal Behaviour Science, 85*, 319–334.

Serpell, J. A. (2005). People in disguise: Anthropomorphism and the human–pet relationship. In L. Daston, & G. Mitmann, (Eds.), *Thinking with Animals: New Perspectives on Anthropomorphism* (pp. 121–136). New York: Columbia University Press.

Shiverdecker, M. D., Schiml, P. A., & Hennessy, M. B. (2013). Human interaction moderates plasma cortisol and behavioral responses of dogs to shelter housing. *Physiology & Behavior, 109*, 75–79.

Shore, E. R., Riley, M. L., & Douglas, D. K. (2006). Pet owner behaviors and attachment to yard versus house dogs. *Anthrozoos, 19*, 325–334.

Simmel, E.C. (1979). *Early experiences and early behaviour: Implications for social development*. New York: Academic Press.

Siniscalchi, M., McFarlane, J. R., Kauter, K. G., Quaranta, A., & Rogers, L. J. (2013). Cortisol levels in hair reflect behavioural reactivity of dogs to acoustic stimuli. *Research in Veterinary Science, 94*, 49–54.

Slabbert, J. M. & Odendaal, J. S. J. (1999). Early prediction of adult police dog efficiency—a longitudinal study. *Applied Animal Behaviour Science, 64*, 269–288.

Soares, G. M., Pereira, J. T., & Paixao, R. L. (2010). Exploratory study of separation anxiety syndrome in apartment dogs. *Ciencia Rural, 40*, 548–553.

Stafford, K. (2007). *The welfare of dogs, Dordrecht*. The Netherlands: Springer.

Steiger, A., Stucki, E., Peyer, N., & Keller, P. (2008). Assessment of animal welfare aspects in extreme breeds of dogs and cats. *Schweizer Archiv Fur Tierheilkunde, 150*, 217–225.

Summers, J. F., Diesel, G., Asher, L., McGreevy, P. D., & Collins, L. M. (2010). Inherited defects in pedigree dogs. Part 2: Disorders that are not related to breed standards. *The Veterinary Journal, 183*, 39–45.

Svartberg, K. (2005). A comparison of behaviour in test and in everyday life: Evidence of three consistent boldness-related personality traits in dogs. *Applied Animal Behaviour Science, 91*, 103–128.

Tami, G., Barone, A., & Diverio, S. (2008). Relationship between management factors and dog behavior in a sample of Argentine Dogos in Italy. *Journal of Veterinary Behavior-Clinical Applications and Research, 3*, 59–73.

Tami, G., & Gallagher, A. (2009). Description of the behaviour of domestic dog (Canis familiaris) by experienced and inexperienced people. *Applied Animal Behaviour Science, 120*, 159–169.

Taylor, K. D., & Mills, D. S. (2007). The effect of the kennel environment on canine welfare: A critical review of experimental studies. *Animal Welfare, 16*, 435–447.

Taylor, P. (2003). Pain management in dogs and cats—More causes and locations to contemplate. *Veterinary Journal, 165*, 186–187.

Terlouw, E. M. C., Schouton, W. G. P., & Ladewig, P. (1997). Physiology. In M. C. Appleby & B. O. Hughes (Eds.), *Animal Welfare*. Wallingford: CAB International.

The Monks of New Skete (1978). *How to be your dog's best friend: A training manual for dog owners*. Boston: Little, Brown & Co.

The Monks of New Skete (1991). *The art of raising a puppy*. Boston: Little, Brown & Co.

Titulaer, M., Blackwell, E. J., Mendl, M., & Casey, R. A. (2013). Cross sectional study comparing behavioural, cognitive and physiological indicators of welfare between short and long term kennelled domestic dogs. *Applied Animal Behaviour Science, 147*, 149–158.

Totton, S. C., Wandeler, A. I., Ribble, C. S., Rosatte, R. C., & McEwen, S. A. (2011). Stray dog population health in Jodhpur, India in the wake of an animal birth control (ABC) program. *Preventive Veterinary Medicine, 98*, 215–220.

Tuber, D. S., Sanders, S., Hennessy, M. B., & Miller, J. A. (1996). Behavioral and glucocorticoid responses of adult domestic dogs (Canis familiaris) to companionship and social separation. *Journal of Comparative Psychology, 110*, 103–108.

Valsecchi, P., Pattacini, O., Beretta, V., Bertozzi J., Zannoni, S., & Viggiani, R. et al. (2007). Effects of a human social enrichment program on behavior and welfare of sheltered dogs. *Journal of Veterinary Behavior: Clinical Applications and Research, 2*, 88–89.

Walker, D. B., Walker, J. C., Cavnar, P. J., Taylor, J. L., Pickel, D. H., & Hall, S. B., et al. (2006). Naturalistic quantification of canine olfactory sensitivity. *Applied Animal Behaviour Science*, *97*, 241–254.

Wells, D. L. (2004). A review of environmental enrichment for kennelled dogs, Canis familiaris. *Applied Animal Behaviour Science*, *85*, 307–317.

Wickens, S. M. (2007). An overview of developments in the regulation of those treating behavioral disorders in animals in the United Kingdom. *Journal of Veterinary Behavior-Clinical Applications and Research*, *2*, 29–34.

Yeates, J. W. (2012). Maximising canine welfare in veterinary practice and research: A review. *The Veterinary Journal*, *192*, 272–278.

Yong, M. H., & Ruffman, T. (2013). Dogs' cortisol and behavioural response to crying human infant behaviour. *Proceedings of the Joint Meeting of the 33rd International Ethological Conference (IEC) & the Association for the Study of Animal Behaviour (ASAB)*.

Printed by Publishers' Graphics LLC
ICISO140221.15.18.6